A PRIMER OF ANALYTIC NUMBER THEORY

This undergraduate introduction to analytic number theory develops analytic skills in the course of a study of ancient questions on polygonal numbers, perfect numbers, and amicable pairs. The question of how the primes are distributed among all integers is central in analytic number theory. This distribution is determined by the Riemann zeta function, and Riemann's work shows how it is connected to the zeros of his function and the significance of the Riemann Hypothesis.

Starting from a traditional calculus course and assuming no complex analysis, the author develops the basic ideas of elementary number theory. The text is supplemented by a series of exercises to further develop the concepts and includes brief sketches of more advanced ideas, to present contemporary research problems at a level suitable for undergraduates. In addition to proofs, both rigorous and heuristic, the book includes extensive graphics and tables to make analytic concepts as concrete as possible.

Jeffrey Stopple is Professor of Mathematics at the University of California, Santa Barbara.

T0243075

A PRIMER OF ANALYTIC NUMBER THEORY

A PRIMER OF ANALYTIC NUMBER THEORY

NUMBER THEORY

From Pythagoras to Riemann

JEFFREY STOPPLE

University of California, Santa Barbara

CAMBRIDGE
UNIVERSITY PRESS

CAMBRIDGE UNIVERSITY PRESS
Cambridge, New York, Melbourne, Madrid, Cape Town, Singapore,
São Paulo, Delhi, Dubai, Tokyo

Cambridge University Press
The Edinburgh Building, Cambridge CB2 8RU, UK

Published in the United States of America by Cambridge University Press, New York

www.cambridge.org
Information on this title: www.cambridge.org/9780521012539

First published 2003

A catalogue record for this publication is available from the British Library

Library of Congress Cataloguing in Publication data
Stopple, Jeffrey, 1958–
A primer of analytic number theory : from Pythagoras to
Riemann / Jeffrey Stopple.
p. cm.
Includes bibliographical references and index.
ISBN 0-521-81309-3 – ISBN 0-521-01253-8 (pb.)
1. Number theory. I. Title.
QA241 .S815 2003
512′.7 – dc21   2002041263

ISBN 978-0-521-81309-9 Hardback
ISBN 978-0-521-01253-9 Paperback

Transferred to digital printing 2009

This book is dedicated to all the former students
who let me practice on them.

Contents

Preface

Good evening. Now, I'm no mathematician but I'd like to talk about
just a couple of numbers that have really been bothering me lately . . .

Laurie Anderson

Number theory is a subject that is so old, no one can say when it started.
That also makes it hard to describe what it is. More or less, it is the study of
interesting properties of integers. Of course, what is interesting depends on
your taste. This is a book about how analysis applies to the study of prime
numbers. Some other goals are to introduce the rich history of the subject and
to emphasize the active research that continues to go on.

History. In the study of right triangles in geometry, one encounters triples
of integers x, y, z such that $x^2 + y^2 = z^2$. For example, $3^2 + 4^2 = 5^2$. These
are called Pythagorean triples, but their study predates even Pythagoras. In
fact, there is a Babylonian cuneiform tablet (designated Plimpton 322 in the
archives of Columbia University) from the nineteenth century B.C. that lists
fifteen very large Pythagorean triples; for example,

$$12709^2 + 13500^2 = 18541^2.$$

The Babylonians seem to have known the theorem that such triples can be
generated as

$$x = 2st, \quad y = s^2 - t^2, \quad z = s^2 + t^2$$

for integers s, t. This, then, is the oldest theorem in mathematics. Pythagoras
and his followers were fascinated by mystical properties of numbers, believing
that numbers constitute the nature of all things. The Pythagorean school of
mathematics also noted this interesting example with sums of cubes:

$$3^3 + 4^3 + 5^3 = 216 = 6^3.$$

This number, 216, is the Geometrical Number in Plato's *Republic*.[1]

The other important tradition in number theory is based on the *Arithmetica* of Diophantus. More or less, his subject was the study of *integer* solutions of equations. The story of how Diophantus' work was lost to the Western world for more than a thousand years is sketched in Section 12.2. The great French mathematician Pierre de Fermat was reading Diophantus' comments on the Pythagorean theorem, mentioned above, when he conjectured that for an exponent $n > 2$, the equation

$$x^n + y^n = z^n$$

has *no* integer solutions x, y, z (other than the trivial solution when one of the integers is zero). This was called "Fermat's Last Theorem," although he gave no proof; Fermat claimed that the margin of the book was too small for it to fit. For more than 350 years, Fermat's Last Theorem was considered the hardest open question in mathematics, until it was solved by Andrew Wiles in 1994. This, then, is the most recent major breakthrough in mathematics.

I have included some historical topics in number theory that I think are interesting, and that fit in well with the material I want to cover. But it's not within my abilities to give a complete history of the subject. As much as possible, I've chosen to let the players speak for themselves, through their own words. My point in including this material is to try to convey the vast timescale on which people have considered these questions.

The Pythagorean tradition of number theory was also the origin of numerology and much number mysticism that sounds strange today. It is my intention neither to endorse this mystical viewpoint nor to ridicule it, but merely to indicate how people thought about the subject. The true value of the subject is in the mathematics itself, not the mysticism. This is perhaps what Françoise Viète meant in dedicating his *Introduction to the Analytic Art* to his patron the princess Catherine de Parthenay in 1591. He wrote very colorfully:

The metal I produced appears to be that class of gold others have desired for so long. It may be alchemist's gold and false, or dug out and true. If it is alchemist's gold, then it will evaporate into a cloud of smoke. But it certainly is true,... with much vaunted labor drawn from those mines, inaccessible places, guarded by fire breathing dragons and noxious serpents....

---

[1] If you watch the movie *Pi* closely, you will see that, in addition to $\pi = 3.14159\ldots$, the number 216 plays an important role, as a tribute to the Pythagoreans. Here's another trivia question: What theorem from this book is on the blackboard during John Nash's Harvard lecture in the movie *A Beautiful Mind*?

Analysis. There are quite a few number theory books already. However, they all cover more or less the same topics: the algebraic parts of the subject. The books that do cover the analytic aspects do so at a level far too high for the typical undergraduate. This is a shame. Students take number theory after a couple of semesters of calculus. They have the basic tools to understand some concepts of analytic number theory, if they are presented at the right level. The prerequisites for this book are two semesters of calculus: differentiation and integration. Complex analysis is specifically not required. We will gently review the ideas of calculus; at the same time, we can introduce some more sophisticated analysis in the context of specific applications. Joseph-Louis Lagrange wrote,

I regard as quite useless the reading of large treatises of pure analysis: too large a number of methods pass at once before the eyes. It is in the works of applications that one must study them; one judges their ability there and one apprises the manner of making use of them.

(Among the areas Lagrange contributed to are the study of Pell's equation, Chapter 11, and the study of binary quadratic forms, Chapter 13.)

This is a good place to discuss what constitutes a proof. While some might call it heresy, a proof is an argument that is convincing. It, thus, depends on the context, on who is doing the proving and who is being convinced. Because advanced books on this subject already exist, I have chosen to emphasize readability and simplicity over absolute rigor. For example, many proofs require comparing a sum to an integral. A picture alone is often quite convincing. In this, it seems Lagrange disagreed, writing in the Preface to *Mécanique Analytique*,

[T]he reader will find no figures in this work. The methods which I set forth do not require . . . geometrical reasonings: but only algebraic operations, subject to a regular and uniform rule of procedure.

In some places, I point out that the argument given is suggestive of the truth but has important details omitted. This is a trade-off that must be made in order to discuss, for example, Riemann's Explicit Formula at this level.

Research. In addition to having the deepest historical roots of all of mathematics, number theory is an active area of research. The Clay Mathematics Institute recently announced seven million-dollar "Millennium Prize Problems," see http://www.claymath.org/prizeproblems/ Two of the seven problems concern number theory, namely the Riemann Hypothesis and the Birch Swinnerton-Dyer conjecture. Unfortunately, without introducing analysis, one can't understand what these problems are about. A couple of years ago, the National Academy of Sciences published a report on the current state of mathematical research. Two of the three important research areas in number theory they named were, again, the Riemann Hypothesis and the Beilinson conjectures (the Birch Swinnerton-Dyer conjecture is a small portion of the latter).

Very roughly speaking, the Riemann Hypothesis is an outgrowth of the Pythagorean tradition in number theory. It determines how the prime numbers are distributed among all the integers, raising the possibility that there is a hidden regularity amid the apparent randomness. The key question turns out to be the location of the zeros of a certain function, the Riemann zeta function. Do they all lie on a straight line? The middle third of the book is devoted to the significance of this. In fact, mathematicians have already identified the next interesting question after the Riemann Hypothesis is solved. What is the distribution of the spacing of the zeros along the line, and what is the (apparent) connection to quantum mechanics? These question are beyond the scope of this book, but see the expository articles Cipra, 1988; Cipra, 1996; Cipra, 1999; and Klarreich, 2000.

The Birch Swinnerton-Dyer conjecture is a natural extension of beautiful and mysterious infinite series identities, such as

$$\frac{1}{1} + \frac{1}{4} + \frac{1}{9} + \frac{1}{16} + \frac{1}{25} + \frac{1}{36} + \cdots = \frac{\pi^2}{6},$$

$$\frac{1}{1} - \frac{1}{3} + \frac{1}{5} - \frac{1}{7} + \frac{1}{9} - \frac{1}{11} + \frac{1}{13} - \cdots = \frac{\pi}{4}.$$

Surprisingly, these are connected to the Diophantine tradition of number theory. The second identity above, Gregory's series for $\pi/4$, is connected to Fermat's observations that no prime that is one less than a multiple of four (e.g., 3, 7, and 11) is a hypotenuse of a right triangle. And every prime that is one more than a multiple of four *is* a hypotenuse, for example 5 in the (3, 4, 5) triangle, 13 in the (5, 12, 13), and 17 in the (8, 15, 17). The last third of the book is devoted to the arithmetic significance of such infinite series identities.

Advice. The Pythagoreans divided their followers into two groups. One group, the μαθηματικη, learned the subject completely and understood all the details. From them comes, our word "mathematician," as you can see for yourself if you know the Greek alphabet (mu, alpha, theta, eta, . . .). The second group, the ακουσματιοκι, or "acusmatics," kept silent and merely memorized the master's words without understanding. The point I am making here is that if you want to be a mathematician, you have to participate, and that means doing the exercises. Most have solutions in the back, but you should at least make a serious attempt before reading the solution. Many sections later in the book refer back to earlier exercises. You will, therefore, want to keep them in a permanent notebook. The exercises offer lots of opportunity to do calculations, which can become tedious when done by hand. Calculators typically do arithmetic with floating point numbers, not integers. You will get a lot more out of the exercises if you have a computer package such as Maple, *Mathematica*, or PARI.

1. Maple is simpler to use and less expensive. In Maple, load the number theory package using the command `with(numtheory);` Maple commands end with a semicolon.
2. *Mathematica* has more capabilities. Pay attention to capitalization in *Mathematica*, and if nothing seems to be happening, it is because you pressed the "return" key instead of "enter."
3. Another possible software package you can use is called PARI. Unlike the other two, it is specialized for doing number theory computations. It is free, but not the most user friendly. You can download it from `http://www.parigp-home.de/`

To see the movies and hear the sound files I created in *Mathematica* in the course of writing the book, or for links to more information, see my home page: `http://www.math.ucsb.edu/~stopple/`

Notation. The symbol $\exp(x)$ means the same as e^x. In this book, $\log(x)$ *always* means natural logarithm of x; you might be more used to seeing $\ln(x)$. If any other base of logarithms is used, it is specified as $\log_2(x)$ or $\log_{10}(x)$. For other notations, see the index.

Acknowledgments. I'd like to thank Jim Tattersall for information on Gerbert, Zack Leibhaber for the Viète translation, Lily Cockerill and David Farmer for reading the manuscript, Kim Spears for Chapter 13, and Lynne Walling for her enthusiastic support.

I still haven't said precisely what number theory – the subject – is. After a Ph.D. and fifteen further years of study, I think I'm only just beginning to figure it out myself.

Chapter 1

Sums and Differences

I met a traveller from an antique land
Who said: Two vast and trunkless legs of stone
Stand in the desert. Near them, on the sand,
Half sunk, a shattered visage lies . . .

Percy Bysshe Shelley

1.1. Polygonal Numbers

The Greek word *gnomon* means the pointer on a sundial, and also a carpenter's square or L-shaped bar. The Pythagoreans, who invented the subject of polygonal numbers, also used the word to refer to consecutive odd integers: 1, 3, 5, 7, ... The *Oxford English Dictionary's* definition of gnomon offers the following quotation, from Thomas Stanley's *History of Philosophy* in 1687 (Stanley, 1978):

Odd Numbers they called Gnomons, because being added to Squares, they keep the same Figures; so Gnomons do in Geometry.

In more mathematical terms, they observed that n^2 is the sum of the first n consecutive odd integers:

$$1 = 1^2,$$
$$1 + 3 = 2^2,$$
$$1 + 3 + 5 = 3^2,$$
$$1 + 3 + 5 + 7 = 4^2,$$
$$\vdots$$

Figure 1.1 shows a geometric proof of this fact; observe that each square is constructed by adding an odd number (the black dots) to the preceding square. These are the gnomons the quotation refers to.

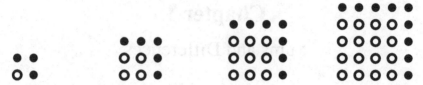

Figure 1.1. A geometric proof of the gnomon theorem.

But before we get to squares, we need to consider triangles. The TRIAN-
GULAR NUMBERS, t_n, are the number of circles (or dots, or whatever) in a
triangular array with n rows (see Figure 1.2).

Since each row has one more than the row above it, we see that

$$t_n = 1 + 2 + \cdots + n - 1 + n.$$

A more compact way of writing this, without the ellipsis, is to use the "Sigma"
notation,

$$t_n = \sum_{k=1}^{n} k.$$

The Greek letter $\sum$ denotes a sum; the terms in the sum are indexed by integers
between 1 and n, generically denoted k. And the thing being summed is the
integer k itself (as opposed to some more complicated function of k.)

Of course, we get the same number of circles (or dots) no matter how we
arrange them. In particular we can make right triangles. This leads to a clever
proof of a "closed-form" expression for t_n, that is, one that does not require
doing the sum. Take two copies of the triangle for t_n, one with circles and one
with dots. They fit together to form a rectangle, as in Figure 1.3. Observe that
the rectangle for two copies of t_n in Figure 1.3 has $n + 1$ rows and n columns,
so $2t_n = n(n + 1)$, or

$$1 + 2 + \cdots + n = t_n = \frac{n(n + 1)}{2}. \tag{1.1}$$

This is such a nice fact that, we will prove it two more times. The next proof
is more algebraic and has a story. The story is that Gauss, as a young student,
was set the task of adding together the first hundred integers by his teacher,
with the hope of keeping him busy and quiet for a while. Gauss immediately
came back with the answer $5050 = 100 \cdot 101/2$, because he saw the following

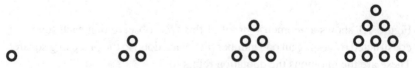

Figure 1.2. The triangular numbers are $t_1 = 1, t_2 = 3, t_3 = 6, t_4 = 10, \ldots$.

Figure 1.3. $2t_1 = 2 \cdot 1, 2t_2 = 3 \cdot 2, 2t_3 = 4 \cdot 3, 2t_4 = 5 \cdot 4, \ldots$.

trick, which works for any n. Write the sum defining t_n twice, once forward and once backward:

$$1+ \ 2+ \quad \cdots + n - 1+n,$$
$$n+n - 1+ \quad \cdots +2 \quad +1.$$

Now, add vertically; each pair of terms sums to $n + 1$, and there are n terms, so $2t_n = n(n + 1)$ or $t_n = n(n + 1)/2$.

The third proof uses mathematical induction. This is a method of proof that works when there are infinitely many theorems to prove, for example, one theorem for each integer n. The first case $n = 1$ must be proven and then it has to be shown that each case follows from the previous one. Think about a line of dominoes standing on edge. The $n = 1$ case is analogous to knocking over the first domino. The inductive step, showing that case $n - 1$ implies case n, is analogous to each domino knocking over the next one in line. We will give a proof of the formula $t_n = n(n + 1)/2$ by induction. The $n = 1$ case is easy. Figure 1.2 shows that $t_1 = 1$, which is equal to $(1 \cdot 2)/2$. Now we get to assume that the theorem is already done in the case of $n - 1$; that is, we can assume that

$$t_{n-1} = 1 + 2 + \cdots + n - 1 = \frac{(n - 1)n}{2}.$$

So

$$t_n = 1 + 2 + \cdots + n - 1 + n = t_{n-1} + n$$
$$= \frac{(n - 1)n}{2} + n = \frac{(n - 1)n}{2} + \frac{2n}{2} = \frac{(n + 1)n}{2}.$$

We have already mentioned the SQUARE NUMBERS, s_n. These are just the number of dots in a square array with n rows and n columns. This is easy; the formula is $s_n = n^2$. Nonetheless, the square numbers, s_n, are more interesting than one might think. For example, it is easy to see that the sum of two consecutive triangular numbers is a square number:

$$t_{n-1} + t_n = s_n. \tag{1.2}$$

Figure 1.4 shows a geometric proof.

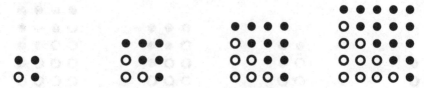

Figure 1.4. Geometric proof of Eq. (1.2).

It is also easy to give an algebraic proof of this same fact:

$$t_{n-1} + t_n = \frac{(n-1)n}{2} + \frac{n(n+1)}{2} = \frac{(n-1+n+1)n}{2} = n^2 = s_n.$$

Figure 1.1 seems to indicate that we can give an inductive proof of the identity

$$1 + 3 + 5 + \cdots + (2n-1) = n^2. \tag{1.3}$$

For the $n = 1$ case we just have to observe that $1 = 1^2$. And we have to show that the $n - 1$st case implies the nth case. But

$$1 + 3 + 5 + \cdots + (2n-3) + (2n-1)$$
$$= \{1 + 3 + 5 + \cdots + (2n-3)\} + 2n - 1.$$

So, by the induction hypothesis, it simplifies to

$$(n-1)^2 + 2n - 1$$
$$= n^2 - 2n + 1 + 2n - 1 = n^2.$$

Exercise 1.1.1. Since we know that $t_{n-1} + t_n = s_n$ and that $1 + 3 + \cdots + (2n-1) = s_n$, it is certainly true that

$$1 + 3 + \cdots + (2n-1) = t_{n-1} + t_n.$$

Give a *geometric* proof of this identity. That is, find a way of arranging the two triangles for t_{n-1} and t_n so that you see an array of dots in which the rows all have an odd number of dots.

Exercise 1.1.2. Give an algebraic proof of Plutarch's identity

$$8t_n + 1 = s_{2n+1}$$

using the formulas for triangular and square numbers. Now give a geometric proof of this same identity by arranging eight copies of the triangle for t_n, plus one extra dot, into a square.

Exercise 1.1.3. Which triangular numbers are also squares? That is, what conditions on m and n will guarantee that $t_n = s_m$? Show that if this happens, then we have

$$(2n + 1)^2 - 8m^2 = 1,$$

a solution to Pell's equation, which we will study in more detail in Chapter 11.

The philosophy of the Pythagoreans had an enormous influence on the development of number theory, so a brief historical diversion is in order.

Pythagoras of Samos (560–480 B.C.). Pythagoras traveled widely in Egypt and Babylonia, becoming acquainted with their mathematics. Iamblichus of Chalcis, in his *On the Pythagorean Life* (Iamblichus, 1989), wrote of Pythagoras' journey to Egypt:

From there he visited all the sanctuaries, making detailed investigations with the utmost zeal. The priests and prophets he met responded with admiration and affection, and he learned from them most diligently all that they had to teach. He neglected no doctrine valued in his time, no man renowned for understanding, no rite honored in any region, no place where he expected to find some wonder.... He spent twenty-two years in the sacred places of Egypt, studying astronomy and geometry and being initiated ... into all the rites of the gods, until he was captured by the expedition of Cambyses and taken to Babylon. There he spent time with the Magi, to their mutual rejoicing, learning what was holy among them, acquiring perfected knowledge of the worship of the gods and reaching the heights of their mathematics and music and other disciplines. He spent twelve more years with them, and returned to Samos, aged by now about fifty-six.

(Cambyses, incidentally, was a Persian emperor who invaded and conquered Egypt in 525 B.C., ending the twenty-fifth dynasty. According to Herodotus in *The Histories*, Cambyses did many reprehensible things against Egyptian religion and customs and eventually went mad.)

The Pythagorean philosophy was that the essence of all things is numbers. Aristotle wrote in *Metaphysics* that

[t]hey thought they found in numbers, more than in fire, earth, or water, many resemblances to things which are and become.... Since, then, all other things seemed in their whole nature to be assimilated to numbers, while numbers seemed to be the first things in the whole of nature, they supposed the elements of numbers to be the elements of all things, and the whole heaven to be a musical scale and a number.

Musical harmonies, the sides of right triangles, and the orbits of different planets could all be described by ratios. This led to mystical speculations about the properties of special numbers. In astronomy the Pythagoreans had the concept of the "great year." If the ratios of the periods of the planets

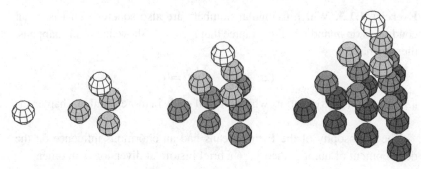

Figure 1.5. The tetrahedral numbers $T_1 = 1, T_2 = 4, T_3 = 10, T_4 = 20, \ldots$.

are integers, then after a certain number of years (in fact, the least common multiple of the ratios), the planets will return to exactly the same positions again. And since astrology says the positions of the planets determine events, according to Eudemus,

... then I shall sit here again with this pointer in my hand and tell you such strange things.

The TETRAHEDRAL NUMBERS, T_n, are three-dimensional analogs of the triangular numbers, t_n. They give the number of objects in a tetrahedral pyramid, that is, a pyramid with triangular base, as in Figure 1.5.

The kth layer of the pyramid is a triangle with t_k objects in it; so, by definition,

$$T_n = t_1 + t_2 + \cdots + t_{n-1} + t_n = \sum_{k=1}^{n} t_k. \qquad (1.4)$$

Here, we use Sigma notation to indicate that the kth term in the sum is the kth triangular number, t_k.

What is the pattern in the sequence of the first few tetrahedral numbers: $1, 4, 10, 20, \ldots$? What is the formula for T_n for general n? It is possible to give a three-dimensional geometric proof that $T_n = n(n + 1)(n + 2)/6$. It helps to use cubes instead of spheres. First shift the cubes so they line up one above the other, as we did in two dimensions. Then try to visualize six copies of the cubes, which make up T_n filling up a box with dimensions n by $n + 1$ by $n + 2$. This would be a three- dimensional analog of Figure 1.3.

If this makes your head hurt, we will give another proof that is longer but not so three dimensional. In fact you can view the following explanation as a two-dimensional analog of Gauss' one dimensional proof that $t_n = n(n + 1)/2$.

We will do this in the case of $n = 5$ for concreteness. From Eq. (1.4) we want to sum all the numbers in a triangle:

$$
\begin{array}{c}
1 \\
1+2 \\
1+2+3 \\
1+2+3+4 \\
1+2+3+4+5
\end{array}
$$

The kth row is the triangular number t_k. We take *three* copies of the triangle, each one rotated by $120°$:

$$
\begin{array}{ccc}
1 & 1 & 5 \\
1+2 & 2+1 & 4+4 \\
1+2+3 & 3+2+1 & 3+3+3 \\
1+2+3+4 & 4+3+2+1 & 2+2+2+2 \\
1+2+3+4+5 & 5+4+3+2+1 & 1+1+1+1+1
\end{array}
$$

The rearranged triangles still have the same sum. This is the analog of Gauss taking a second copy of the sum for t_n written backward. Observe that if we add the left and center triangles together, in each row the sums are constant:

$$
\begin{array}{ccccc}
1 & + & 1 & = & 2 \\
1+2 & + & 2+1 & = & 3+3 \\
1+2+3 & + & 3+2+1 & = & 4+4+4 \\
1+2+3+4 & + & 4+3+2+1 & = & 5+5+5+5 \\
1+2+3+4+5 & + & 5+4+3+2+1 & = & 6+6+6+6+6
\end{array}
$$

In row k, all the entries are $k + 1$, just as Gauss found. In the third triangle, all the entries in row k are the same; they are equal to $n - k + 1$, and $k + 1$ plus $n - k + 1$ is $n + 2$.

$$
\begin{array}{ccccc}
2 & + & 5 & = & 7 \\
3+3 & + & 4+4 & = & 7+7 \\
4+4+4 & + & 3+3+3 & = & 7+7+7 \\
5+5+5+5 & + & 2+2+2+2 & = & 7+7+7+7 \\
6+6+6+6+6 & + & 1+1+1+1+1 & = & 7+7+7+7+7
\end{array}
$$

We get a triangle with t_n numbers in it, each of which is equal to $n + 2$. So,

$$3T_n = t_n(n + 2) = n(n + 1)(n + 2)/2,$$

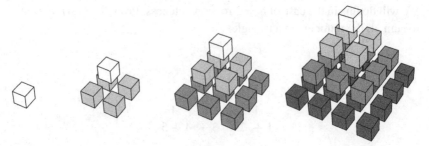

Figure 1.6. The pyramidal numbers $P_1 = 1$, $P_2 = 5$, $P_3 = 14$, $P_4 = 30, \ldots$.

and therefore,

$$T_n = n(n+1)(n+2)/6. \tag{1.5}$$

Exercise 1.1.4. Use mathematical induction to give another proof of Eq. (1.5), with T_n defined by Eq. (1.4).

The PYRAMIDAL NUMBERS, P_n, give the number of objects in a pyramid with a square base, as in Figure 1.6. The kth layer of the pyramid is a square with $s_k = k^2$ objects in it; so, by definition,

$$P_n = 1^2 + 2^2 + 3^2 + \cdots + n^2 = \sum_{k=1}^{n} k^2.$$

Since we know a relationship between square numbers and triangular numbers, we can get a formula for P_n in terms of the formula for T_n, as follows. From Eq. (1.2) we have $t_k + t_{k-1} = k^2$ for every k. This even works for $k = 1$ if we define $t_0 = 0$, which makes sense. So,

$$P_n = \sum_{k=1}^{n} k^2 = \sum_{k=1}^{n} \{t_k + t_{k-1}\}$$

$$= \sum_{k=1}^{n} t_k + \sum_{k=1}^{n} t_{k-1} = T_n + T_{n-1}.$$

According to Eq. (1.5) this is just

$$P_n = n(n+1)(n+2)/6 + (n-1)n(n+1)/6$$
$$= n(n+1)(2n+1)/6.$$

The formulas

$$1 + 2 + \cdots + n = n(n + 1)/2, \qquad (1.6)$$

$$1^2 + 2^2 + \cdots + n^2 = n(n + 1)(2n + 1)/6 \qquad (1.7)$$

are beautiful. Can we generalize them? Is there a formula for sums of cubes? In fact there is, due to Nicomachus of Gerasa. Nicomachus observed the interesting pattern in sums of odd numbers:

$$
\begin{array}{ll}
1 & = 1^3, \\
3 + 5 & = 2^3, \\
7 + 9 + 11 & = 3^3, \\
13 + 15 + 17 + 19 & = 4^3, \\
21 + 23 + 25 + 27 + 29 & = 5^3, \\
\quad\vdots & \quad\vdots
\end{array}
$$

This seems to indicate that summing consecutive cubes will be the same as summing consecutive odd numbers.

$$1 + 3 + 5 = 1^3 + 2^3,$$

$$1 + 3 + 5 + 7 + 9 + 11 = 1^3 + 2^3 + 3^3,$$

$$\vdots$$

But how many odd numbers do we need to take? Notice that 5 is the third odd number, and $t_2 = 3$. Similarly, 11 is the sixth odd number, and $t_3 = 6$. We guess that the pattern is that the sum of the first n cubes is the sum of the first t_n odd numbers. Now Eq. (1.3) applies and this sum is just $(t_n)^2$. From Eq. (1.1) this is $(n(n + 1)/2)^2$. So it seems as if

$$1^3 + 2^3 + \cdots + n^3 = n^2(n + 1)^2/4. \qquad (1.8)$$

But the preceding argument was mostly inspired guessing, so a careful proof by induction is a good idea. The base case $n = 1$ is easy because $1^3 = 1^2 \cdot 2^2/4$. Now we can assume that the $n - 1$ case

$$1^3 + 2^3 + \cdots + (n - 1)^3 = (n - 1)^2 n^2/4$$

Table 1.1. *Another proof of*
Nicomachus identity

1	2	3	4	5	...
2	4	6	8	10	...
3	6	9	12	15	...
4	8	12	16	20	...
5	10	15	20	25	...
⋮	⋮	⋮	⋮	⋮	

is true and use it to prove the next case. But

$$1^3 + 2^3 + \cdots + (n-1)^3 + n^3$$
$$= \{1^3 + 2^3 + \cdots + (n-1)^3\} + n^3$$
$$= \frac{(n-1)^2 n^2}{4} + n^3$$

by the induction hypothesis. Now, put the two terms over the common de-nominator and simplify to get $n^2(n+1)^2/4$.

Exercise 1.1.5. Here's another proof that

$$1^3 + 2^3 + 3^3 + \cdots + n^3 = n^2(n+1)^2/4, \tag{1.9}$$

with the details to be filled in. The entries of the multiplication table are shown in Table 1.1. Each side of the equation can be interpreted as a sum of all the entries in the table. For the left side of Eq. (1.9), form "gnomons" starting from the upper-left corner. For example, the second one is 2, 4, 2. The third one is 3, 6, 9, 6, 3, and so on.

What seems to be the pattern when you add up the terms in the kth gnomon? To prove your conjecture, consider the following questions:

1. What is the common factor of all the terms in the kth gnomon?
2. If you factor this out, can you write what remains in terms of triangular numbers?
3. Can you write what remains in terms of squares?
4. Combine these ideas to prove the conjecture you made.

The right side of Eq. (1.9) is t_n^2. Why is the sum of the n^2 entries in the first n rows and n columns equal to $t_n \cdot t_n$?

1.2. The Finite Calculus

The results in the previous sections are beautiful, but some of the proofs are almost *too* clever. In this section we will see some structure that simplifies things. This will build on skills you already have from studying calculus.

For example, if we want to go beyond triangular numbers and squares, the next step is pentagonal numbers. But the pictures are hard to draw because of the fivefold symmetry of the pentagon. Instead, consider what we've done so far:

$$
\begin{array}{lllllll}
n: & 1 & 2 & 3 & 4 & 5 & \ldots, \\
t_n: & 1 & 3 & 6 & 10 & 15 & \ldots, \\
s_n: & 1 & 4 & 9 & 16 & 25 & \ldots.
\end{array}
$$

In each row, consider the differences between consecutive terms:

$$
\begin{array}{lllllll}
(n+1) - n: & 1 & 1 & 1 & 1 & 1 & \ldots, \\
t_{n+1} - t_n: & 2 & 3 & 4 & 5 & 6 & \ldots, \\
s_{n+1} - s_n: & 3 & 5 & 7 & 9 & 11 & \ldots.
\end{array}
$$

There is nothing new here; in the third row, we are just seeing that each square is formed by adding an odd number (gnomon) to the previous square. If we now compute the differences again, we see

$$
\begin{array}{lllllll}
0 & 0 & 0 & 0 & 0 & \ldots, \\
1 & 1 & 1 & 1 & 1 & \ldots, \\
2 & 2 & 2 & 2 & 2 & \ldots.
\end{array}
$$

In each case, the second differences are constant, and the constant increases by one in each row.

For convenience we will introduce the DIFFERENCE OPERATOR, Δ, on functions $f(n)$, which gives a new function, $\Delta f(n)$, defined as $f(n+1) - f(n)$. This is an analog of derivative. We can do it again,

$$
\begin{aligned}
\Delta^2 f(n) &= \Delta(\Delta f)(n) \\
&= (\Delta f)(n+1) - (\Delta f)(n) \\
&= f(n+2) - 2f(n+1) + f(n),
\end{aligned}
$$

in an analogy with the second derivative. Think of the triangular numbers and square numbers as functions and not sequences. So,

$$s(n) = n^2,$$
$$\Delta s(n) = (n+1)^2 - n^2$$
$$= n^2 + 2n + 1 - n^2 = 2n + 1,$$
$$\Delta^2 s(n) = (2(n+1)+1) - (2n+1) = 2.$$

Based on the pattern of second differences, we expect that the pentagonal numbers, $p(n)$, should satisfy $\Delta^2 p(n) = 3$ for all n. This means that $\Delta p(n) = 3n + C$ for some constant C, since

$$\Delta(3n + C) = (3(n+1) + C) - (3n + C) = 3.$$

What about $p(n)$ itself? To correspond to the $+C$ term, we need a term, $Cn + D$ for some other constant D, since

$$\Delta(Cn + D) = (C(n+1) + D) - (Cn + D) = C.$$

We also need a term whose difference is $3n$. We already observed that for the triangular numbers, $\Delta t(n) = n + 1$. So, $\Delta t(n-1) = n$ and $\Delta(3t(n-1)) = 3n$. So,

$$p(n) = 3t(n-1) + Cn + D = 3(n-1)n/2 + Cn + D$$

for some constants C and D. We expect $p(1) = 1$ and $p(2) = 5$, because they are pentagonal numbers; so, plugging in, we get

$$0 + C + D = 1,$$
$$3 + 2C + D = 5.$$

Solving, we get that $C = 1$ and $D = 0$, so

$$p(n) = 3(n-1)n/2 + n = n(3n-1)/2.$$

This seems to be correct, since it gives

$$\begin{aligned}
p(n): \quad & 1 \quad 5 \quad 12 \quad 22 \quad 35 \quad \ldots, \\
\Delta p(n): \quad & 4 \quad 7 \quad 10 \quad 13 \quad 16 \quad \ldots, \\
\Delta^2 p(n): \quad & 3 \quad 3 \quad 3 \quad 3 \quad 3 \quad \ldots.
\end{aligned}$$

Exercise 1.2.1. Imitate this argument to get a formula for the hexagonal numbers, $h(n)$.

The difference operator, Δ, has many similarities to the derivative d/dx in calculus. We have already used the fact that

$$\Delta(f + g)(n) = \Delta f(n) + \Delta g(n) \quad \text{and} \quad \Delta(c \cdot f)(n) = c \cdot \Delta f(n)$$

in an analogy with the corresponding rules for derivatives. But the rules are not exactly the same, since

$$\frac{d}{dx}x^2 = 2x \quad \text{but} \quad \Delta n^2 = 2n + 1, \text{ not } 2n.$$

What functions play the role of powers x^m? It turns out to be the FACTORIAL POWERS

$$n^{\underline{m}} = \underbrace{n(n - 1)(n - 2) \cdots (n - (m - 1))}_{m \text{ consecutive integers}}.$$

An empty product is 1 by convention, so

$$n^{\underline{0}} = 1, \quad n^{\underline{1}} = n, \quad n^{\underline{2}} = n(n - 1), \quad n^{\underline{3}} = n(n - 1)(n - 2), \ldots. \quad (1.10)$$

Observe that

$$\Delta(n^{\underline{m}}) = (n + 1)^{\underline{m}} - n^{\underline{m}}$$
$$= [(n + 1) \cdots (n - (m - 2))] - [n \cdots (n - (m - 1))].$$

The last $m - 1$ factors in the first term and the first $m - 1$ factors in the second term are both equal to $n^{\underline{m-1}}$. So we have

$$\Delta(n^{\underline{m}}) = [(n + 1) \cdot n^{\underline{m-1}}] - [n^{\underline{m-1}} \cdot (n - (m - 1))]$$
$$= \{(n + 1) - (n - (m - 1))\} \cdot n^{\underline{m-1}}$$
$$= m \cdot n^{\underline{m-1}}.$$

What about negative powers? From Eq. (1.10) we see that

$$n^{\underline{2}} = \frac{n^{\underline{3}}}{n - 2}, \quad n^{\underline{1}} = \frac{n^{\underline{2}}}{n - 1}, \quad n^{\underline{0}} = \frac{n^{\underline{1}}}{n - 0}.$$

It makes sense to define the negative powers so that the pattern continues:

$$n^{\underline{-1}} = \frac{n^{\underline{0}}}{n - -1} = \frac{1}{n + 1},$$
$$n^{\underline{-2}} = \frac{n^{\underline{-1}}}{n - -2} = \frac{1}{(n + 1)(n + 2)},$$
$$n^{\underline{-3}} = \frac{n^{\underline{-2}}}{n - -3} = \frac{1}{(n + 1)(n + 2)(n + 3)},$$
$$\vdots$$

One can show that for any m, positive or negative,

$$\Delta(n^{\underline{m}}) = m \cdot n^{\underline{m-1}}. \tag{1.11}$$

Exercise 1.2.2. Verify this in the case of $m = -2$. That is, show that $\Delta(n^{\underline{-2}}) = -2 \cdot n^{\underline{-3}}$.

The factorial powers combine in a way that is a little more complicated than ordinary powers. Instead of $x^{m+k} = x^m \cdot x^k$, we have that

$$n^{\underline{m+k}} = n^{\underline{m}}(n - m)^{\underline{k}} \quad \text{for all } m, k. \tag{1.12}$$

Exercise 1.2.3. Verify this for $m = 2$ and $k = -3$. That is, show that $n^{\underline{-1}} = n^{\underline{2}}(n - 2)^{\underline{-3}}$.

The difference operator, Δ, is like the derivative d/dx, and so one might ask about the operation that undoes Δ the way an antiderivative undoes a derivative. This operation is denoted Σ:

$$\Sigma f(n) = F(n), \quad \text{if } F(n) \text{ is a function with } \Delta F(n) = f(n).$$

Don't be confused by the symbol Σ; we are not computing any sums. $\Sigma f(n)$ denotes a *function*, not a number. As in calculus, there is more than one possible choice for $\Sigma f(n)$. We can add a constant C to $F(n)$, because $\Delta(C) = C - C = 0$. Just as in calculus, the rule (1.11) implies that

$$\Sigma n^{\underline{m}} = \frac{n^{\underline{m+1}}}{m + 1} + C \quad \text{for } m \neq -1. \tag{1.13}$$

Exercise 1.2.4. We were already undoing the difference operator in finding pentagonal and hexagonal numbers. Generalize this to polygonal numbers with a sides, for any a. That is, find a formula for a function $f(n)$ with

$$\Delta^2 f(n) = a - 2, \quad \text{with } f(1) = 1 \text{ and } f(2) = a.$$

In calculus, the point of antiderivatives is to compute definite integrals. Geometrically, this is the area under curves. The Fundamental Theorem of Calculus says that if

$$F(x) = \int f(x)dx, \quad \text{then} \quad \int_a^b f(x)dx = F(b) - F(a).$$

We will think about this more carefully in Interlude 1, but for now the important point is the finite analog. We can use the operator Σ on functions to compute actual sums.

Theorem (Fundamental Theorem of Finite Calculus, Part I). *If*

$$\Sigma f(n) = F(n), \quad then \quad \sum_{a \le n < b} f(n) = F(b) - F(a).$$

Proof. The hypothesis $\Sigma f(n) = F(n)$ is just another way to say that $f(n) = \Delta F(n)$. The sum on the left is

$$\sum_{a \le n < b} f(n) = f(a) + f(a+1) + \cdots + f(b-2) + f(b-1)$$

$$= \Delta F(a) + \Delta F(a+1) + \cdots + \Delta F(b-2) + \Delta F(b-1)$$
$$= (F(a+1) - F(a)) + (F(a+2) - F(a+1)) + \cdots$$
$$\cdots + (F(b-1) - F(b-2)) + (F(b) - F(b-1))$$
$$= -F(a) + F(b).$$

$\square$

Notice that it does not matter which choice of constant C we pick, because $(F(b) + C) - (F(a) + C) = F(b) - F(a)$.

As an application, we can use the fact that $\Sigma n^{\underline{1}} = \frac{n^{\underline{2}}}{2}$ to say that

$$1 + 2 + \cdots + n = \sum_{0 \le k < n+1} k^{\underline{1}} = \frac{(n+1)^{\underline{2}}}{2} - \frac{0^{\underline{2}}}{2} = \frac{n(n+1)}{2}.$$

This is formula (1.6) for triangular numbers.

Here is another example. Because

$$n^{\underline{1}} + n^{\underline{2}} = n + n(n-1) = n^2,$$

we can say that

$$\Sigma n^2 = \Sigma(n^{\underline{1}} + n^{\underline{2}}) = \frac{n^{\underline{2}}}{2} + \frac{n^{\underline{3}}}{3}.$$

So,

$$\sum_{0 \le k < n+1} k^2 = \left(\frac{(n+1)^{\underline{2}}}{2} + \frac{(n+1)^{\underline{3}}}{3} \right) - \left(\frac{0^{\underline{2}}}{2} + \frac{0^{\underline{3}}}{3} \right)$$

$$= \frac{(n+1)n}{2} + \frac{(n+1)n(n-1)}{3}$$

$$= \frac{n(n+1)(2n+1)}{6}.$$

This is just Eq. (1.7) again.

Exercise 1.2.5. First, verify that

$$n^{\underline{1}} + 3n^{\underline{2}} + n^{\underline{3}} = n^3.$$

Now use this fact to find formulas for

$$\sum_{0 \le k < n+1} k^3.$$

Your answer should agree with formula (1.8).

In fact, one can do this for any exponent m. We will see that there are integers called STIRLING NUMBERS, $\left\{{m \atop k}\right\}$, which allow you to write ordinary powers in terms of factorial powers:

$$n^m = \sum_{k=0}^{m} \left\{{m \atop k}\right\} n^{\underline{k}}. \tag{1.14}$$

In the preceding example, we saw that

$$\left\{{2 \atop 0}\right\} = 0, \left\{{2 \atop 1}\right\} = 1, \left\{{2 \atop 2}\right\} = 1.$$

In the first part of Exercise 1.2.5, you verified that

$$\left\{{3 \atop 0}\right\} = 0, \left\{{3 \atop 1}\right\} = 1, \left\{{3 \atop 2}\right\} = 3, \left\{{3 \atop 3}\right\} = 1.$$

Exercise 1.2.6. Use the Stirling numbers

$$\left\{{4 \atop 0}\right\} = 0, \left\{{4 \atop 1}\right\} = 1, \left\{{4 \atop 2}\right\} = 7, \left\{{4 \atop 3}\right\} = 6, \left\{{4 \atop 4}\right\} = 1$$

to show that

$$1^4 + 2^4 + \cdots + n^4 = n(n+1)(2n+1)(3n^2+3n-1)/30. \tag{1.15}$$

The Stirling numbers are sort of like the binomial coefficients $\binom{m}{k}$. Binomial coefficients are found in Pascal's triangle, which you have probably seen:

$$
\begin{array}{c}
1 \\
1 \ 1 \\
1 \ 2 \ 1 \\
1 \ 3 \ 3 \ 1 \\
1 \ 4 \ 6 \ 4 \ 1
\end{array}
$$

The first and last entry in each row is always 1; the rest are computed by adding the two binomial coefficients on either side in the previous row. Suppose we make a similar triangle for the Stirling numbers. The Stirling number $\left\{ {m \atop k} \right\}$ is the kth entry in row m here:

$$
\begin{array}{ccccccccc}
 & & & & 1 & & & & \\
 & & & 1 & & 1 & & & \\
 & & 1 & & 3 & & 1 & & \\
 & 1 & & 7 & & 6 & & 1 & \\
1 & & 15 & & 25 & & 10 & & 1
\end{array}
$$

Exercise 1.2.7. Try to find the pattern in this triangle, similar to Pascal's. Here's a hint, but don't read it unless you're really stuck. The 3 is computed from the 1 and the second entry, also a 1, above it. The 7 is computed from the 1 and the second entry, a 3, above it. The 6 is computed from the 3 and the third entry, a 1, above it. What is the pattern?

Fill in the next row of Stirling's triangle.

In fact, if we make this a little more precise, we can prove the theorem now. First, though, we need to define

$$
\left\{ {m \atop 0} \right\} = \begin{cases} 1, & \text{if } m = 0, \\ 0, & \text{if } m > 0, \end{cases} \quad \text{and } \left\{ {m \atop k} \right\} = 0, \quad \text{if } k > m \text{ or } k < 0.
$$

Theorem. *If we now* define *the Stirling numbers by the recursion you discovered, that is,*

$$
\left\{ {m \atop k} \right\} = k \left\{ {m-1 \atop k} \right\} + \left\{ {m-1 \atop k-1} \right\},
$$

then Eq. (1.14) is true.

Notice that we have switched our point of view; the recursion is now the definition and the property (1.14) that we are interested in is a theorem. This is perfectly legal, as long as we make it clear that is what is happening. You may have indexed things slightly differently; make sure your recursion is equivalent to this one.

Proof. We can prove Eq. (1.14) by induction. The case of $m = 1$ is already done. From the boundary conditions ($k > m$ or $k < 0$) defined earlier, we can write (1.14) more easily as a sum over all k:

$$n^m = \sum_k \left\{ {m \atop k} \right\} n^{\underline{k}}.$$

The extra terms are 0. For the inductive step, we can assume that

$$n^{m-1} = \sum_k \left\{ {m-1 \atop k} \right\} n^{\underline{k}}$$

in order to prove (1.14). But

$$\sum_k \left\{ {m \atop k} \right\} n^{\underline{k}} = \sum_k \left(k \left\{ {m-1 \atop k} \right\} + \left\{ {m-1 \atop k-1} \right\} \right) n^{\underline{k}}$$

by the recursion for Stirling numbers. Thus,

$$\sum_k \left\{ {m \atop k} \right\} n^{\underline{k}} = \sum_k k \left\{ {m-1 \atop k} \right\} n^{\underline{k}} + \sum_k \left\{ {m-1 \atop k-1} \right\} n^{\underline{k}}.$$

We need to notice that Eq. (1.12) implies

$$n^{\underline{k+1}} = n \cdot n^{\underline{k}} - k \cdot n^{\underline{k}},$$

so that

$$k \cdot n^{\underline{k}} = n \cdot n^{\underline{k}} - n^{\underline{k+1}}.$$

Plug this in to see that

$$\sum_k \left\{ {m \atop k} \right\} n^{\underline{k}} = \sum_k n \cdot \left\{ {m-1 \atop k} \right\} n^{\underline{k}} - \sum_k \left\{ {m-1 \atop k} \right\} n^{\underline{k+1}} + \sum_k \left\{ {m-1 \atop k-1} \right\} n^{\underline{k}}.$$

The last two sums cancel; they are secretly equal since the factorial power is always one more than the lower parameter in the Stirling number. So,

$$\sum_k \left\{ {m \atop k} \right\} n^{\underline{k}} = n \cdot \sum_k \left\{ {m-1 \atop k} \right\} n^{\underline{k}} = n \cdot n^{m-1} = n^m$$

by the induction hypothesis.   □

Exercise 1.2.8. You now know enough to compute sums of any mth power in closed form. Show that

$$1^5 + 2^5 + \cdots + n^5 = (2n^2 + 2n - 1)(n+1)^2 n^2 / 12. \qquad (1.16)$$

You can find out more about Stirling numbers in Graham, Knuth, and Patashnik, 1994.

As with the polygonal numbers, once we have a closed-form expression, there seems to be nothing left to say. But notice that the rule (1.13) misses one case. There is no factorial power whose difference is $n^{\underline{-1}}$. In other words, $\Sigma n^{\underline{-1}}$ is not a factorial power. (This is the finite analog of the calculus fact that no power of x has derivative $1/x$.) So we make a definition instead, defining the nth HARMONIC NUMBER to be

$$H_n = \sum_{1 \leq k \leq n} \frac{1}{k} = 1 + \frac{1}{2} + \cdots + \frac{1}{n}. \qquad (1.17)$$

Notice that after changing the variable slightly, we can also write

$$H_n = \sum_{0 \leq k < n} \frac{1}{k+1}.$$

What is ΔH_n? We compute

$$H_{n+1} - H_n = \left(1 + \frac{1}{2} + \cdots + \frac{1}{n} + \frac{1}{n+1}\right) - \left(1 + \frac{1}{2} + \cdots + \frac{1}{n}\right)$$

$$= \frac{1}{n+1} = n^{\underline{-1}}.$$

So, the Harmonic numbers are the finite analog of logarithms in that

$$\Delta H_n = n^{\underline{-1}}$$

is true. Harmonic numbers are interesting, as shown in Eq. (1.17), which provides a generalization of the formulas (1.6), (1.7), (1.8), (1.15), and (1.16). In some sense they are even more interesting, because there is no closed-form expression for them as for the formulas mentioned earlier.

Actually, we can do this same procedure for *any* $f(n)$, not just $n^{\underline{-1}}$.

Theorem (Fundamental Theorem of Finite Calculus, Part II). *If a new function $F(n)$ is defined by*

$$F(n) = \sum_{0 \leq k < n} f(n) \quad \text{for some} \quad f(n),$$

then

$$\Delta F(n) = f(n), \quad \text{so} \quad F(n) = \Sigma f(n).$$

Proof. This proof is exactly the same as the proof for the Harmonic numbers.   □

Exercise 1.2.9. Suppose that $f(n) = 2^n$ (ordinary exponent, not factorial). Show that $\Delta f(n) = f(n)$ and $f(0) = 1$. What function in calculus are we imitating? Use the Fundamental Theorem, Part I, to show that

$$1 + 2 + 2^2 + \cdots + 2^n = \sum_{0 \le k < n+1} 2^k = 2^{n+1} - 1.$$

Exercise 1.2.10. More generally, suppose that $f(n) = x^n$. Here $x \ne 1$ is a constant, and n is still the variable. Show that $\Delta f(n) = (x - 1)f(n)$, and therefore $\Sigma f(n) = f(n)/(x - 1)$. Use this to show that

$$1 + x + x^2 + \cdots + x^n = \sum_{0 \le k < n+1} x^k = \frac{x^{n+1} - 1}{x - 1}.$$

This sum is called the GEOMETRIC SERIES.

Exercise 1.2.11. The Rhind papyrus is the oldest known mathematical document: 14 sheets of papyrus from the fifteenth dynasty, or about 1700 B.C. Problem 79 says, "There are seven houses. Each house has seven cats. Each cat catches seven mice. Each mouse eats seven ears of spelt [a grain related to wheat]. Each ear of spelt produces seven hekats [a bulk measure]. What is the total of all of these?" Use the Geometric series to answer this, the oldest known mathematical puzzle.

Archimedes, too, knew of the Geometric series.

Archimedes (287–212 B.C.). Archimedes is better known for his beautiful theorems on area and volume in geometry than for his work in number theory. However, the Geometric series and other series, as we will see, are vital in number theory. Archimedes used the Geometric series in his work *Quadrature of the Parabola*. He approximated the area below a parabola using a collection of congruent triangles. The sum of the areas was a Geometric series. Archimedes' works were not widely studied until the Byzantines wrote commentaries in the sixth century A.D. Thabit ibn Qurra wrote commentaries in the ninth century. From these texts, Archimedes' work became known in the west. Nicole Oresme quoted at length from Archimedes, as did Leonardo of Pisa.

Accounts of his death by Livy, Plutarch, and others all more or less agree that he was killed by a Roman soldier in the sack of Syracuse (in Sicily) in 212 B.C., while he was doing some mathematics. His grave was marked by a cylinder circumscribing a sphere, to commemorate his theorem in solid geometry: that the ratio of the volumes is 3:2. Cicero, as *Quaestor* of Sicily in 75 B.C., described his search for the site (Cicero, 1928):

I shall call up from the dust on which he drew his figures an obscure, insignificant person, Archimedes. I tracked out his grave . . . and found it enclosed all round and covered with brambles and thickets . . . . I noticed a small column rising a little above the bushes, on which there was a figure of a sphere and a cylinder . . . . Slaves were sent in with sickles and when a passage to the place was opened we approached the pedestal; the epigram was traceable with about half of the lines legible, as the latter portion was worn away.

Cicero goes on to add,

Who in all the world, who enjoys merely some degree of communion with the Muses, . . . is there who would not choose to be the mathematician rather than the tyrant?

The most useful trick in calculus for finding antiderivatives is "u substitution." This does not translate very well to finite calculus, except for very simple changes of variables involving translation. That is, if $\Delta f(k) = g(k)$ and a is any constant, then $\Delta(f(k+a)) = g(k+a)$.

Exercise 1.2.12. Use this and the fact that $2(k-1)^{\underline{-2}} = 1/t_k$ to find the sum of the reciprocals of the first n triangular numbers

$$\frac{1}{t_1} + \frac{1}{t_2} + \cdots + \frac{1}{t_n}.$$

Can you compute

$$\frac{1}{T_1} + \frac{1}{T_2} + \cdots + \frac{1}{T_n},$$

the sum of the reciprocals of the first n tetrahedral numbers?

Toward the end of this book, we will need one more tool based on this finite analog of calculus. If you are just casually skimming, you may skip the rest of this chapter. In calculus, another useful method of finding antiderivatives is integration by parts. This is exactly the same thing as the product rule for derivatives, just written in antiderivative notation. That is, if you have functions $u(x)$ and $v(x)$, then

$$(u(x)v(x))' = u(x)'v(x) + u(x)v(x)';$$

so,

$$u(x)v(x)' = (u(x)v(x))' - u(x)'v(x).$$

If we take antiderivatives of both sides of the equation and use the fact that $\int (u(x)v(x))' dx = u(x)v(x)$, we get

$$\int u(x)v(x)' dx = u(x)v(x) - \int u(x)'v(x)dx.$$

If we suppress mention of the variable x and use the abbreviations $u(x)'dx = du$ and $v(x)'dx = dv$, then this is the formula for integration by parts you know and love (or at least know):

$$\int u\,dv = uv - \int v\,du.$$

For a finite analog, it seems we should start by applying the difference operator, Δ, to a product of two functions, for an analog of the product rule. This gives

$$\Delta(u(n)v(n)) = u(n+1)v(n+1) - u(n)v(n).$$

We can add and subtract a term $u(n)v(n+1)$ to get

$$\begin{aligned} \Delta(u(n)v(n)) &= u(n+1)v(n+1) - u(n)v(n+1) \\ &\quad + u(n)v(n+1) - u(n)v(n) \\ &= (u(n+1) - u(n))v(n+1) \\ &\quad + u(n)(v(n+1) - v(n)) \\ &= \Delta u(n)v(n+1) + u(n)\Delta v(n). \end{aligned}$$

This is not exactly what you might expect. The function v is shifted by one so that $v(n+1)$ appears. We will denote this shift operator on functions by E, so $Ef(n) = f(n+1)$. Then the product rule in this setting says

$$\Delta(uv) = \Delta u \cdot Ev + u \cdot \Delta v$$

when the variable n is suppressed. As in the derivation of the integration-by-parts formula, we rearrange the terms to say

$$u \cdot \Delta v = \Delta(uv) - \Delta u \cdot Ev.$$

Applying the Σ operator, which undoes Δ, we get that

$$\Sigma(u \cdot \Delta v) = uv - \Sigma(\Delta u \cdot Ev).$$

This identity is called SUMMATION BY PARTS. Remember that so far it is just an identity between functions.

Suppose we want to use Summation by Parts to compute

$$\sum_{0 \le k < n} k^{\underline{1}} H_k.$$

First we need to find the function $\Sigma(k^{\underline{1}} H_k)$. Let $u(k) = H_k$, so $\Delta u(k) = k^{\underline{-1}}$. Then $k^{\underline{1}} = \Delta v(k)$, so we can choose $v(k) = k^{\underline{2}}/2$. Summation by Parts says that

$$\Sigma(k^{\underline{1}} H_k) = H_k \cdot \frac{k^{\underline{2}}}{2} - \Sigma\left(E\left(\frac{k^{\underline{2}}}{2}\right)k^{\underline{-1}}\right).$$

Now $k^{\underline{2}}/2 = k(k-1)/2$, so $E(k^{\underline{2}}/2) = (k+1)k/2$, and then $E(k^{\underline{2}}/2)k^{\underline{-1}}$ is equal to $k/2 = k^{\underline{1}}/2$. Thus,

$$\begin{aligned}
\Sigma k^{\underline{1}} H_k &= H_k \cdot \frac{k^{\underline{2}}}{2} - \Sigma\frac{k^{\underline{1}}}{2} \\
&= H_k \cdot \frac{k^{\underline{2}}}{2} - \frac{k^{\underline{2}}}{4} \\
&= \frac{k^{\underline{2}}}{2}\left(H_k - \frac{1}{2}\right).
\end{aligned}$$

Remember, this is just saying that

$$\Delta\left(\frac{k^{\underline{2}}}{2}\left(H_k - \frac{1}{2}\right)\right) = k^{\underline{1}} H_k.$$

Now the Fundamental Theorem, Part I, says that

$$\begin{aligned}
\sum_{0 \le k < n} k^{\underline{1}} H_k &= \left(\frac{n^{\underline{2}}}{2}\left(H_n - \frac{1}{2}\right)\right) - \left(\frac{0^{\underline{2}}}{2}\left(H_0 - \frac{1}{2}\right)\right) \\
&= \frac{n^{\underline{2}}}{2}\left(H_n - \frac{1}{2}\right).
\end{aligned}$$

Exercise 1.2.13. Use Summation by Parts and the Fundamental Theorem to compute $\sum_{0 \le k < n} H_k$. (Hint: You can write $H_k = H_k \cdot 1 = H_k \cdot k^{\underline{0}}$.) Your answer will have Harmonic numbers in it, of course.

Exercise 1.2.14. Use Summation by Parts and the Fundamental Theorem to compute $\sum_{0 \le k < n} k 2^k$. (Hint: You need the first part of Exercise 1.2.9.)

Chapter 2

Products and Divisibility

I am one who becomes two
I am two who becomes four
I am four who becomes eight
I am the one after that
  Egyptian hieroglyphic inscription from the 22nd dynasty (Hopper, 2000)

2.1. Conjectures

Questions about the divisors, d, of an integer n are among the oldest in mathematics. The DIVISOR FUNCTION $\tau(n)$ counts how many divisors n has. For example, the divisors of 8 are $1, 2, 4$, and 8, so $\tau(8) = 4$. The divisors of 12 are $1, 2, 3, 4, 6$, and 12, so $\tau(12) = 6$. The SIGMA FUNCTION $\sigma(n)$ is defined as the sum of the divisors of n. So,

$$\sigma(8) = 1 + 2 + 4 + 8 = 15,$$
$$\sigma(12) = 1 + 2 + 3 + 4 + 6 + 12 = 28.$$

In the Sigma notation of Chapter 1,

$$\sigma(n) = \sum_{d\mid n} d.$$

The difference here is that we are summing not over a set of *consecutive* integers but only those d which divide n, as the subscript $d \mid n$ indicates. Similarly,

$$\tau(n) = \sum_{d\mid n} 1.$$

Here, we add to our count a 1, not d, for each divisor d of n.

Exercise 2.1.1. Isaac Newton computed how many divisors 60 has in his 1732 work *Arithmetica Universalis*. What is $\tau(60)$?

This section consists mostly of exercises in which you will try to make conjectures about the functions $\tau(n)$ and $\sigma(n)$. Girolamo Cardano was the first person to do this, in his 1537 book *Practica Arithmetica*. Cardano is more famous now for his work on solutions to the cubic equation $x^3 + px + q = 0$. He was famous in his own time also, as a physician and astrologer, and in fact was imprisoned during the Inquisition for casting the horoscope of Christ.

Exercise 2.1.2. Compute values of $\tau(n)$ and $\sigma(n)$ for integers n that are powers of a single prime number; for example,

$$n = 2, 4, 8, 16, \ldots,$$
$$n = 3, 9, 27, 81, \ldots,$$
$$n = 5, 25, 125, 625, \ldots.$$

Try to phrase a precise conjecture for integers n which are of the form p^k for prime p. At some point you will need the general fact that for any x and k,

$$(1 - x)(1 + x + x^2 + x^3 + \cdots + x^k) = 1 - x^{k+1}. \tag{2.1}$$

A little algebra shows that in the product, many terms cancel; in fact, all but the first and last do. So,

$$1 + x + x^2 + x^3 + \cdots + x^k = \frac{1 - x^{k+1}}{1 - x}.$$

You already derived this identity in Exercise 1.2.10; it will appear many more times in this book.

We will say that two integers, m and n, are RELATIVELY PRIME if they have no prime factor in common. For example, 10 and 12 are not relatively prime; both are divisible by 2. But 9 and 10 are relatively prime.

Exercise 2.1.3. Choose several pairs of integers m and n and compute $\tau(n)$, $\tau(m)$, and $\tau(mn)$. What relationship is there, if any? Try to make a precise conjecture. Be sure to look at enough examples; your first try at a conjecture may be false. Table 2.1 contains factorizations for integers less than 300.

Exercise 2.1.4. Repeat this process with the function $\sigma(n)$.

Exercise 2.1.5. You should combine the conjectures of the previous exercises to get a conjecture for the general formula: If an integer n factors as

2. Products and Divisibility

Table 2.1. *Factor Table*

$30 = 2^1 \cdot 3^1 \cdot 5^1$	$31 = 31^1$	$32 = 2^5$	$33 = 3^1 \cdot 11^1$	$34 = 2^1 \cdot 17^1$
$35 = 5^1 \cdot 7^1$	$36 = 2^2 \cdot 3^2$	$37 = 37^1$	$38 = 2^1 \cdot 19^1$	$39 = 3^1 \cdot 13^1$
$40 = 2^3 \cdot 5^1$	$41 = 41^1$	$42 = 2^1 \cdot 3^1 \cdot 7^1$	$43 = 43^1$	$44 = 2^2 \cdot 11^1$
$45 = 3^2 \cdot 5^1$	$46 = 2^1 \cdot 23^1$	$47 = 47^1$	$48 = 2^4 \cdot 3^1$	$49 = 7^2$
$50 = 2^1 \cdot 5^2$	$51 = 3^1 \cdot 17^1$	$52 = 2^2 \cdot 13^1$	$53 = 53^1$	$54 = 2^1 \cdot 3^3$
$55 = 5^1 \cdot 11^1$	$56 = 2^3 \cdot 7^1$	$57 = 3^1 \cdot 19^1$	$58 = 2^1 \cdot 29^1$	$59 = 59^1$
$60 = 2^2 \cdot 3^1 \cdot 5^1$	$61 = 61^1$	$62 = 2^1 \cdot 31^1$	$63 = 3^2 \cdot 7^1$	$64 = 2^6$
$65 = 5^1 \cdot 13^1$	$66 = 2^1 \cdot 3^1 \cdot 11^1$	$67 = 67^1$	$68 = 2^2 \cdot 17^1$	$69 = 3^1 \cdot 23^1$
$70 = 2^1 \cdot 5^1 \cdot 7^1$	$71 = 71^1$	$72 = 2^3 \cdot 3^2$	$73 = 73^1$	$74 = 2^1 \cdot 37^1$
$75 = 3^1 \cdot 5^2$	$76 = 2^2 \cdot 19^1$	$77 = 7^1 \cdot 11^1$	$78 = 2^1 \cdot 3^1 \cdot 13^1$	$79 = 79^1$
$80 = 2^4 \cdot 5^1$	$81 = 3^4$	$82 = 2^1 \cdot 41^1$	$83 = 83^1$	$84 = 2^2 \cdot 3^1 \cdot 7^1$
$85 = 5^1 \cdot 17^1$	$86 = 2^1 \cdot 43^1$	$87 = 3^1 \cdot 29^1$	$88 = 2^3 \cdot 11^1$	$89 = 89^1$
$90 = 2^1 \cdot 3^2 \cdot 5^1$	$91 = 7^1 \cdot 13^1$	$92 = 2^2 \cdot 23^1$	$93 = 3^1 \cdot 31^1$	$94 = 2^1 \cdot 47^1$
$95 = 5^1 \cdot 19^1$	$96 = 2^5 \cdot 3^1$	$97 = 97^1$	$98 = 2^1 \cdot 7^2$	$99 = 3^2 \cdot 11^1$
$100 = 2^2 \cdot 5^2$	$101 = 101^1$	$102 = 2^1 \cdot 3^1 \cdot 17^1$	$103 = 103^1$	$104 = 2^3 \cdot 13^1$
$105 = 3^1 \cdot 5^1 \cdot 7^1$	$106 = 2^1 \cdot 53^1$	$107 = 107^1$	$108 = 2^2 \cdot 3^3$	$109 = 109^1$
$110 = 2^1 \cdot 5^1 \cdot 11^1$	$111 = 3^1 \cdot 37^1$	$112 = 2^4 \cdot 7^1$	$113 = 113^1$	$114 = 2^1 \cdot 3^1 \cdot 19^1$
$115 = 5^1 \cdot 23^1$	$116 = 2^2 \cdot 29^1$	$117 = 3^2 \cdot 13^1$	$118 = 2^1 \cdot 59^1$	$119 = 7^1 \cdot 17^1$
$120 = 2^3 \cdot 3^1 \cdot 5^1$	$121 = 11^2$	$122 = 2^1 \cdot 61^1$	$123 = 3^1 \cdot 41^1$	$124 = 2^2 \cdot 31^1$
$125 = 5^3$	$126 = 2^1 \cdot 3^2 \cdot 7^1$	$127 = 127^1$	$128 = 2^7$	$129 = 3^1 \cdot 43^1$
$130 = 2^1 \cdot 5^1 \cdot 13^1$	$131 = 131^1$	$132 = 2^2 \cdot 3^1 \cdot 11^1$	$133 = 7^1 \cdot 19^1$	$134 = 2^1 \cdot 67^1$
$135 = 3^3 \cdot 5^1$	$136 = 2^3 \cdot 17^1$	$137 = 137^1$	$138 = 2^1 \cdot 3^1 \cdot 23^1$	$139 = 139^1$
$140 = 2^2 \cdot 5^1 \cdot 7^1$	$141 = 3^1 \cdot 47^1$	$142 = 2^1 \cdot 71^1$	$143 = 11^1 \cdot 13^1$	$144 = 2^4 \cdot 3^2$
$145 = 5^1 \cdot 29^1$	$146 = 2^1 \cdot 73^1$	$147 = 3^1 \cdot 7^2$	$148 = 2^2 \cdot 37^1$	$149 = 149^1$
$150 = 2^1 \cdot 3^1 \cdot 5^2$	$151 = 151^1$	$152 = 2^3 \cdot 19^1$	$153 = 3^2 \cdot 17^1$	$154 = 2^1 \cdot 7^1 \cdot 11^1$
$155 = 5^1 \cdot 31^1$	$156 = 2^2 \cdot 3^1 \cdot 13^1$	$157 = 157^1$	$158 = 2^1 \cdot 79^1$	$159 = 3^1 \cdot 53^1$
$160 = 2^5 \cdot 5^1$	$161 = 7^1 \cdot 23^1$	$162 = 2^1 \cdot 3^4$	$163 = 163^1$	$164 = 2^2 \cdot 41^1$
$165 = 3^1 \cdot 5^1 \cdot 11^1$	$166 = 2^1 \cdot 83^1$	$167 = 167^1$	$168 = 2^3 \cdot 3^1 \cdot 7^1$	$169 = 13^2$
$170 = 2^1 \cdot 5^1 \cdot 17^1$	$171 = 3^2 \cdot 19^1$	$172 = 2^2 \cdot 43^1$	$173 = 173^1$	$174 = 2^1 \cdot 3^1 \cdot 29^1$
$175 = 5^2 \cdot 7^1$	$176 = 2^4 \cdot 11^1$	$177 = 3^1 \cdot 59^1$	$178 = 2^1 \cdot 89^1$	$179 = 179^1$
$180 = 2^2 \cdot 3^2 \cdot 5^1$	$181 = 181^1$	$182 = 2^1 \cdot 7^1 \cdot 13^1$	$183 = 3^1 \cdot 61^1$	$184 = 2^3 \cdot 23^1$
$185 = 5^1 \cdot 37^1$	$186 = 2^1 \cdot 3^1 \cdot 31^1$	$187 = 11^1 \cdot 17^1$	$188 = 2^2 \cdot 47^1$	$189 = 3^3 \cdot 7^1$
$190 = 2^1 \cdot 5^1 \cdot 19^1$	$191 = 191^1$	$192 = 2^6 \cdot 3^1$	$193 = 193^1$	$194 = 2^1 \cdot 97^1$
$195 = 3^1 \cdot 5^1 \cdot 13^1$	$196 = 2^2 \cdot 7^2$	$197 = 197^1$	$198 = 2^1 \cdot 3^2 \cdot 11^1$	$199 = 199^1$
$200 = 2^3 \cdot 5^2$	$201 = 3^1 \cdot 67^1$	$202 = 2^1 \cdot 101^1$	$203 = 7^1 \cdot 29^1$	$204 = 2^2 \cdot 3^1 \cdot 17^1$
$205 = 5^1 \cdot 41^1$	$206 = 2^1 \cdot 103^1$	$207 = 3^2 \cdot 23^1$	$208 = 2^4 \cdot 13^1$	$209 = 11^1 \cdot 19^1$
$210 = 2^1 \cdot 3^1 \cdot 5^1 \cdot 7^1$	$211 = 211^1$	$212 = 2^2 \cdot 53^1$	$213 = 3^1 \cdot 71^1$	$214 = 2^1 \cdot 107^1$
$215 = 5^1 \cdot 43^1$	$216 = 2^3 \cdot 3^3$	$217 = 7^1 \cdot 31^1$	$218 = 2^1 \cdot 109^1$	$219 = 3^1 \cdot 73^1$
$220 = 2^2 \cdot 5^1 \cdot 11^1$	$221 = 13^1 \cdot 17^1$	$222 = 2^1 \cdot 3^1 \cdot 37^1$	$223 = 223^1$	$224 = 2^5 \cdot 7^1$
$225 = 3^2 \cdot 5^2$	$226 = 2^1 \cdot 113^1$	$227 = 227^1$	$228 = 2^2 \cdot 3^1 \cdot 19^1$	$229 = 229^1$
$230 = 2^1 \cdot 5^1 \cdot 23^1$	$231 = 3^1 \cdot 7^1 \cdot 11^1$	$232 = 2^3 \cdot 29^1$	$233 = 233^1$	$234 = 2^1 \cdot 3^2 \cdot 13^1$
$235 = 5^1 \cdot 47^1$	$236 = 2^2 \cdot 59^1$	$237 = 3^1 \cdot 79^1$	$238 = 2^1 \cdot 7^1 \cdot 17^1$	$239 = 239^1$
$240 = 2^4 \cdot 3^1 \cdot 5^1$	$241 = 241^1$	$242 = 2^1 \cdot 11^2$	$243 = 3^5$	$244 = 2^2 \cdot 61^1$
$245 = 5^1 \cdot 7^2$	$246 = 2^1 \cdot 3^1 \cdot 41^1$	$247 = 13^1 \cdot 19^1$	$248 = 2^3 \cdot 31^1$	$249 = 3^1 \cdot 83^1$
$250 = 2^1 \cdot 5^3$	$251 = 251^1$	$252 = 2^2 \cdot 3^2 \cdot 7^1$	$253 = 11^1 \cdot 23^1$	$254 = 2^1 \cdot 127^1$
$255 = 3^1 \cdot 5^1 \cdot 17^1$	$256 = 2^8$	$257 = 257^1$	$258 = 2^1 \cdot 3^1 \cdot 43^1$	$259 = 7^1 \cdot 37^1$
$260 = 2^2 \cdot 5^1 \cdot 13^1$	$261 = 3^2 \cdot 29^1$	$262 = 2^1 \cdot 131^1$	$263 = 263^1$	$264 = 2^3 \cdot 3^1 \cdot 11^1$
$265 = 5^1 \cdot 53^1$	$266 = 2^1 \cdot 7^1 \cdot 19^1$	$267 = 3^1 \cdot 89^1$	$268 = 2^2 \cdot 67^1$	$269 = 269^1$
$270 = 2^1 \cdot 3^3 \cdot 5^1$	$271 = 271^1$	$272 = 2^4 \cdot 17^1$	$273 = 3^1 \cdot 7^1 \cdot 13^1$	$274 = 2^1 \cdot 137^1$
$275 = 5^2 \cdot 11^1$	$276 = 2^2 \cdot 3^1 \cdot 23^1$	$277 = 277^1$	$278 = 2^1 \cdot 139^1$	$279 = 3^2 \cdot 31^1$
$280 = 2^3 \cdot 5^1 \cdot 7^1$	$281 = 281^1$	$282 = 2^1 \cdot 3^1 \cdot 47^1$	$283 = 283^1$	$284 = 2^2 \cdot 71^1$
$285 = 3^1 \cdot 5^1 \cdot 19^1$	$286 = 2^1 \cdot 11^1 \cdot 13^1$	$287 = 7^1 \cdot 41^1$	$288 = 2^5 \cdot 3^2$	$289 = 17^2$
$290 = 2^1 \cdot 5^1 \cdot 29^1$	$291 = 3^1 \cdot 97^1$	$292 = 2^2 \cdot 73^1$	$293 = 293^1$	$294 = 2^1 \cdot 3^1 \cdot 7^2$
$295 = 5^1 \cdot 59^1$	$296 = 2^3 \cdot 37^1$	$297 = 3^3 \cdot 11^1$	$298 = 2^1 \cdot 149^1$	$299 = 13^1 \cdot 23^1$

$p_1^{k_1} p_2^{k_2} \ldots p_m^{k_m}$, then

$$\tau(n) = \ldots$$

and

$$\sigma(n) = \ldots.$$

We will need one other function related to $\sigma(n)$:

$$s(n) = \sum_{\substack{d|n \\ d<n}} d = \sigma(n) - n.$$

This adds up the divisors of n other than n itself. We say integer n is DEFICIENT if $s(n) < n$, ABUNDANT if $s(n) > n$, and PERFECT if $s(n) = n$. Because $s(8) = 7$ and $s(12) = 16$, 8 is deficient and 12 is abundant. Six is a perfect number, and so is 28. The function $s(n)$ is sometimes called the sum of the ALIQUOT PARTS of n, after an archaic word meaning a proper divisor.

Exercise 2.1.6. Based on Exercise 2.1.3, you might be tempted to assume that there is a nice relationship between $s(n)$, $s(m)$, and $s(mn)$. Is there a relationship?

The concept of perfect numbers goes back at least to Archytas of Tarentum. Abundant and deficient numbers were first defined by Nicomachus of Gerasa in his work *Introduction to Arithmetic*. This was the twilight of Greek mathematics and the first writing on number theory since the time of Euclid; so, another historical diversion is in order.

Nicomachus of Gerasa (100 A.D.). In addition to perfect, abundant, and deficient numbers, Nicomachus also wrote on polygonal numbers. The style is very different from earlier Greek mathematics, though. Nicomachus included no proofs, only examples, and often extrapolated incorrectly from them. He was really interested in the mystical properties he ascribed to these special numbers. Another book he wrote was titled *Theology of Numbers*. Nicomachus also wrote on the theory of music in *Manual of Harmony*, about the Pythagorean philosophy of music based on number and ratio. This work, like Viète's some 1,500 years later, was dedicated to a noblewoman patron. It is speculated in Nicomachus of Gerasa, 1938, that she must have been knowledgeable in mathematics and music to read it. All that remains of Gerasa, in Jordan, are spectacular Roman ruins. In the same style as Nicomachus, later came Iamblichus of Chalcis.

Iamblichus of Chalcis (250 A.D.–330 A.D.). Iamblichus lived in a period when the Roman empire was in serious decline. Almost nothing of his life is known, despite what you might find on the Internet about him being descended from an ancient line of priest–kings of Syria. He wrote on "neo-Pythagorean" philosophy, including arithmetic, geometry, and astronomy. His biography of Pythagoras was quoted in Chapter 1, and he also wrote a commentary on Nicomachus. But he was more of a philosopher and a mystic than a mathematician. To him, individual numbers were symbols of individual gods of the Greek pantheon. For example, seven is the only number among the first ten that is neither divisible by nor a divisor of any of the others. For this reason, it represents Athena, the virgin goddess. One and two were not even considered numbers. One is the "monad," representing unity or the absence of multitude. Two is the "dyad," representing duality. Strange as this may seem, consider that ordinary language conventions still support this. If you speak of "a number of" objects, you don't mean one or two. Even the PARI number theory package treats one and two in a special way. It includes "universal objects" `gun` and `gdeaux`, which do not belong on the stack; other integers are treated on an ad hoc basis.

Not long after Iamblichus, St. Augustine wrote in *DeGenesi ad Litteram,*

[t]he good Christian should beware of mathematicians, and all those who make empty prophecies. The danger already exists that the mathematicians have made a covenant with the devil to darken the spirit and to confine man in the bonds of Hell.

This is a quotation you will see on many a math professor's office door or website. But, in fact, St. Augustine uses the word mathematician here to mean astrologer. It was St. Augustine himself who introduced much number-theory mysticism to Christian theology. In *The City of God*, Book XI, he wrote the following:

We must not despise the science of numbers, which, in many passages of holy scripture, is found to be of eminent service to the careful.

Both Nicomachus and Iamblichus had a great influence on Boethius.

Boethius (480 A.D.–524 A.D.). Boethius came from a prominent Roman family related to emperors and popes. He was executed by Theodoric, the Gothic king of Rome, for suspected disloyalty. He was the last Roman who studied the works of ancient Greece, and he is influential mainly because of the works that he translated and adapted, and which thus became widely known in the medieval world. Books were difficult to obtain in Rome at the start of the Dark

Ages; Boethius did obtain those of Nicomachus and Iamblichus on which his mathematical writings were based.

His *Arithmetic* includes the four Pythagorean mathematical disciplines, arithmetic, astronomy, music, and geometry, that he calls the *quadrivium*. Geometry treats quantities at rest, while astronomy treats quantities in motion. Arithmetic studies numbers abstractly, whereas music studies relations between numbers as harmonies. The term *trivium* came to refer, by analogy, to the three subjects involving language: grammar, logic, and rhetoric. The concept of *university* was based on this division of all knowledge into the *trivium* and the *quadrivium*. A university is a place that teaches all seven topics, and thus everything.

On abundant, deficient, and perfect numbers, Boethius wrote the following (Masi, 1983):

So these numbers, those whose parts added together exceed the total, are seen to be similar to someone who is born with many hands more than nature usually gives, as in the case of a giant who has a hundred hands, or three bodies joined together, such as the triple formed Geryon. The other number, whose parts when totaled are less than the size of the entire number, is like one born with some limb missing, or with an eye missing, like the ugliness of the Cyclops' face. Between these two kinds of number, as if between two elements unequal and intemperate, is put a number which holds the middle place between the extremes like one who seeks virtue. That number is called perfect and it does not extend in a superfluous progression nor is it reduced in a contracted reduction, but it maintains the place of the middle. . . .

(The Cyclops is the one-eyed monster in Homer's *Odyssey*, while Geryon is a three-headed giant slain by Hercules as one of his twelve labors.) "The Neo-Pythagorean theory of number as the very divine essence of the world is the view around which the four sciences of the quadrivium are developed," according to *Dictionary of Scientific Biography*, 1970–1980.

Close to five hundred years later, the Dark Ages began to draw to a close. Mathematics was reintroduced into Europe via Islamic sources in Catalonian Spain. Gerbert d'Aurillac, who had studied at the monastery Santa Maria de Ripoll near Barcelona, reorganized the cathedral school at Rheims around the trivium and the quadrivium. Students had to master these subjects before beginning the study of theology. He was tutor to the Holy Roman Emperor Otto III and gave him an inscribed copy of Boethius's *Arithmetic*. Gerbert was elected Pope Sylvester II on April 2, 999. Gregorovius, in his *History of the City of Rome in the Middle Ages* (Gregorovius, 1971), writes,

Gerbert in Rome is like a solitary torch in the darkness of the night. The century of the grossest ignorance closed strangely enough with the appearance of a renowned genius. . . . If the Romans noticed their aged Pope watching the stars from his observatory in a tower of the Lateran, or surrounded in his study by parchments and drawings of geometrical figures, designing a sundial with his own hand, or studying astronomy on a globe covered with horse's skin, they probably believed him in league with the devil. A second Ptolemy seemed to wear the tiara, and the figure of Sylvester II marks a fresh period in the Middle Ages, that of the scholastics.

Exercise 2.1.7. We know that $6 = 2^1 \cdot 3$ and $28 = 2^2 \cdot 7$ are perfect numbers. The next ones are $496 = 2^4 \cdot 31$ and $8128 = 2^6 \cdot 127$ (check this!). What is the pattern in $3, 7, 31, 127, \ldots$? What is the pattern in the exponents $1, 2, 4, 6, \ldots$? Try to make a conjecture about perfect numbers. Euclid, in his *Elements*, proved a general theorem about perfect numbers around the year 300 B.C.

Exercise 2.1.8. This is a continuation of Exercise 2.1.7.

1. The integer $130816 = 2^8(2^9 - 1)$ factors into primes as $2^8 \cdot 7 \cdot 73$. Compute $s(130816)$ as $\sigma(130816) - 130816$. You will want to use the formula you conjectured for $\sigma(n)$ in Exercise 2.1.5. Is 130816 perfect?
2. The integer $2096128 = 2^{10}(2^{11} - 1)$ factors as $2^{10} \cdot 23 \cdot 89$. Compute $s(2096128)$. Is 2096128 perfect?
3. The integer $33550336 = 2^{12}(2^{13} - 1)$ factors as $2^{12} \cdot 8191$; that is, $2^{13} - 1 = 8191$ already is a prime. Compute $s(33550336)$. Is 33550336 perfect?

Refine the conjecture you made in Exercise 2.1.7 if necessary.

   If $2^p - 1$ is a prime, then the number p is automatically prime; we don't have to assume it. This follows from the polynomial identity (2.1). If p were to factor as $p = ab$, then

$$2^{ab} - 1 = (2^a - 1)(1 + 2^a + 2^{2a} + 2^{3a} + \cdots + 2^{(b-1)a})$$

factors nontrivially as well. Make sure you believe it is the same identity. What is the x here relative to Eq. (2.1)? What is the k? For this reason, we know that $2^3 - 1 = 7$ is a factor of $2^9 - 1 = 511$ without doing any work. This theorem was observed independently by Cataldi and Fermat in the sixteenth century. Before this, it was widely believed that $2^n - 1$ was prime for every odd n, with $511 = 2^9 - 1$ often given as an example (Dickson, 1999, Chap. 1).

Exercise 2.1.9. Find a factor of

$$151115727451828646838271 = 2^{77} - 1.$$

(In fact, you can find more than one.)

The primes 3, 7, 31, and 127 in Exercise 2.1.7 are examples of Mersenne numbers, after Marin Mersenne. Mersenne, a Franciscan friar, had an active correspondence with Fermat and Galileo. Mersenne wrote about his numbers in *Cogitata Physico-Mathematica* in 1644. In general, a MERSENNE NUMBER, M_p, is any integer of the form $2^p - 1$, where p is a prime number. So, $3 = M_2$, $7 = M_3$, $31 = M_5$, and $127 = M_7$ are prime Mersenne numbers, whereas $M_{11} = 2^{11} - 1 = 23 \cdot 89$ is a composite Mersenne number. And $2^9 - 1$ is not a Mersenne number at all because 9 is not a prime. It is still unknown whether there are infinitely many Mersenne numbers that are prime, although this is generally expected to be true. Remarkably, it is also not known whether there are infinitely many composite Mersenne numbers. At least one set must be infinite; probably both sets are.

Exercise 2.1.10. Based on your refined conjecture, can you find another perfect number?

Exercise 2.1.11. Can you prove your conjecture about perfect numbers?

Exercise 2.1.12. There is an interesting theorem about perfect numbers and sums of cubes. For example,

$$28 = 1^3 + 3^3.$$

Try to make a conjecture about what is true. Ignore the first perfect number, 6; it doesn't fit the pattern.

Exercise 2.1.13. Above you made a conjecture about perfect numbers and sum of cubes of odd numbers. In fact, you know enough to prove it. First of all, what does Eq. (1.8) say about a closed-form expression for

$$1^3 + 2^3 + 3^3 + \cdots + N^3 + \cdots + (2N)^3?$$

What about

$$2^3 + 4^3 + 6^3 + \cdots + (2N)^3?$$

(Hint: Factor.) Now subtract to get a closed-form expression for

$$1^3 + 3^3 + 5^3 + \cdots + (2N - 1)^3.$$

What value of N will give you a perfect number?

Exercise 2.1.14. Compute $s(n)$ for as many n as possible. Determine whether n is deficient, perfect, or abundant. Look for patterns when n is odd or even. Is

$s(n)$ odd or even? (The truth is complicated, so don't be afraid to modify your conjecture.) You can compute $s(n)$ easily in *Mathematica* by first defining it as a function

```
s[n_] := DivisorSigma[1, n] - n
```

The input s [8] returns the answer 7. In Maple, after loading the numtheory package (see p. xiii), you can define the function using

```
s:=n->sigma(n)-n;
```

The input s (12) ; returns the answer 16.

Exercise 2.1.15. The number $284 = 2^2 \cdot 71$ isn't perfect; $s(284) = 220 = 2^2 \cdot 5 \cdot 11$. What *is* interesting about $220 = s(284)$? I realize this is very much an open-ended question. Be patient; experiment and do computations.

Pairs of numbers, such as 220 and 284, that have the property you discovered earlier are called AMICABLE PAIRS. Iamblichus of Chalcis ascribed the discovery of these numbers to Pythagoras. In the ninth century, Thabit ibn Qurra wrote about amicable pairs.

Thabit ibn Qurra (836–901). Thabit ibn Qurra belonged to the Sabian sect, descended from Babylonian star worshipers according to *Dictionary of Scientific Biography*, 1970–1980. The Sabians spoke Arabic and took Arabic names but held onto their religion for a long while after the Arab conquest of the region that is now part of Turkey. For this reason, they produced many excellent astronomers and mathematicians. Thabit's great knowledge led to his invitation to the court of the Caliph in Baghdad. He translated many ancient Greek texts into Arabic. All the works of Archimedes that did not survive in the original came to us through his versions. His *Book on the Determination of Amicable Numbers by an Easy Method* contains ten propositions in number theory, including a method for generating more such pairs, described below. He was probably the first to discover the amicable pair $17296 = 2^4 \cdot 23 \cdot 47$, $18416 = 2^4 \cdot 1151$.

Amicable pairs have long been used in astrology, sorcery, and the concoction of love potions. Al Magriti wrote in his grimoire *Aim of the Wise* in 945 that he had put to the test the erotic effect of

giving any one the smaller number 220 to eat, and himself eating the larger number 284.

Ibn Khaldun wrote in *Muqaddimah* that

persons who have concerned themselves with talismans affirm that the amicable numbers
220 and 284 have an influence to establish a union or close friendship between two
individuals.

In the thirteenth and fourteenth century, manuscripts in Hebrew by Samuel
ben Yehuda and others carried the study of amicable pairs from the Islamic
world to the court of Robert of Anjou in Naples. The interest in them was
again motivated by their occult properties (Lévy, 1996). Abraham Azulai
commented in the sixteenth century that, in the "Book of Genesis," Jacob
gives Esau 220 goats:

Our ancestor Jacob prepared his present in a wise way. This number 220 is a hidden
secret, being one of a pair of numbers such that the parts of it are equal to the other one
284, and conversely. And Jacob had this in mind; this has been tried by the ancients in
securing the love of kings and dignitaries.

Exercise 2.1.16. If you found something interesting in Exercise 2.1.15, so
that you know what amicable pairs *are*, show that if m and n form an amicable
pair, then

$$\sigma(m) = m + n = \sigma(n).$$

If you did *not* discover anything in Exercise 2.1.15, you should work backward
now. Suppose m and n are integers such that $\sigma(m) = m + n = \sigma(n)$. What
can you say is true about $s(m)$ and $s(n)$?

According to Mersenne, Fermat told him a rule for generating amicable
pairs, which is equivalent to the one discovered by Thabit ibn Qurra. Write
in a column the powers of two: 2, 4, 8, 16, 32, ... In a column on the
right, write down three times the powers of two: 6, 12, 24, 48, 96, ... In
the next column to the right, enter the number on the left minus one *if* this
is a prime, and if it is composite just leave a blank: $6 - 1 = 5$, $12 - 1 = 11$,
$24 - 1 = 23$, $48 - 1 = 47$, $96 - 1 = 95 = 5 \cdot 19$, ... Finally, the column
on the far right follows the pattern $6 \cdot 12 - 1 = 71$, $12 \cdot 24 - 1 = 287 = 7 \cdot$
41, $24 \cdot 48 - 1 = 1151$, $48 \cdot 96 - 1 = 4607 = 17 \cdot 271 \ldots$ *if* this number
is prime, and the entry is blank otherwise. Look at Table 2.2, and read this
paragraph again.
    Any time we have an L-shape pattern of three primes in the right two
columns, we can build an amicable pair. From 5, 11, and 71 we get the pair $m =$
$5 \cdot 11 \cdot 4 = 220$, $n = 71 \cdot 4 = 284$. From 23, 47, and 1151 we get the pair
$m = 23 \cdot 47 \cdot 16 = 17296$, $n = 1151 \cdot 16 = 18416$, and so forth. Notice the

Table 2.2. *Thabit ibn Qurra's Algorithm*

2	6	5	—
4	12	11	71
8	24	23	
16	48	47	1151
32	96		
64	192	191	
128	384	383	73727
⋮	⋮	⋮	⋮

power of 2 comes from the first column. In words, the algorithm says that if k is an exponent such that $p = 3 \cdot 2^{k-1} - 1$, $q = 3 \cdot 2^k - 1$, and $r = 9 \cdot 2^{2k-1} - 1$ are all primes, then $m = p \cdot q \cdot 2^k$ and $n = r \cdot 2^k$ form an amicable pair. In the table, p and q form the vertical part of the L, with p above and q below. The horizontal part is r.

Exercise 2.1.17. Show that this is true. That is, for p, q, and r as above, show that the property of Exercise 2.1.16 holds:

$$\sigma(p \cdot q \cdot 2^k) = p \cdot q \cdot 2^k + r \cdot 2^k = \sigma(r \cdot 2^k).$$

Use the formula you conjectured for σ in Exercise 2.1.5, which is also proved in Eq. (2.3).

Exercise 2.1.18. Show that for p, q, and r as above, $q = 2p + 1$ and $r = 2p^2 + 4p + 1$. In particular, this means that p is an example of a Sophie Germain prime, a prime number p such that $2p + 1$ is also prime. We will see these again in Section 7.3.

Mersenne was also apparently the first to conjecture that there are infinitely many amicable pairs. But, after all this attention, only three such pairs of numbers were known until 1750, when the Swiss mathematician Euler (pronounced "oil-er," not "you-ler") found *fifty-nine* new amicable pairs, including the pair

$$3^5 \cdot 7^2 \cdot 13 \cdot 19 \cdot 53 \cdot 6959 = 1084730902983,$$
$$3^5 \cdot 7^2 \cdot 13 \cdot 19 \cdot 179 \cdot 2087 = 1098689026617.$$

The second smallest amicable pair,

$$1184 = 2^5 \cdot 37 \quad \text{and} \quad 1210 = 2 \cdot 5 \cdot 11^2,$$

was overlooked until 1866, when it was discovered by a sixteen year old, Nicolo Paganini. As this book is being written, 2,683,135 amicable pairs are known.

Exercise 2.1.19. This is another exercise where you need *Mathematica* or Maple. Pick a random integer $n \leq 300$ and compute $s(n)$, $s(s(n))$, $s(s(s(n)))$, and so forth. Make a conjecture about what happens. Pick another n and do it again. But don't pick $n = 276$; if you do, don't say I didn't warn you.

Exercise 2.1.20. Which integers m do you think can be written as $m = s(n)$ for some n? (In mathematical language, when is m in the range of the function s?) If m is in the range of s, how many different n have $m = s(n)$? Is the number finite or infinite?

Exercise 2.1.21. Let $D(k)$ denote the smallest integer such that $D(k)$ has exactly k divisors, i.e., the least integer such that $\tau(D(k)) = k$. For example, since 6 is the smallest integer with four divisors, $D(4) = 6$. Find $D(8)$, $D(12)$, $D(16)$, $D(24)$. In 1644, Mersenne asked his correspondents to find a number with 60 divisors. Can you find $D(60)$? Notice that this is not computing $\tau(60)$; you did that in Exercise 2.1.1.

Exercise 2.1.22. With the values 11, then 12, then 13, then 14 for n, compute

$$\left(\sum_{k|n} \tau(k) \right)^2 \quad \text{and compare it to} \quad \sum_{k|n} \tau(k)^3.$$

Make a conjecture. Is your conjecture true for other values of n? Can you prove it?

2.2. Theorems

In this section, we will prove some of the conjectures made earlier. First we need a lemma.

Lemma. *Suppose two integers, m and n, are relatively prime. Then, every integer d that divides the product mn can be written in exactly one way, $d = bc$, with b dividing m and c dividing n.*

Recall that we defined relatively prime in the last section to mean that m and n have no factor in common. To help understand what the lemma says before we actually prove it, we will see an example of how it fails to be true without this hypothesis. For example, $m = 12$ and $n = 10$ have the prime factor 2 in common. $d = 4$ divides the product 120, but we can write this two ways: as $4 \cdot 1$, with $b = 4$ dividing 12 and $c = 1$ dividing 10; or as $2 \cdot 2$, with $b = 2$ dividing 12 and $c = 2$ dividing 10.

Proof. We will show the contrapositive. That is, "P implies Q" is logically the same as "not Q implies not P." Suppose that d divides mn. We can write $d = bc$ and also $d = b'c'$, with both b and b' dividing m and both c and c' dividing n. We need to show that m and n are not relatively prime. We can write the integer $(mn)/d$ in two ways:

$$\frac{m}{b}\frac{n}{c} = \frac{mn}{d} = \frac{m}{b'}\frac{n}{c'}.$$

So, cross multiplying,

$$\frac{b'c'}{bc} = \frac{mn}{mn} = 1.$$

Since $b \neq b'$, some prime p dividing b does not divide b' (or vice versa); it must divide c' or it won't cancel out to give 1 on the right side. So, p divides b, which divides m, and p divides c', which divides n; so, m and n are both divisible by p. $\square$

For completeness, we will include the following, which you discovered in Exercise 2.1.2.

Lemma. *If $n = p^k$ is a power of a prime, then $\tau(n) = k + 1$ and $\sigma(n) = (p^{k+1} - 1)/(p - 1)$.*

Proof. This is easy; the divisors d of p^k are $1, p, p^2, \ldots, p^k$. There are $k + 1$ of them. Their sum is

$$1 + p + p^2 + \cdots + p^k = \frac{1 - p^{k+1}}{1 - p} = \frac{p^{k+1} - 1}{p - 1},$$

according to Exercise 1.2.10. $\square$

Lemma. *If m and n are relatively prime, then $\tau(mn) = \tau(m)\tau(n)$.*

Proof. Imagine a sheet of paper with the divisors of m listed at the top, starting with 1 and ending with m. Imagine the divisors of n listed down the left side, starting with 1 and ending with n. We fill in a rectangular table by forming products: In the column that has a divisor b of m, and in the row that has a divisor c of n, we put the number bc. So, for example, the upper left entry of the table is $1 \cdot 1$. Each bc divides mn, and every divisor of mn occurs exactly once according to the lemma above. The table has $\tau(m)$ columns and $\tau(n)$ rows, so there are $\tau(m)\tau(n)$ entries in all. □

Lemma. *If m and n are relatively prime, then $\sigma(mn) = \sigma(m)\sigma(n)$.*

Exercise 2.2.1. The proof is a modification of the proof for τ. The number $\sigma(mn)$ is the sum of all of the entries in the table. Suppose that we look at a single row in the table, indexed by some divisor c of n. The entries in the row are $1 \cdot c, \ldots, m \cdot c$. What is the sum of all the entries in this particular row? Now, use this expression to add up all the row sums. (If you're stuck, write out the whole table for $m = 10$ and $n = 21$.)

Theorem. *If n factors as $p_1^{k_1} p_2^{k_2} \ldots p_t^{k_t}$, then*

$$\tau(n) = (k_1 + 1)(k_2 + 1) \ldots (k_t + 1), \tag{2.2}$$

$$\sigma(n) = \frac{p_1^{k_1+1} - 1}{p_1 - 1} \frac{p_2^{k_2+1} - 1}{p_2 - 1} \ldots \frac{p_t^{k_t+1} - 1}{p_t - 1}. \tag{2.3}$$

Proof. This is just a combination of the previous lemmas. Because $p_1^{k_1}$, $p_2^{k_2}, \ldots, p_t^{k_t}$ are all relatively prime,

$$\tau(n) = \tau\left(p_1^{k_1}\right)\tau\left(p_2^{k_2}\right) \ldots \tau\left(p_t^{k_t}\right) = (k_1 + 1)(k_2 + 1) \ldots (k_t + 1),$$

and similarly for $\sigma(n)$. □

Theorem (Euclid). *If p is a prime number such that $M_p = 2^p - 1$ is also a prime, then $n = 2^{p-1} \cdot M_p$ is a perfect number.*

Proof. To show $s(n) = n$, we show that $\sigma(n) = s(n) + n$ is just $2n$. By the previous theorem,

$$\sigma(n) = \sigma(2^{p-1}M_p) = \sigma(2^{p-1})\sigma(M_p)$$

$$= \frac{2^p - 1}{2 - 1} \frac{M_p^2 - 1}{M_p - 1} = (2^p - 1)(M_p + 1),$$

but $2^p - 1$ is just M_p and therefore $M_p + 1$ is just 2^p. So,

$$\sigma(n) = M_p 2^p = 2 \cdot M_p 2^{p-1} = 2n.$$

$\square$

The philosopher and mathematician René Descartes ("I think, therefore I am") told Mersenne in 1638 he could prove that every even perfect number is of the form Euclid described. This is a partial converse to Euclid's theorem. Descartes never wrote down a proof. Euler did; it was published after his death and some 2,000 years after Euclid.

Theorem (Euler). *If n is a perfect number that is even, then $n = 2^{p-1} M_p$, with $M_p = 2^p - 1$ being a prime.*

It is conjectured that there are no odd perfect numbers, and there are various theorems that show that they must be very scarce. There is a theorem, for example, that says that any odd perfect number must be bigger than 10^{300}.

Proof. Since n is even, we can write it as a power of 2 times an odd number: $n = 2^{p-1} m$ with m odd and $p > 1$. Eventually, it will turn out that p is a prime, but for now it is just some positive integer. Because 2^{p-1} and m are relatively prime,

$$\sigma(n) = \sigma(2^{p-1} m) = \sigma(2^{p-1})\sigma(m) = (2^p - 1)\sigma(m).$$

On the other hand, because n is perfect by hypothesis,

$$\sigma(n) = 2n = 2^p m.$$

Setting these expressions equal, we have

$$(2^p - 1)\sigma(m) = 2^p m.$$

Because $2^p - 1$ is odd, $\sigma(m)$ must be divisible by 2^p; i.e., $\sigma(m) = 2^p q$ for some q, so

$$(2^p - 1)2^p q = 2^p m \quad \text{or} \quad (2^p - 1)q = m.$$

This implies that q divides m but is not equal to m, which we will need in a moment. Also, multiplying and rearranging, we have

$$2^p q = m + q \quad \text{or} \quad \sigma(m) = m + q.$$

But $\sigma(m)$ is the sum of *all* the divisors of m. Certainly m and 1 are divisors, and we just showed that q is a divisor of m. The only solution is that $q = 1$, and because these are the only divisors of m, we conclude that m must be a prime. But, a few steps back, we showed that $m = (2^p - 1)q = 2^p - 1$. Because m is a prime, Exercise 2.1.9 says that p is a prime and $m = M_p$ is a Mersenne prime. So our n is of the form $n = 2^{p-1}M_p$, with M_p being a prime. $\square$

It should be pointed out that there is no analog of this theorem for amicable pairs. That is, Thabit ibn Qurra's algorithm is not the only way to find amicable pairs; they can take many different forms. For example, Paganini's pair $2^5 \cdot 37$, $2 \cdot 5 \cdot 11^2$, mentioned earlier, does not arise from Thabit ibn Qurra's algorithm.

Exercise 2.2.2. Use Euler's theorem and Eq. (1.1) to show that every even perfect number is a triangular number.

2.3. Structure

The functions $\tau(n)$ and $\sigma(n)$ are examples of functions defined on the set of natural numbers

$$\mathbb{N} = \{1, 2, 3, 4 \dots \}.$$

From a more modern viewpoint, it is often helpful to consider such objects more generally, and the relations between them. So we will define an ARITH-METICAL FUNCTION $f(n)$ as any function whose domain is the natural numbers. Sometimes it is more convenient not to specify the variable and to just write, for example, f. Some very simple arithmetical functions are

$$
\begin{aligned}
u(n) &= 1, &&\text{for all } n, \\
N(n) &= n, &&\text{for all } n, \\
e(n) &= \begin{cases} 1, & \text{if } n = 1, \\ 0, & \text{if } n > 1. \end{cases}
\end{aligned}
$$

So, u is a function which always gives 1, whereas N is a function that always returns the input value unchanged.

We can combine two arithmetical functions, f and g, with an operation called CONVOLUTION. The new function is denoted $f * g$ (pronounced "f splat g"):

$$f * g(n) = \sum_{d|n} f(d)g(n/d).$$

So, for example,

$$N * u(n) = \sum_{d|n} N(d)u(n/d) = \sum_{d|n} d \cdot 1 = \sigma(n),$$

or, omitting mention of the variables, $N * u = \sigma$.

Exercise 2.3.1. What well-loved function is $u * u$?

The divisors of an integer n come in pairs: For every d dividing n, the integer $c = n/d$ is another divisor. For that reason, we can write the definition of convolution in a more symmetric way,

$$f * g(n) = \sum_{\substack{c,d \\ c \cdot d = n}} f(d)g(c),$$

and from this it is clear that convolution is commutative, that is, that $f * g = g * f$. One can also show that convolution is associative. That is, if you have three functions, f, g, and h, then,

$$(f * g) * h(n) = \sum_{\substack{b,c,d \\ b \cdot c \cdot d = n}} f(d)g(c)h(b) = f * (g * h)(n).$$

Exercise 2.3.2. Convolution of functions is not ordinary multiplication, where multiplication by the number 1 leaves things unchanged. Convince yourself that that role is played here by the function e, that is, that $f * e = f$ for every function f.

The Möbius μ function (the Greek letter mu, pronounced "mew") has a more complicated definition. We define $\mu(1) = 1$, and if the input value $n > 1$ factors into primes as

$$n = p_1^{a_1} p_2^{a_2} \cdots p_k^{a_k},$$

then

$$\mu(n) = \begin{cases} 0, & \text{if any exponent } a_i > 1, \\ (-1)^k, & \text{if every exponent } a_i = 1. \end{cases}$$

So, $\mu(2) = \mu(3) = \mu(5) = -1$, whereas $\mu(4) = 0$ and $\mu(6) = 1$.

Exercise 2.3.3. Compute $\mu * u(n)$ for $n = 4, 6, 10, 15, 30, 60,$ and 210. You might begin to see a pattern. In your calculations for $n = 30$ and 210, single out one of the primes dividing n, say the largest one, and call it q. Group the divisors d of n according to whether or not the d is divisible by q, and compare values of the Möbius function. Can you prove your conjecture now?

The Möbius function is important despite its funny-looking definition, because of the following.

Theorem. *The convolution* $\mu * u = e$. *In other words,*

$$\sum_{d\mid n} \mu(d) = \begin{cases} 1, & \text{if } n = 1, \\ 0, & \text{if } n > 1. \end{cases}$$

Proof. Proving that $\mu * u(1) = 1$ is easy; the only divisor of 1 is $d = 1$ and $\mu(1) = 1$. Suppose that $n > 1$, and factor it again as $n = p_1^{a_1} p_2^{a_2} \ldots p_k^{a_k}$. The only divisors d that contribute nonzero terms to the sum are those made from products of distinct primes. As in the exercise above, single out one of the primes dividing n, the largest perhaps, and call it q. For every d that is not divisible by q, there is a unique divisor that is, namely dq. And $\mu(d) = -\mu(dq)$ because one has an even number of primes, the other an odd number. □

The preceding theorem looks like an isolated curiosity, but in fact it leads to a very useful, general result.

Theorem (Möbius Inversion). *Suppose that f and g are arithmetical functions. If*

$$f(n) = \sum_{d\mid n} g(d), \quad \text{then} \quad g(n) = \sum_{d\mid n} f(d)\mu(n/d),$$

and conversely.

Proof. In the notation of convolution, the theorem just claims that

$$f = g * u \Leftrightarrow g = f * \mu.$$

Be sure you believe this. To prove the $\Rightarrow$ half, suppose that $f = g * u$. Then, $f * \mu = (g * u) * \mu$. By associativity, this is $g * (u * \mu)$. By commutativity, this is equal to $g * (\mu * u)$. According to the previous theorem, this is $g * e$, which is g according to Exercise 2.3.2.

Exercise 2.3.4. Write down the $\Leftarrow$ half of the proof. That is, suppose that $g = f * \mu$. Show that $g * u = f$.

$\square$

From this theorem, we get identities that look complicated but are easy. We've already noticed that $\sigma = N * u$. So, by using Möbius inversion, $N = \sigma * \mu$. Written out, this says that

$$\sum_{d|n} \sigma(d)\mu(n/d) = n \quad \text{for all } n.$$

Exercise 2.3.5. Show that for all n,

$$\sum_{d|n} \tau(d)\mu(n/d) = 1.$$

Chapter 3

Order of Magnitude

In his *Introduction to Arithmetic*, Nicomachus wrote the following:

It comes about that even as fair and excellent things are few and easily numerated, while ugly and evil ones are widespread, so also the abundant and deficient numbers are found in great multitude and irregularly placed – for the method of their discovery is irregular – but the perfect numbers are easily enumerated and arranged with suitable order; for only one is found among the units, 6, only one among the tens, 28, and a third in the rank of the hundreds, 496 alone, and a fourth within the limits of the thousands, that is, below ten thousand, 8128.

Nicomachus is clearly implying that the nth perfect number has n digits. We already know this is wrong; we discovered in Chapter 2 that the fifth perfect number is 33550336. According to Dickson, 1999, Iamblichus in his *Commentary* on Nicomachus states this even more explicitly, and the mistake was subsequently repeated by Boethius in the fifth century. In the twelfth century, Abraham ben Meir ibn Ezra made the same claim in his commentary to the Pentateuch. In the fourteenth century, Thomas Bradwardine, mathematician and physicist, repeated the claim in his book *Arithmetica Speculativa*. Bradwardine became Archbishop of Canterbury but died shortly after of the Black Death in 1349.

Despite being wrong, this claim by Nicomachus is important because it is the very first of its kind. It examines the distribution of perfect numbers among all the integers. Because of Euler's theorem (see Chapter 2), we know that even perfect numbers correspond to prime Mersenne numbers. The question is, then, for a prime number p, how often is $M_p = 2^p - 1$ also prime?

Similarly, Mersenne's conjecture that there are infinitely many amicable pairs depends ultimately on relationships between prime numbers, as in Exercise 2.1.18. A much simpler question, still unsolved, is for a prime number p: how often is $2p + 1$ also a prime?

These are problems that require not algebra but analysis. We will shed some light on these particular questions in Sections 7.3 and 7.4 of Chapter 7.

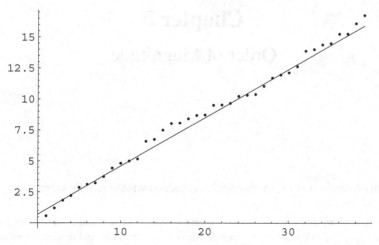

Figure 3.1. A log log plot of the first 39 perfect numbers.

But, for now, we will merely say that Nicomachus was not *completely* wrong in thinking that the perfect numbers are "arranged with suitable order." Figure 3.1 is a log log plot of the first known 39 perfect numbers (that is, the points in the plane $(1, \log(\log(6)))$, $(2, \log(\log(28)))$, $(3, \log(\log(496)))$, and so forth). The line $y = \log(2)\exp(-\gamma)x + \log(2)$ is also shown.

3.1. Landau Notation

Analytic number theory uses the techniques of calculus to answer such questions as how many, how big, and how often about the arithmetic functions we've considered so far. In order to do this, we need a language to compare functions.

For example, how does $7x^3 - 12x + 1$ compare to x^3? Generally, we will want to replace complicated functions with simpler ones. So, in comparing two functions, $f(x)$ and $h(x)$, we define a new relation, $\ll$ (pronounced "less than less than"). We say

$$f(x) \ll h(x) \quad \text{as} \quad x \to \infty$$

if we can find some constants C and x_0 so that

$$f(x) \le Ch(x) \quad \text{when} \quad x > x_0.$$

For example, we can say

$$7x^3 - 12x + 9 \ll x^3 \quad \text{as} \quad x \to \infty.$$

The constant $C = 8$ will work as

$$7x^3 - 12x + 9 \leq 8x^3 \quad \text{exactly when}$$
$$0 \leq x^3 + 12x - 9.$$

The cubic polynomial $x^3 + 12x - 9$ is not always positive but it is positive for $x = 1$. The derivative of $x^3 + 12x - 9$ is $3x^2 + 12$, which is always positive. So, $x^3 + 12x - 9$ is always increasing, and thus it is positive for $x \geq 1$. So, $x_0 = 1$ satisfies the definition in this case.

Here's a useful trick for dealing with inequalities: If $F(x)$ is an increasing function, then $a \leq b$ is true exactly when $F(a) \leq F(b)$ is true. This is just the definition of "increasing." So, we can show, for example, that

$$\exp(\sqrt{\log(x)}) \ll x \quad \text{as} \quad x \to \infty$$

will be true, because $\log(\ )$ and squaring are both increasing functions (calculus again). So,

$$
\begin{aligned}
\exp(\sqrt{\log(x)}) \leq x & \quad \Leftrightarrow \quad \text{(take logarithm)}, \\
\sqrt{\log(x)} \leq \log(x) & \quad \Leftrightarrow \quad \text{(square both sides)}, \\
\log(x) \leq (\log(x))^2 & \quad \Leftrightarrow \quad \text{(substitute } y = \log(x)\text{)}, \\
y \leq y^2, & \quad \quad \text{which is true for } y > 1 \text{ or } x > e.
\end{aligned}
$$

So, the relation is true with $C = 1$.

The point of $\ll$ notation is to simplify complicated expressions, and to suppress constants we don't care about. For that reason, there's no point in worrying about the smallest choice of C. Sometimes we will say just

$$\exp(\sqrt{\log(x)}) \ll x$$

when it is clear from the context that we mean as $x \to \infty$. Also, we don't have to call the variable x; for example, the Mersenne numbers $M_p = 2^p - 1$ satisfy

$$M_p \ll e^p.$$

The nth triangular number, $t_n = n(n+1)/2$, satisfies

$$t_n \ll n^2.$$

(You should check this.)

The preceding examples were done pretty carefully; we won't always include that much detail.

Exercise 3.1.1. Try to decide which of the following are true as $x \to \infty$:

$$2x + 1 \ll x; \qquad 10x + 100 \ll \exp(x); \qquad 2 + \sin(x) \ll 1;$$

$$\exp(-x) \ll \frac{1}{x}; \qquad \log(e^3 x) \ll x; \qquad \log(x) + 1 \ll \log(x).$$

Exercise 3.1.2. Show that $\sigma(n) = \sum_{d|n} d$ satisfies

$$\sigma(n) \ll n^2.$$

(Hint: Compare $\sigma(n)$ to the triangular numbers, t_n.)

Exercise 3.1.3. Show that the function $\tau(n) = \sum_{d|n} 1$ satisfies

$$\tau(n) \leq 2\sqrt{n} \quad \text{so} \quad \tau(n) \ll \sqrt{n}.$$

(Hint: For each divisor d of n, n/d is also a divisor of n.)

The preceding relation $\ll$ behaves like an ordering. It is reflexive: $f(x) \ll f(x)$ is always true. And it is transitive: If $f(x) \ll g(x)$ and $g(x) \ll h(x)$, then $f(x) \ll h(x)$. It is not symmetric: $f(x) \ll g(x)$ does not mean $g(x) \ll f(x)$. So, we need a new concept for two functions, $f(x)$ and $g(x)$, that are about the same size, up to some error term or fudge factor of size $h(x)$. We say

$$f(x) = g(x) + O(h(x)) \qquad \text{if} \qquad |f(x) - g(x)| \ll h(x).$$

This is pronounced "$f(x)$ is $g(x)$ plus Big Oh of $h(x)$." For example,

$$(x + 1)^2 = x^2 + O(x) \quad \text{as} \quad x \to \infty,$$

because

$$|(x + 1)^2 - x^2| = |2x + 1| \ll x.$$

For a fixed choice of $h(x)$, the relation $f(x) = g(x) + O(h(x))$ really is an equivalence relation between $f(x)$ and $g(x)$. That is, it is reflexive, symmetric, and transitive. If $f(x) - g(x)$ and $h(x)$ are both positive, we don't need the absolute values in the definition, and we will be able to ignore them.

Here is an example with an integer parameter n instead of x: The nth triangular number, t_n, satisfies

$$t_n = \frac{n^2}{2} + O(n),$$

because $t_n - n^2/2 = n(n+1)/2 - n^2/2 = n/2$.

Exercise 3.1.4. Show that as $x \to \infty$,

$$\frac{x}{x+1} = 1 + O\left(\frac{1}{x}\right),$$

$$\cosh(x) = \exp(x)/2 + O(\exp(-x)),$$

where the hyperbolic cosine function $\cosh(x)$ is $(\exp(x) + \exp(-x))/2$.

Exercise 3.1.5. Show that the sum of the squares of the first n integers is

$$\sum_{k=1}^{n} k^2 = \frac{n^3}{3} + O(n^2).$$

If we have a pair of functions that satisfies $f(x) \ll h(x)$, then from the definitions it is certainly true that $f(x) = 0 + O(h(x))$. Because adding 0 never changes anything, we might write

$$f(x) = O(h(x)) \quad \text{if} \quad f(x) \ll h(x).$$

Many books do this, but it can be confusing for beginners, because "$= O(\ )$" is not an equivalence.

3.2. Harmonic Numbers

In Exercises 3.1.3, 3.1.2, and 3.1.5 you proved some simple estimates of the functions $\tau(n)$ and $\sigma(n)$. We next consider the Harmonic numbers, H_n, which seem less connected to number theory. Their definition requires only division, not divisibility. Nonetheless, the estimate we will make in this section is fundamental.

Lemma. *For all $n > 1$,*

$$H_n - 1 < \log(n) < H_{n-1}. \tag{3.1}$$

Proof. The basic idea is geometric. We know that $\log(n)$ in calculus is a definite integral,

$$\log(n) = \int_1^n \frac{1}{x} \, dx,$$

so it is the area under the curve $y = 1/x$ between $x = 1$ and $x = n$. First, we show that $H_n - 1 < \log(n)$. We know that $H_n - 1 = 1/2 + 1/3 + \cdots + 1/n$, and that the $n - 1$ rectangles with width 1 and heights $1/2, 1/3, \ldots, 1/n$ have total area $H_n - 1$. The diagram on the top in Figure 3.2 shows the example

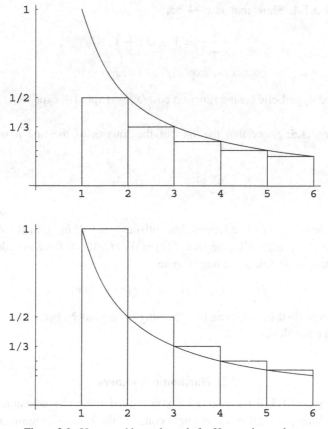

Figure 3.2. Upper and lower bounds for Harmonic numbers.

of $n = 6$. The horizontal and vertical scales are not the same. Because all the rectangles fit below $y = 1/x$, the area of the rectangles is less than the area under the curve, so $H_n - 1 < \log(n)$. The other inequality is just as easy. We know that $H_{n-1} = 1 + 1/2 + \cdots + 1/(n-1)$ and that the $n-1$ rectangles with width 1 and heights 1, $1/2, \ldots, 1/(n-1)$ have total area H_{n-1}. The case of $n = 6$ is on the bottom in Figure 3.2. Now, the curve fits under the rectangles instead of the other way around, so $\log(n) < H_{n-1}$. □

In Big Oh notation, this says

Lemma.

$$H_n = \log(n) + O(1). \tag{3.2}$$

Proof. This is easy. Since $H_{n-1} < H_n$, we have from (3.1)

$$H_n - 1 < \log(n) < H_n.$$

Subtract H_n from both sides, then multiply by -1 to get

$$0 < H_n - \log(n) < 1.$$

$\square$

Exercise 3.2.1. Use this proof to show that

$$\log(n) < H_n < \log(n) + 1.$$

So, the Harmonic number, H_n, is about the same size as $\log(n)$. In fact, not only is the difference between them bounded in size, it actually has a limiting value.

Theorem. *There is a real number* γ, *called Euler's constant, such that*

$$H_n = \log(n) + \gamma + O(1/n). \tag{3.3}$$

Euler's constant γ is about $0.57721566490153286061\ldots$. This is the most important number that you've never heard of before.

Proof. Consider again the bottom of Figure 3.2, which shows that $\log(n) < H_{n-1}$. The difference between H_{n-1} and $\log(n)$ is the area above $y = 1/x$ and below all the rectangles. This is the shaded region shown on the top in Figure 3.3. For each n, let E_n (E for error) denote the area of this region; so, numerically, $E_n = H_{n-1} - \log(n)$. On the bottom in Figure 3.3, we've moved all the pieces horizontally to the left, which does not change the area. Because they all fit into the rectangle of height 1 and width 1, we see that $E_n \leq 1$. Be sure that you believe this; we are going to use this trick a lot.

Because this is true for every n, infinitely many times, we see that the area of *all infinitely many* pieces is some finite number less than 1, which we will denote γ.

Now that we're sure γ exists, consider $\gamma - (H_{n-1} - \log(n))$. This is just $\gamma - E_n$, the total area of all except the first n of the pieces. The first n fit into the rectangle between height 1 and height $1/n$. This is just the bottom of Figure 3.3 again. So all the rest fit into a rectangle between height $1/n$ and 0, which has area $1/n$. This means that

$$0 < \gamma - (H_{n-1} - \log(n)) < 1/n.$$

Multiply by -1 to reverse all the inequalities:

$$-1/n < H_{n-1} - \log(n) - \gamma < 0.$$

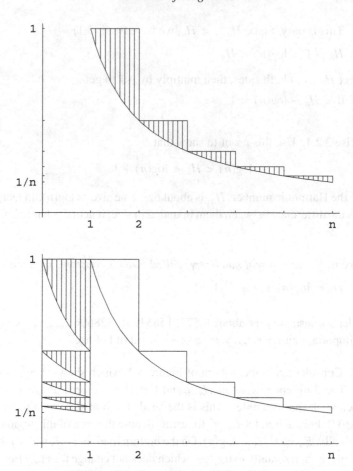

Figure 3.3. Geometric proof of eq. (3.3).

Add $1/n$ to both inequalities to see that

$$0 < H_n - \log(n) - \gamma < 1/n.$$

This is implies that

$$H_n = \log(n) + \gamma + O(1/n).$$

$\square$

We actually proved something a little stronger than the statement of the theorem: The error is actually less than $1/n$, not just a constant times $1/n$.

Table 3.1 shows the Harmonic numbers for some multiples of 10. Observe that even though the numbers are small in decimal notation, as an exact

Table 3.1. *Harmonic Numbers*

n	H_n	$\log(n) + \gamma$
10	$\dfrac{7381}{2520} = 2.92897\ldots$	$2.8798\ldots$
20	$\dfrac{55835135}{15519504} = 3.59774\ldots$	$3.57295\ldots$
30	$\dfrac{9304682830147}{2329089562800} = 3.99499\ldots$	$3.97841\ldots$
40	$\dfrac{2078178381193813}{485721041551200} = 4.27854\ldots$	$4.2661\ldots$
50	$\dfrac{13943237577224054960759}{3099044504245996706400} = 4.49921\ldots$	$4.48924\ldots$
60	$\dfrac{15117092380124150817026911}{3230237388259077233637600} = 4.67987\ldots$	$4.67156\ldots$
70	$\dfrac{42535343474848157886823113473}{8801320137209899102584580800} = 4.83284\ldots$	$4.82571\ldots$
80	$\dfrac{4880292608058024066886120358155997}{982844219842241906412811281988800} = 4.96548\ldots$	$4.95924\ldots$
90	$\dfrac{36531827789907675893960153728753282855861}{7187667549454894553044722570650750294400} = 5.08257\ldots$	$5.07703\ldots$
100	$\dfrac{14466636279520351160221518043104131447711}{2788815009188499086581352357412492142272} = 5.18738\ldots$	$5.18239\ldots$

fraction they involve a very large number of digits. As expected, the difference between H_{10} and $\log(10) + \gamma$ is less than 0.1; the difference between H_{100} and $\log(100) + \gamma$ is less than 0.01.

Exercise 3.2.2. H_{1000} is a fraction whose numerator is 433 digits long and whose denominator is 432 digits. Use the theorem to estimate H_{1000}, accurate to three decimal places.

H_{10000} has a numerator of 4345 digits and a denominator of 4344 digits. Use the theorem to estimate H_{10000}, accurate to four decimal places. Estimate H_{100000}, accurate to five decimal places.

This is somewhat paradoxical: The *larger* n is, the *better* approximation to H_n we get.

Exercise 3.2.3. Table 3.2 compares Harmonic numbers for some very large n to a function more complicated than $\log(n) + \gamma$. Examine the data and try to

Table 3.2. *Numerical Evidence for a Conjecture on*
Harmonic Numbers

n	H_n	$\log(n) + \gamma + 1/(2n)$
10	2.9289682539682539683	2.9298007578955785446
10^2	5.1873775176396202608	5.1873858508896242286
10^3	7.4854708605503449127	7.4854709438836699127
10^4	9.7876060360443822642	9.7876060368777155967
10^5	12.090146129863427947	12.090146129871761281

make a conjecture. Give yourself extra credit if you can state your conjectures
in Big Oh notation.

Exercise 3.2.4. Table 3.3 compares harmonic numbers for some very large
n to a function still more complicated. Examine the data and try to make a
conjecture in Big Oh notation.

3.3. Factorials

For a little more practice with Big Oh notation, we will try to get an estimate
for $n!$. The basic idea is the same as before: comparing a sum to an integral.
To get a sum from $n!$, use logarithms:

$$\log(n!) = \sum_{k=1}^{n} \log(k).$$

The relevant integral for comparison purposes will be

$$\int_{1}^{n} \log(x)dx = (x\log(x) - x)\big|_1^n = n\log(n) - n + 1.$$

(This is integration by parts: $u = \log(x)$, $dv = dx$, etc.)

Table 3.3. *Numerical Evidence for a Stronger Conjecture on*
Harmonic Numbers

n	H_n	$\log(n) + \gamma + 1/(2n) - 1/(12n^2)$
10	2.9289682539682539683	2.9289674245622452113
10^2	5.1873775176396202608	5.1873775175562908953
10^3	7.4854708605503449127	7.4854708605503365793
10^4	9.7876060360443822642	9.7876060360443822633
10^5	12.090146129863427947	12.090146129863427947

Lemma.

$$\log(n!) = n \log(n) - n + O(\log(n)). \tag{3.4}$$

Proof. Suppose k is an integer and x satisfies

$$k - 1 \le x \le k,$$

then

$$\log(k - 1) \le \log(x) \le \log(k),$$

because log is an increasing function. We can integrate between $x = k - 1$ and $x = k$:

$$\int_{k-1}^{k} \log(k - 1)dx \le \int_{k-1}^{k} \log(x)dx \le \int_{k-1}^{k} \log(k)dx. \tag{3.5}$$

The first and last integrals are constant in x, so this says

$$\log(k - 1) \le \int_{k-1}^{k} \log(x)dx \le \log(k).$$

Multiply by -1 to reverse all inequalities and add $\log(k)$ to each term to get

$$0 \le \log(k) - \int_{k-1}^{k} \log(x)dx \le \log(k) - \log(k - 1). \tag{3.6}$$

View this as $n - 1$ inequalities, with $k = 2, 3, \ldots, n$, and add them together. The sum of integrals combines as

$$\sum_{k=2}^{n} \int_{k-1}^{k} \log(x)dx =$$

$$\int_{1}^{2} \log(x)dx + \int_{2}^{3} \log(x)dx + \cdots + \int_{n-1}^{n} \log(x)dx =$$

$$\int_{1}^{n} \log(x)dx,$$

whereas the last sum on the right side of (3.6) "telescopes" down to

$$\sum_{k=2}^{n} (\log(k) - \log(k - 1)) =$$

$$(\log(2) - \log(1)) + (\log(3) - \log(2)) + \cdots + (\log(n) - \log(n - 1)) =$$
$$\log(n) - \log(1) = \log(n).$$

We get

$$0 \le \sum_{k=2}^{n} \log(k) - \int_{1}^{n} \log(x)dx \le \log(n).$$

We have already calculated the integral above, so

$$0 \le \log(n!) - (n \log(n) - n) \le \log(n) + 1 \ll \log(n).$$

$\square$

Exercise 3.3.1. We wrote out all the inequalities for that lemma, just for the practice. In fact, the proof is much easier than we made it seem. Give a geometric proof of the lemma, analogous to the way we proved (3.3). The relevant diagrams are in Figure 3.4. You will still need to know $\int_{1}^{n} \log(x)dx = n \log(n) - n + 1$.

Exercise 3.3.2. Use (3.4) to show that

$$n! \ll n \left(\frac{n}{e}\right)^{n}.$$

This looks like we replaced the simple expression $n!$ with a more complicated one, contrary to our philosophy of what $\ll$ is good for. The point is that even though $n!$ looks simple, it is defined recursively. To understand what this means, compute 20! and $20(20/e)^{20}$. Later, we will get a better estimate.

3.4. Estimates for Sums of Divisors

In Chapter 2, we introduced the divisor function, $\tau(n)$, which counts the number of divisors, the sigma function, $\sigma(n)$, which sums the divisors, and $s(n)$, which is the sum of the proper divisors, that is, those less than n. How big can these functions be? In the classification of integers as deficient, perfect, or abundant, how abundant can an integer be? Exercises 3.1.3 and 3.1.2 proved the estimates

$$\tau(n) \ll n^{1/2} \quad \text{and} \quad \sigma(n) \ll n^{2}.$$

In this section and the next we will get estimates that are better, that is, closer to the true size of these functions.

Theorem. $\sigma(n)$ is $\ll n \log(n)$.

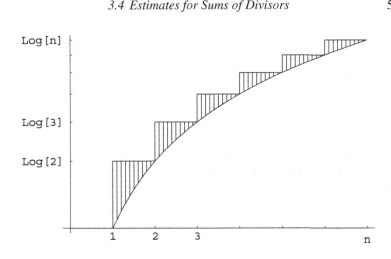

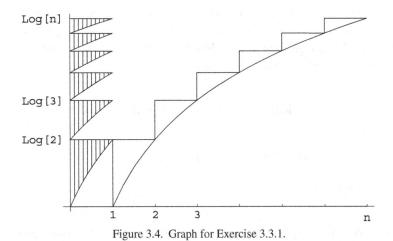

Figure 3.4. Graph for Exercise 3.3.1.

Proof. In fact, we will show that

$$\sigma(n) \le n\log(n) + n \quad \text{for all } n. \tag{3.7}$$

Exercise 3.4.1. Show that $n\log(n) + n \ll n\log(n)$.

To prove (3.7) we will use the same method that we used in Exercise 3.1.3, that is, that the divisors of n come in pairs. Whenever d divides n, so does n/d. So,

$$\sigma(n) = \sum_{d|n} d = \sum_{d|n} \frac{n}{d}.$$

The second sum above is the same as the first, but with the terms written in a different order. If you're not convinced, write out both explicitly for some small n, such as $n = 12$. We can now write

$$\frac{\sigma(n)}{n} = \sum_{d|n} \frac{1}{d} \leq \sum_{d=1}^{n} \frac{1}{d} = H_n \leq \log(n) + 1.$$

The second sum above includes *all* integers $d \leq n$, not just those that divide n, so it is bigger. The last inequality comes from Exercise 3.2.1. Multiply both sides by n to get (3.7). $\square$

Exercise 3.4.2. From the theorem you can deduce an estimate for $s(n)$. What is it?

Just as $\log(n)$ is much smaller than n, so $n \log(n)$ is much smaller than n^2. Both Exercise 3.1.2 and the theorem used the fact that the divisors of n are a subset of all integers less than n. But the theorem used the deeper relation between Harmonic numbers and logarithms.

3.5. Estimates for the Number of Divisors

Theorem. *The divisor function is bounded by*

$$\tau(n) \ll n^{1/3}.$$

Proof. Here it is helpful to write $n = p_1^{t_1} \dots p_k^{t_k}$ as a product of prime powers and to use the fact that $\tau(n)$ is multiplicative. Consider first the case of powers of a fixed prime: $n = p^t$. How does $\tau(p^t) = t + 1$ compare to $p^{t/3}$? It should be smaller as t increases, because $p^{t/3} = \exp(t \log(p)/3)$ grows exponentially in t. So,

$$t + 1 \ll t \quad \text{and } t \ll \exp(t \log(p)/3), \quad \text{so } t + 1 \ll \exp(t \log(p)/3).$$

We should be able to multiply these inequalities for various p to get the result for general n, right? Well, not exactly. The $\ll$ notation hides a constant C, and we have to worry about how fast C grows with p. For example, if $C = p$, our bound has exponent $4/3$, not $1/3$.

To get around this, suppose first that $p > e^3 = 20.855\dots$ is fixed, so $\log(p) > 3$. How does $t + 1$ compare to $\exp(t \log(p)/3)$? The two are both equal to 1 when $t = 0$. To see which grows faster, compare derivatives at $t = 0$:

$$\frac{d}{dt} (t + 1)|_{t=0} = 1,$$

$$\frac{d}{dt} (\exp(t \log(p)/3))|_{t=0} = \left(\frac{\log(p)}{3} \exp(t \log(p)/3) \right)\Big|_{t=0},$$

$$= \frac{\log(p)}{3} > 1.$$

The exponential is already increasing faster at $t = 0$, so

$$\tau(p^t) = t + 1 \leq p^{t/3} \quad \text{for all } t,$$

and for all primes $p \geq 23$. So, $\tau(n) \leq n^{1/3}$ as long as n is divisible only by primes $p \geq 23$.

This still leaves primes $p = 2, 3, 5, \ldots, 19$. For each of these primes, we determine by calculus that the function $(t + 1)p^{-t/3}$ has a maximum at $t = 3/\log(p) - 1$; the maximum value is some constant $C(p)$. (These graphs are shown in Figure 3.5.) So,

$$t + 1 \leq C(p)p^{t/3} \quad \text{for all } t, \quad \text{for } p = 2, 3, \ldots, 19.$$

Set $C(p) = 1$ for $p > 19$ and let C be the product of all the constants $C(p)$. Now, we can safely multiply the inequalities: For n, which factors

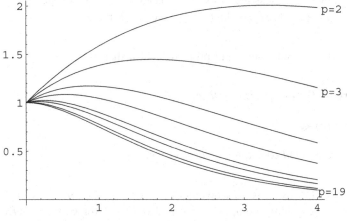

Figure 3.5. $(t + 1)p^{-t/3}$ for $p = 2, 3, 5, \ldots, 19$.

as $n = \prod_i p_i^{t_i}$, we see

$$\tau\left(\prod_i p_i^{t_i}\right) = \prod_i (t_i + 1) \le \prod_i C(p_i) p_i^{t_i/3} \le C \prod_i p_i^{t_i/3} = Cn^{1/3}.$$

$\square$

Exercise 3.5.1. Verify my calculation that the maximum of the function $(t+1)p^{-t/3} = (t+1)\exp(-t\log(p)/3)$ occurs at $t = 3/\log(p) - 1$. For the eight primes $p = 2, 3, 5, \ldots, 19$, plug this t value back into find the constant $C(p)$. Multiply them together to show that $C \le 4$. So,

$$\tau(n) \le 4n^{1/3} \quad \text{for all } n.$$

Exercise 3.5.2. The improvement in the bound on $\tau(n)$, from exponent $1/2$ to exponent $1/3$, is much less than we got for $\sigma(n)$ in the previous theorem. On the other hand, there was nothing special about the exponent $1/3$. If we want to prove

$$\tau(n) \ll n^{1/4},$$

how does the proof need to be modified? That is, how big does a prime number p have to be so that

$$\tau(p^t) = t + 1 \le p^{t/4} \quad \text{for all } t \ge 0?$$

How many primes less than this bound need to be treated separately? Does the function $(t+1)\exp(-t\log(p)/4)$ have a maximum?

In fact, this same method of proof will show that

Theorem. *For any positive integer k,*

$$\tau(n) \ll n^{1/k}$$

where the implied constant C depends on k.

We are starting to see the advantage of the $\ll$ notation. It was a painful exercise to explicitly compute the fact that $C \le 4$ for exponent $1/3$. You certainly don't want to compute the constant for exponent $1/4$.

3.6. Very Abundant Numbers

We have improved the bound on the sigma function, first getting $\sigma(n) \ll n^2$, then $\sigma(n) \ll n \log(n)$. Perhaps we can do better. Is it true that $\sigma(n) \ll n$? In fact, it is *not* true; we can prove the opposite.

Before we do this, we should think about negating the definition of $\ll$. In slightly more formal language than we used in Section 3.1, $f(n) \ll h(n)$ if there exists a pair of constants C and N_0 such that for all $n \geq N_0$, we have $f(n) \leq Ch(n)$. The shorthand symbols $\exists$, "there exists," and $\forall$, "for all," are useful:

$$\exists \, C, N_0 \quad \text{such that} \quad \forall \, n > N_0, \quad f(n) \leq Ch(n).$$

The algebra of negation is easy. We apply the "not" to the whole expression by moving it left to right, negating each portion in turn. The negation of $\exists$ is $\forall$ and vice versa. For example, the negation of the statement that "there exists a word that rhymes with orange" is the statement that "every word does not rhyme with orange." The negation of $f(n) \ll h(n)$ is that

$$\forall \, C, N_0, \quad \exists \, n > N_0 \quad \text{such that} \quad f(n) > Ch(n).$$

We can rephrase this to simplify it. The function $f(n)$ is not $\ll h(n)$ if for all C there exist infinitely many n such that $f(n) > Ch(n)$. (Think about why this is the same.)

Theorem. *The function $\sigma(n)$ is not $\ll n$. That is, for every constant C, there are infinitely many integers n such that $\sigma(n) > Cn$.*

Proof. Given any constant C, we need to produce infinitely many integers n such that $\sigma(n) > Cn$. We let N be any integer larger than $\exp(C)$, so $\log(N) > C$. We can choose $n = N!$ so the integers $d = 1, 2, 3, \ldots, N$ are a subset of all the divisors of n. So,

$$\frac{\sigma(n)}{n} = \sum_{d \mid n} \frac{1}{d} \geq \sum_{d=1}^{N} \frac{1}{d} = H_N > \log(N) > C.$$

The first equality comes from the proof of (3.7), whereas the first $\geq$ comes from the preceding remark. We know that $H_N > \log(N)$ from Exercise 3.2.1, whereas we know that $\log(N) > C$ from the way we chose N. $\square$

Exercise 3.6.1. The theorem for $\sigma(n)$ implies a corresponding result for $s(n)$. What is it? For a given constant C, if $\sigma(n) > Cn$, what inequality holds for $s(n)$?

Exercise 3.6.2. The inequalities in this proof are not very "sharp." That is, typically a factorial $N!$ has many more divisors than just $1, 2, 3, \ldots, N$. So,

$$\sum_{d|N!} \frac{1}{d} \quad \text{is much larger than} \quad \sum_{d=1}^{N} \frac{1}{d}.$$

As a result, our counterexample n is much larger than it needs to be. To see this, let $C = 2$ and compute explicitly the smallest N and n of the theorem. This integer will be abundant, but it is certainly not the first, as $s(12) = 16$. Repeat with $C = 3$ to find an integer n with $s(n) > 2n$. (You might not be able to compute the factorial without a computer.) The smallest example is $s(180) = 366$. On the other hand, a computer search doesn't turn up any n with $s(n) > 10n$. But the theorem tells me that

$$s(n) > 10n \quad \text{for } n = 59875! \approx 9.5830531968 \times 10^{260036}.$$

The theorem above implies that factorial integers $N!$ tend to be very abundant. In Exercise 2.1.14, you may have conjectured that the odd integers were all deficient. Jordanus de Nemore claimed to have actually proved this in 1236. In fact, a generalization of the preceding argument will show the opposite. But factorials tend to be very even, so we'll fix this by introducing the double factorial:

$$n!! = n \cdot n - 2 \cdot n - 4 \cdot n - 6 \ldots.$$

So, for example, $5!! = 5 \cdot 3 \cdot 1 = 15$. Don't confuse this function with an iterated factorial $(n!)!$, which would be the factorial of $n!$. If n is odd, so is $n!!$.

Exercise 3.6.3. For $n = 7, 9, 11$, and 13, compute $n!!$ and $\sigma(n!!)$. Use this to get $s(n!!)$.

With this new function, we can prove

Theorem. *The odd integers can also be as abundant as we like. That is, for any constant C, there are infinitely many odd integers n such that $\sigma(n) > Cn$.*

Proof. This will be similar to the previous theorem. We need to make $\sigma(n) > Cn$. Given a constant C, we pick N to be any integer larger than $\exp(2C)$, so $\log(N)/2 > C$. We pick $n = 2N + 1!!$. Then, as before,

$$\frac{\sigma(2N + 1!!)}{2N + 1!!} = \sum_{\substack{d|2N+1!!}} \frac{1}{d} > \sum_{\substack{d \leq 2N+1 \\ d \text{ odd}}} \frac{1}{d}.$$

That is, the odd integers below $2N + 1$ are a subset of all the divisors of $2N + 1!!$. How can we relate this awkward sum to a Harmonic number? Well,

$$\sum_{\substack{d \leq 2N+1 \\ d \text{ odd}}} \frac{1}{d} = \sum_{d=1}^{2N+1} \frac{1}{d} - \sum_{d=1}^{N} \frac{1}{2d}.$$

That is, we take the sum of all the integers and subtract the sum of the even ones. These are exactly the integers of the form $2d$, for all $d \leq N$. (If you have doubts, compare the two sides explicitly for the case of $N = 4$.) This is, by definition and according to Exercise 3.2.1, equal to

$$H_{2N+1} - \frac{1}{2} H_N > \log(2N + 1) - \frac{1}{2}(\log(N) + 1).$$

(Since H_N is being subtracted, we need to replace it with something larger.) This is greater than

$$\log(2N) - \frac{1}{2}(\log(N) + 1),$$

which is equal to

$$\log(2) + \log(N) - \frac{1}{2}\log(N) - \frac{1}{2} > \frac{\log(N)}{2} > C,$$

because $\log(2) - 1/2 > 0$, and because of the way N was chosen. So, for $n = 2N + 1!!$, we have $\sigma(n) > Cn$, or $s(n) > (C - 1)n$. $\qquad \square$

Exercise 3.6.4. With $C = 2$, compute the smallest $2N + 1$ of the theorem, so that its double factorial is odd and abundant. Don't try to calculate $2N + 1!!$ without a computer; it is about $3.85399 \cdots \times 10^{90}$. Compare this with your answer to Exercise 3.6.3.

Charles de Bouvelles, in 1509, was the first person to find an odd abundant number, namely, $45045 = 5 \cdot 7 \cdot 9 \cdot 11 \cdot 13 = 13!!/3$. The smallest odd abundant number, $945 = 9!!$, was first noticed by Bachet about a century later.

3.7. Highly Composite Numbers

In this section, we will think about analogous results for the divisor function $\tau(n)$. Can we find examples of integers n with "lots" of divisors? We already know that $\tau(n) \ll n^{1/k}$ for any k. So, we might try to show that $\tau(n)$ can sometimes be bigger than $\log(n)$ or powers of $\log(n)$. We will approach this using some lemmas and exercises.

Lemma. $\tau(n)$ *is not* $\ll 1$.

Proof. This is not hard. From our formula for $\tau(n)$, we know that if we take $n = 2^m$ to be any power of 2, then

$$\tau(n) = m + 1.$$

It is clear, then, that for this sequence of powers of 2, the divisor function is unbounded; i.e., $\tau(n)$ is not $\ll 1$. To help us generalize the lemma later, we will explicitly relate the size of n to $\tau(n)$ when $n = 2^m$. Then, $\log(n) = m \log(2)$, or $m = \log(n)/\log(2)$. So,

$$\tau(n) = \frac{\log(n)}{\log(2)} + 1.$$

In words, when n is a power of 2, $\tau(n)$ is about the size of $\log(n)$.   □

Exercise 3.7.1. The lemma says that for any choice of constant C, there exist infinitely many integers with $\tau(n) > C$. Make this explicit for $C = 100$; that is, find an infinite set of integers with $\tau(n) > 100$.

Exercise 3.7.2. What, if anything, is special about the number 2 in the lemma?

Lemma. $\tau(n)$ *is not* $\ll \log(n)$.

Proof. This is similar to the lemma above; instead of taking powers of a single prime, we consider n to be of the form $2^m \cdot 3^m$; then,

$$\tau(n) = (m + 1)^2.$$

Now, $n = 6^m$, so $m = \log(n)/\log(6)$ and $\tau(n) = (\log(n)/\log(6) + 1)^2$. We want to show that this function is not $\ll \log(n)$. To simplify, change the variables with $x = \log(n)$. We must show that for any C, the inequality

$$\left(\frac{x}{\log(6)} + 1\right)^2 \leq Cx$$

is eventually false. Multiplying this out, we get the equivalent inequality:

$$\frac{x^2}{\log(6)^2} + \left(\frac{2}{\log(6)} - C\right)x + 1 \leq 0.$$

The function on the left is a polynomial with positive lead term; not only is it eventually positive, it goes to infinity.   □

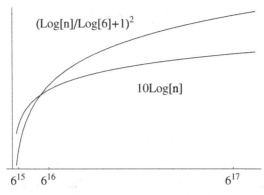

Figure 3.6. Graph for Exercise 3.7.3.

Exercise 3.7.3. The lemma says that for any choice of constant C, there exist infinitely many integers with $\tau(n) > C \log(n)$. Make this explicit for $C = 10$; that is, find an infinite set of integers with $\tau(n) > 10 \log(n)$. You need to think about the proof of the lemma and refer to Figure 3.6. What choices of m work?

Exercise 3.7.4. Imitate the previous two lemmas to show that

$$\tau(n) \text{ is not } \ll (\log(n))^2.$$

Which integer might you consider powers of?

This same method of proof works for any exponent k.

Theorem. *No matter how big k is,*

$$\tau(n) \text{ is never } \ll (\log(n))^k.$$

Proof. Write out the details of this if you like. □

Chapter 4

Averages

So far, we've seen that $\tau(n)$ is "less than less than" any root of n but sometimes bigger than any power of $\log(n)$. Part of the difficulty is that $\tau(n)$ is very irregular. For example, $\tau(p) = 2$ for any prime p, whereas $\tau(2^m) = m + 1$. So, the function jumps from 2 at $n = 127$ to 8 at $n = 128$. Figure 4.1 show a plot of data points $(\log(n), \tau(n))$ for all n below 1000. For consecutive integers, there does not seem to be much correlation between one value of the divisor function and the next. (The appearance of horizontal lines is an artifact; $\tau(n)$ takes on only integer values.)

One way of smoothing out random fluctuations is by averaging. If you go bowling or play golf, your score changes from game to game, but your average changes more slowly. In this chapter, we will take another look at the size of arithmetic functions by forming averages.

We will need a little more terminology. We will say a function $F(n)$ is ASYMPTOTIC to $G(n)$ as n goes to infinity if the limit of $F(n)/G(n)$ is 1. We write this as $F(n) \sim G(n)$. If you are worried that you don't know enough about limits, you can look ahead to Section I1.2, where we talk about limits more carefully.

For example, recall (3.3), which said that

$$H_n = \log(n) + \gamma + O(1/n).$$

Subtract the $\log(n)$ and multiply by γ^{-1} to get

$$\frac{H_n - \log(n)}{\gamma} = 1 + O(1/n).$$

(The γ^{-1} is absorbed by the implicit constant in the definition of Big Oh.) This says that

$$H_n - \log(n) \sim \gamma,$$

because the sequence $1/n \to 0$.

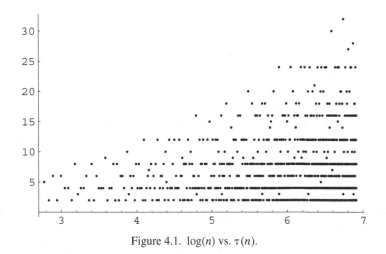

Figure 4.1. $\log(n)$ vs. $\tau(n)$.

Exercise 4.0.1. It is also true that $H_n \sim \log(n)$. Show this.

Exercise 4.0.2. Show that the nth triangular number satisfies

$$t_n \sim \frac{n^2}{2}.$$

Exercise 4.0.3. Use (3.4) to show that $\log(n!) \sim n\log(n)$ or, equivalently, $\log(n!)/\log(n) \sim n$. (If necessary, convert (3.4) to $\ll$ notation.) Try to interpret this geometrically, by plotting pairs of points $(n, \log(n!)/\log(n))$ for some small n.

Typically the arithmetic functions we like, such as $\tau(n)$, are too complicated to be asymptotic to any simple function. That is where the idea of averaging comes in. Starting with a complicated function, $f(n)$, we seek a simpler function, $g(n)$, such that the sums of the first n values of f and g are asymptotic as $n \to \infty$:

$$f(1) + f(2) + \cdots + f(n) \sim g(1) + g(2) + \cdots + g(n). \qquad (4.1)$$

Notice that the role of the variable n here is not simply to be plugged into the function f or g, but rather to tell us how many terms to take in the sum. Equation (4.1) is true if and only if the two sides, when divided by n, are asymptotic; so, it really is a statement about averages.

4.1. Divisor Averages

To make a conjecture about the average order of $\tau(n)$, you will do some calculations that will give an idea of what is likely to be true. The point of view is geometric. Consider, for example, computing $\tau(8)$. For every divisor d of 8, we know that $c = 8/d$ is another divisor of 8, so we have a pair of integers (c, d) such that $cd = 8$. So, we have a point on the hyperbola $xy = 8$ with *integer* coordinates. Any point in the plane with integer coordinates is called a LATTICE POINT; a divisor of 8 gives a lattice point that lies on the hyperbola $xy = 8$.

Exercise 4.1.1. Compute $\tau(8)$, and also identify the lattice points corresponding to the divisors of 8 in Figure 4.2. Repeat this for $k = 6, 4, 7, 5, 3, 2, 1$. That is, compute $\tau(k)$, identify the lattice points (c, d) corresponding to each of the divisors of k, and draw in the hyperbola $xy = k$ on Figure 4.2. Make sure it goes through the relevant lattice points.

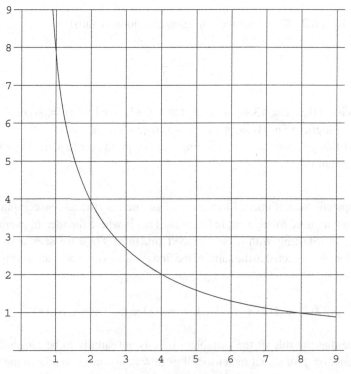

Figure 4.2. Graph for exercise 4.4.1.

Observe that every lattice point under the hyperbola $xy = 8$ is on one of the hyperbolas $xy = k$ with $k \leq 8$, except for those on the coordinate axes $x = 0$ and $y = 0$. Count these lattice points.

Exercise 4.1.2. Now, compute

$$\sum_{k=1}^{8} \tau(k) = \tau(1) + \tau(2) + \cdots + \tau(8)$$

and compare that answer to your answer to the previous exercise.

Exercise 4.1.3. There is nothing special about $n = 8$ in the previous exercise. How do you think that $\sum_{k=1}^{n} \tau(k)$ relates to the number of lattice points under the hyperbola $xy = n$?

Exercise 4.1.4. Each of these lattice points is the upper left hand corner of a square of area 1; so, the number of lattice points is the area of the region the squares cover. Identify these squares in the example of $n = 8$ of Figure 4.2.

We expect that the area of the region should be "approximately" the area under the hyperbola $y = n/x$ between $x = 1$ and $x = n$. So,

$$\sum_{k=1}^{n} \tau(k) \quad \text{is about size} \quad \int_{1}^{n} \frac{n}{x} \, dx.$$

Exercise 4.1.5. Compute this integral. Treat n as a constant; x is the variable.

We say that a function $f(n)$ has AVERAGE ORDER $g(n)$ if (4.1) is true.

Exercise 4.1.6. Exercise 4.0.3 showed that $\log(n!) \sim n \log(n)$. Use this to make a conjecture for the average order of $\tau(n)$.

We arranged the definition of average order so that it is an equivalence relation. In particular, it is true that $f(n)$ has average order $f(n)$. It is also true that if $f(n)$ has average order $g(n)$, then $g(n)$ has average order $f(n)$. However, it emphatically does not say that $g(n)$ is the average of $f(n)$. It would be more appropriate to say that $f(n)$ and $g(n)$ have the same average, but we are stuck with this terminology. An example will help clarify. Remember the function $N(n)$ from Section 2.3: $N(n) = n$ for all positive integers n (a very simple function). We know a closed-form expression for the sum of the first *n* values,

$$N(1) + N(2) + \cdots + N(n) = t_n = \frac{n^2}{2} + O(n) \sim \frac{n^2}{2};$$

so, the average of the first *n* values is

$$\frac{1}{n}(N(1) + N(2) + \cdots + N(n)) = \frac{t_n}{n} = \frac{n}{2} + O(1) \sim \frac{n}{2}.$$

The average of the first *n* values of N is $N(n)/2$. We don't see this paradox with the function $\log(n)$. It grows more slowly, so it is closer to being a constant function. So, $\log(n)$ actually is asymptotic to the average of the first *n* values of log; that is what Exercise 4.0.3 says.

We will now prove the conjecture you made previously. In fact, we can prove something a little better.

Theorem. *The sum of the first n values of the divisor function is*

$$\sum_{k=1}^{n} \tau(k) = n \log(n) + O(n). \tag{4.2}$$

Proof. The idea of the proof is exactly the same as the preceding exercises; we just need to be more precise about what "approximately" means in Exercise 4.1.3. To summarize, the divisors d of any $k \leq n$ are exactly the lattice points $(c = k/d, d)$ on the hyperbola $xy = k$. And, conversely, every lattice point (c, d) under $xy = n$ lies on the hyperbola $xy = k$ with $k = cd \leq n$, so it corresponds to a divisor d of k.

So, $\sum_{k=1}^{n} \tau(k)$ is the number of lattice points on and under the hyperbola $xy = n$. Each lattice point is the upper-left corner of a square of area 1. Meanwhile, the area under the hyperbola $xy = n$ between $x = 1$ and $x = n$ is $n \log(n)$, by integration. We need to see that the area covered by the squares differs from the area under the hyperbola by some constant C times n. In fact, $C = 1$ will do.

Recall how we estimated Harmonic numbers and factorials geometrically in Chapter 3. Figure 4.3 shows a typical example, $n = 8$. On the left are shown all the lattice points on and under $xy = n$ and the corresponding squares, which have lattice points in the upper-left corners. The error is shown on the right. The new twist is that the squares are neither all below nor all above the curve; there is a mixture of both. Squares that extend above the hyperbola are shaded in vertical stripes. These represent a quantity by which $\sum_{k=1}^{n} \tau(k)$ *exceeds* $n \log(n)$. As before, slide them over against the y axis; their area is less than that of a $1 \times n$ rectangle, which is n.

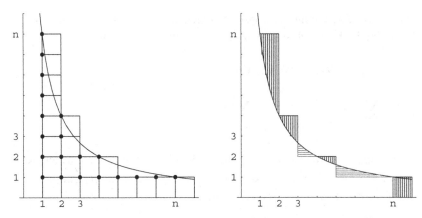

Figure 4.3. Geometric proof of Eq. (4.2).

Shaded in horizontal stripes is the area under $xy = n$ that is not covered by any squares. This represents a quantity by which $\sum_{k=1}^{n} \tau(k)$ falls short of $n \log(n)$. So, it should be subtracted from, not added to, the total error. Again, this area is less than n. (Imagine sliding them down to the x axis.) The difference of two numbers, each less than n, is less than n in absolute value. □

Now, we will give another proof with equations instead of geometry. To do this we need to make our Big Oh notation more flexible. The basic idea was that $f(n) = g(n) + O(h(n))$ meant that we could replace a complicated expression $f(n)$ with a simpler one $g(n)$ if we were willing to tolerate an error of no more than (a constant times) $h(n)$. We will now allow more than one simplification to take place. We may have more than one Big Oh per line, as long as we eventually combine the errors in a consistent way. So, for example, (3.2) says that after multiplying by $\sqrt{n}$,

$$\sqrt{n} H_n = \sqrt{n} \log(n) + O(\sqrt{n}).$$

That is, if we start with an error bounded by a constant independent of the variable, and then multiply by $\sqrt{n}$, we have an error bounded by a constant times $\sqrt{n}$. (Write out (3.2) in $\ll$ notation and multiply by $\sqrt{n}$ if you have doubts.) Adding this to (3.4) gives

$$\sqrt{n} H_n + \log(n!) = \sqrt{n} \log(n) + O(\sqrt{n}) + n \log(n) - n + O(\log(n))$$
$$= \sqrt{n} \log(n) + n \log(n) - n + O(\sqrt{n}).$$

Since $\log(n) \ll \sqrt{n}$, the $O(\log(n))$ error can be absorbed into the $O(\sqrt{n})$

error by making the hidden constant bigger. This, too, can be made explicit by writing both Big Oh statements in $\ll$ notation and adding.

Second Proof of (4.2). We know that

$$\sum_{k=1}^{n} \tau(k) = \sum_{k=1}^{n} \sum_{d|k} 1 = \sum_{\substack{c,d \\ cd \leq n}} 1.$$

This is the same idea as before: that all the divisors of all the integers less than or equal n are exactly the same as all pairs of integers (c, d) with $cd \leq n$. But if $cd \leq n$, then $d \leq n$ and $c \leq n/d$; so,

$$\sum_{k=1}^{n} \tau(k) = \sum_{d \leq n} \sum_{c \leq n/d} 1.$$

The inner sum counts how many integers c are less than or equal to n/d. This is $[n/d]$, the integer part of the rational number n/d. And we know that rounding down a number changes it by an error less than 1, so $[n/d] = n/d + O(1)$. We have

$$\sum_{k=1}^{n} \tau(k) = \sum_{d \leq n} [n/d] = \sum_{d \leq n} \{n/d + O(1)\}$$

$$= \sum_{d \leq n} n/d + \sum_{d \leq n} O(1) = nH_n + O(n),$$

where the first term comes from the definition of Harmonic numbers. For the second, notice we make an error of at most 1 for each $d \leq n$, that is, n different errors. We get an error of at most n, and

$$\sum_{k=1}^{n} \tau(k) = n \{\log(n) + O(1)\} + O(n) = n \log(n) + O(n)$$

according to (3.2).   □

Corollary. *The average order of $\tau(n)$ is $\log(n)$.*

Proof. Divide both sides of (4.2) by $n \log(n)$ to get

$$\frac{\sum_{k=1}^{n} \tau(k)}{n \log(n)} = 1 + O\left(\frac{1}{\log(n)}\right).$$

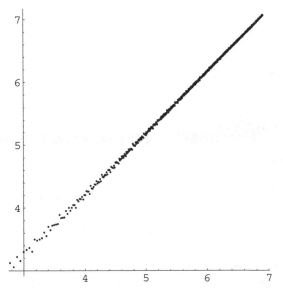

Figure 4.4. $\log(n)$ vs. the average of $\tau(n)$.

Because $1/\log(n) \to 0$, as $n \to \infty$, this says that

$$\sum_{k=1}^{n} \tau(k) \sim n \log(n), \quad \text{which is} \quad \sim \log(n!) = \sum_{k=1}^{n} \log(k)$$

according to Exercise 4.0.3. $\qquad\qquad\qquad\qquad\qquad\qquad\qquad\qquad\qquad\qquad\qquad \square$

Because $n \log(n) \sim \sum_{k=1}^{n} \tau(k)$, we see that $\log(n) \sim (\sum_{k=1}^{n} \tau(k))/n$, which really looks like an average. Figure 4.4 shows a plot of points of the form

$$(\log(n), \frac{1}{n}\sum_{k=1}^{n} \tau(k))$$

for all n below 1000. You should compare this to Figure 4.1, which compares $\log(n)$ to $\tau(n)$ without averaging. (The vertical scales in these two plots are not the same.)

4.2. Improved Estimate

In Chapter 3, we made an approximation (3.2) to the Harmonic numbers, H_n, and then refined it with (3.3). Exercise 3.2.3 of Chapter 3 showed (without

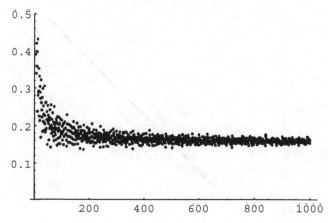

Figure 4.5. Graphical evidence for an improved conjecture on the average size of $\tau(n)$.

proof) that still-more-accurate approximations may be possible. Perhaps we can do the same with the average size of $\tau(n)$? After dividing both sides of (4.2) by n, we see that

$$\frac{1}{n}\sum_{k=1}^{n}\tau(k) - \log(n)$$

is a bounded sequence. Now, look at Figure 4.5, which shows this difference for all n below 1000. It seems possible that this sequence may be not merely bounded but actually convergent.

To prove such a theorem, we will need some lemmas. We already understand triangular numbers, $\sum_{k=1}^{n} k = n(n+1)/2$, when n is an integer. We will need a formula for $\sum_{k \leq t} k$, the sum over positive integers less than a *real* number t. Because $k \leq t$ exactly when $k \leq [t]$, the integer part of t, this is still a triangular number; it is $[t]([t]+1)/2$. The problem is that this is too exact. Because it is an integer-valued function of a real variable, it is not continuous, just as $[t]$ is not. For purposes of analysis, we prefer to have a continuous function and let the Big Oh absorb the discontinuity.

Lemma. *For a real number t,*

$$\sum_{k \leq t} k = \frac{t^2}{2} + O(t). \tag{4.3}$$

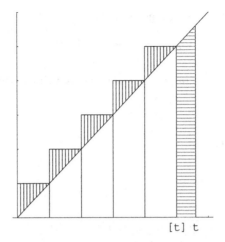

Figure 4.6. Graph for exercise 4.2.1.

Exercise 4.2.1. We know that

$$\frac{t^2}{2} = \int_0^t x\,dx$$

is the area of the large right triangle in Figure 4.6. The amount by which $\sum_{k \le t} k$ exceeds this is the area shaded in vertical stripes. On the other hand, the horizontal stripes show the area of the triangle not covered by any of the rectangles; this error has the opposite sign.

Use Figure 4.6 to give a geometric proof of (4.3).

Exercise 4.2.2. Alternately, prove (4.3) by writing

$$\frac{[t]([t]+1)}{2} = \frac{(t + O(1))(t + O(1) + 1)}{2}$$

and simplifying.

We will similarly extend Harmonic numbers to real variables. That is, we define

$$H_t = \sum_{k \le t} \frac{1}{k}.$$

As with the preceding triangular numbers, we know an exact formula; that is, $H_t = H_{[t]}$.

Exercise 4.2.3. By viewing $\log(t) - \log([t])$ as an integral, show that

$$\log([t]) = \log(t) + O(1/[t]).$$

These two equations imply that our previous estimate still holds.

Lemma.

$$H_t = \log(t) + \gamma + O(1/t). \tag{4.4}$$

Proof.

$$H_t = H_{[t]} = \log([t]) + \gamma + O(1/[t])$$

according to (3.3),

$$= \log(t) + +O(1/[t]) + \gamma + O(1/[t])$$

according to the previous exercise, and

$$= \log(t) + \gamma + O(1/t)$$

because $1/[t] \ll 1/t$. (Verify this.) □

We are now ready to improve estimate (4.2).

Theorem.

$$\sum_{k=1}^{n} \tau(k) = n\log(n) + (2\gamma - 1)n + O(\sqrt{n}), \tag{4.5}$$

where γ is Euler's constant (see Chapter 3).

The numerical value of $2\gamma - 1$ is about 0.154431. Compare this to the height of the horizontal asymptote in Figure 4.5. Dividing both sides of (4.5) by n says that

$$\frac{1}{n}\sum_{k=1}^{n} \tau(k) - \log(n) = 2\gamma - 1 + O\left(\frac{1}{\sqrt{n}}\right),$$

and because $1/\sqrt{n} \to 0$, the sequence on the left really does have the limit $2\gamma - 1$.

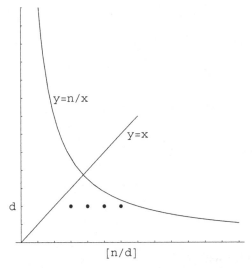

Figure 4.7. Diagram for the proof of Eq. (4.5).

Proof. To improve on (4.2), we will make use of symmetry. The left side of Figure 4.3 is symmetric about the line $y = x$. Every lattice point that we want to count is either on the line $y = x$ or is one of a pair of lattice points situated symmetrically about the line. This is just a restatement of the fact that if (c, d) is a lattice point under the hyperbola, so is (d, c).

The lattice points on the line $y = x$ are clearly just $(1, 1)$, $(2, 2)$, ..., $([\sqrt{n}], [\sqrt{n}])$. There are exactly $[\sqrt{n}]$ of them, and this number is already smaller than the error $O(\sqrt{n})$ that the theorem allows, so we may ignore these points.

It remains to count the lattice points on a line at height d that lie between the line $y = x$ and the hyperbola $y = n/x$. There are $[n/d] - d$ such points. (Look at Figure 4.7 until you believe this.) We must do this for each of the horizontal lines $d = 1, d = 2, \ldots, d = [\sqrt{n}]$, then multiply by 2. So,

$$\sum_{k=1}^{n} \tau(k) = 2 \sum_{d \leq \sqrt{n}} \{[n/d] - d\} + O(\sqrt{n})$$

$$= 2 \sum_{d \leq \sqrt{n}} \{n/d + O(1) - d\} + O(\sqrt{n})$$

$$= 2 \sum_{d \leq \sqrt{n}} n/d + 2 \sum_{d \leq \sqrt{n}} O(1) - 2 \sum_{d \leq \sqrt{n}} d + O(\sqrt{n}).$$

The $\sqrt{n}$ different errors of size $O(1)$ accumulate to be as big as $O(\sqrt{n})$; so, we have

$$\sum_{k=1}^{n} \tau(k) = 2n \sum_{d \leq \sqrt{n}} \frac{1}{d} - 2 \sum_{d \leq \sqrt{n}} d + O(\sqrt{n})$$

$$= 2n H_{\sqrt{n}} - 2 \left\{ \frac{n}{2} + O(\sqrt{n}) \right\} + O(\sqrt{n})$$

according to (4.3), with $t = \sqrt{n}$. According to (4.4), with $t = \sqrt{n}$,

$$\sum_{k=1}^{n} \tau(k) = 2n \{ \log(\sqrt{n}) + \gamma + O(1/\sqrt{n}) \} - n + O(\sqrt{n}).$$

This gives

$$\sum_{k=1}^{n} \tau(k) = n \log(n) + (2\gamma - 1)n + O(\sqrt{n}),$$

because $\log(\sqrt{n}) = \log(n)/2$, and n errors of size $O(1/\sqrt{n})$ accumulate only to size $O(\sqrt{n})$.   □

Dirichlet proved this theorem in 1849. Mathematicians are still working on decreasing the error term in the estimate. Voronoi obtained $O(x^{1/3})$ in 1904, Van der Corput proved $O(x^{27/82})$ in 1928, and Chih got $O(x^{15/46})$ in 1950. The best-known estimate, by Kolesnik, is slightly worse than $O(x^{12/37})$. On the other hand, Hardy and Landau showed in 1915 that the error is at least as big as $O(x^{1/4})$.

Exercise 4.2.4. Get your calculator out and compute $1/3, 27/82, 15/46$, and $12/37$ as decimals to see how small these improvements are (and thus how hard the problem is).

4.3. Second-Order Harmonic Numbers

Before we can compute averages of the function $\sigma(n)$, we will need to know about the Second-order Harmonic numbers. These are a generalization of H_n defined by

$$H_n^{(2)} = \sum_{k=1}^{n} \frac{1}{k^2}.$$

Exercise 4.3.1. Compute the Second-order Harmonic numbers $H_1^{(2)}, H_2^{(2)}$, and $H_3^{(2)}$ exactly, as rational numbers.

This is obviously tedious. As with ordinary Harmonic numbers, the numerators and denominators get big quickly. In fact,

$$H_{50}^{(2)} = \frac{3121579929551692678469635660835626209661709}{1920815367859463099600511526151929560192000}.$$

However, unlike ordinary Harmonic numbers, we will show that

Theorem. *There is a real number $\zeta(2)$ such that*

$$H_n^{(2)} = -\frac{1}{n} + \zeta(2) + O\left(\frac{1}{n^2}\right). \tag{4.6}$$

The numerical value of $\zeta(2)$ is about 1.6449340668482264365 .... This is the analog for Second-order Harmonic numbers of (3.3) in Chapter 3. The constant $\zeta(2)$ is the analog of Euler's constant γ. The notation for this constant is a little funny looking. In Exercise 4.3.3 you will look at a generalization to Third-order Harmonic numbers, and more generally kth-order Harmonic numbers. The constants $\zeta(2), \zeta(3), \ldots, \zeta(k), \ldots$ are values of the RIEMANN ZETA FUNCTION, which we will be very interested in soon.

Proof. The area under $y = 1/x^2$ between $x = 1$ and $x = \infty$ is given by the improper integral

$$\int_1^\infty \frac{1}{x^2}dx = \lim_{B\to\infty} \int_1^B \frac{1}{x^2}dx = \lim_{B\to\infty} \frac{-1}{x}\bigg|_1^B = \lim_{B\to\infty} 1 - \frac{1}{B} = 1.$$

The *area* under the curve is finite, even though it stretches infinitely far to the right. The top half of Figure 4.8 shows the infinitely many rectangles of height $1/k^2$, for $k = 2, 3, \ldots$, fit under this curve. In particular, their area is also finite, and in fact less than 1. We can define $\zeta(2)$ as this number plus 1, to include the first rectangle with height 1.

Now that we have defined the number $\zeta(2)$, we can prove the theorem, which after rearranging the terms claims that

$$\zeta(2) - H_n^{(2)} = \frac{1}{n} + O\left(\frac{1}{n^2}\right).$$

The number on the left is the area of all infinitely many rectangles except the first n. The bottom half of Figure 4.8 shows (on a different scale) that this is

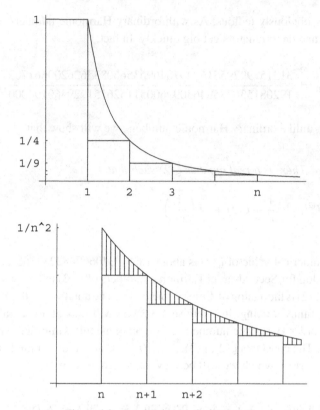

Figure 4.8. Geometric proof of Eq. (4.6).

approximated by the area under the curve from n to ∞, which is

$$\int_n^\infty \frac{1}{x^2}dx = \lim_{B\to\infty} \int_n^B \frac{1}{x^2}dx = \lim_{B\to\infty} \frac{1}{n} - \frac{1}{B} = \frac{1}{n}.$$

The error in making this approximation is the shaded area of Figure 4.8. As usual, we see that all these pieces will fit into a rectangle of width 1 and height $1/n^2$, so the error is less than $1/n^2$.   □

Exercise 4.3.2. Table 4.1 compares the decimal expansion of $H_n^{(2)}$ to $\zeta(2) - 1/n$ for some powers of 10. Check that the error in this approximation seems to be about size $1/n^2$, as predicted by (4.6).

Exercise 4.3.3. What do you think the definition of the Third-order Harmonic numbers $H_n^{(3)}$ should be? Prove a theorem similar to (4.6). (The numerical value of $\zeta(3)$ is about $1.2020569031595942854\ldots$.) In fact, you can just as

Table 4.1. *Second-Order Harmonic Numbers*

n	$H_n^{(2)}$	$\zeta(2) - 1/n$
10	1.5497677311665406904...	1.5449340668482264365...
10^2	1.6349839001848928651...	1.6349340668482264365...
10^3	1.6439345666815598031...	1.6439340668482264365...
10^4	1.6448340718480597698...	1.6448340668482264365...
10^5	1.6449240668982262698...	1.6449240668482264365...
10^6	1.6449330668487264363...	1.6449330668482264365...

easily define kth-order harmonic numbers, $H_n^{(k)}$, and constants $\zeta(k)$ for any positive integer k.

Exercise 4.3.4. If your answer to the previous exercise is correct, numerical evidence should confirm it. Some of the Third-order Harmonic numbers are listed in Table 4.2. Fill in the rest of the table with the estimate from your theorem, and compare to see how big the error is.

4.4. Averages of Sums

Now, we have the tools we need to think about averages of $\sigma(n)$.

Theorem. *For all n,*

$$\sum_{k=1}^{n} \sigma(k) = \zeta(2)\frac{n^2}{2} + O(n \log(n)). \tag{4.7}$$

Proof. We can view this, like the theorem about $\tau(n)$, in terms of lattice points. But, now, we are not counting the number of points; instead, we are

Table 4.2. *Third-Order Harmonic Numbers*

n	$H_n^{(3)}$
10	1.1975319856741932517...
10^2	1.2020074006596776104...
10^3	1.2020564036593442855...
10^4	1.2020568981600942604...
10^5	1.2020569031095947854...
10^6	1.2020569031590942859...

adding up the "y-coordinates" d of the lattice points (c, d). Because of this, it will be easier to imitate the second proof of (4.2). We have

$$\sum_{k=1}^{n} \sigma(k) = \sum_{k=1}^{n} \sum_{d|k} d = \sum_{\substack{c,d \\ cd \leq n}} d$$

$$= \sum_{c \leq n} \sum_{d \leq n/c} d = \sum_{c \leq n} \left\{ \frac{1}{2} \frac{n^2}{c^2} + O\left(\frac{n}{c}\right) \right\}$$

according to (4.3), with $t = n/c$, and

$$= \frac{n^2}{2} \sum_{c \leq n} \frac{1}{c^2} + O\left(n \sum_{c \leq n} \frac{1}{c}\right).$$

Here, the Second-order Harmonic numbers make their appearance:

$$= \frac{n^2}{2} \left\{ \zeta(2) - \frac{1}{n} + O\left(\frac{1}{n^2}\right) \right\} + O(nH_n)$$

according to (4.6), and

$$= \zeta(2)\frac{n^2}{2} - \frac{n}{2} + O(1) + O(n \log(n)).$$

But, the error $O(n \log(n))$ is already bigger than the $O(1)$ and the exact term $-n/2$. So, this is the same as

$$\sum_{k=1}^{n} \sigma(k) = \zeta(2)\frac{n^2}{2} + O(n \log(n)).$$

Maybe you objected to the preceding claim that

$$\sum_{c \leq n} O\left(\frac{n}{c}\right) = O\left(n \sum_{c \leq n} \frac{1}{c}\right).$$

This just says that the sum of n errors, each bounded by a constant K times n/c, is in fact bounded by $Kn \sum 1/c$.   □

Corollary. *The average order of $\sigma(n)$ is $\zeta(2)n$.*

You might have expected the average order to be $\zeta(2)n/2$. If so, go back and look at the example with $N(n)$, at the beginning of this chapter.

Proof. Divide both sides of (4.7) by $\zeta(2)n^2/2$, and use the fact that

$n \log(n)/n^2 = \log(n)/n \to 0$ as $n \to \infty$ to see that

$$\sum_{k=1}^{n} \sigma(k) \sim \zeta(2)\frac{n^2}{2}.$$

Meanwhile, we can multiply the triangular numbers t_n by $\zeta(2)$ to see that

$$\sum_{k=1}^{n} \zeta(2)k = \zeta(2)t_n = \zeta(2)\frac{n^2}{2} + O(n).$$

So,

$$\sum_{k=1}^{n} \zeta(2)k \sim \zeta(2)\frac{n^2}{2}$$

as well, and $\sim$ is an equivalence relation. □

In the following series of exercises, we will use this to find the average order of $s(n) = \sigma(n) - n$.

Exercise 4.4.1. Write out what (4.7) means in $\ll$ notation. Now, using the fact that $|a| < b$ means that $-b < a < b$, write out what this means in term of actual inequalities, i.e., without $\ll$ symbols.

Exercise 4.4.2. What do we already know about triangular numbers that

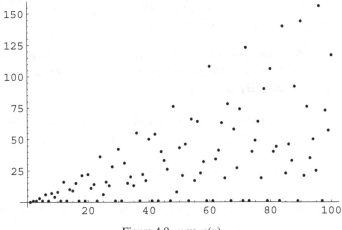

Figure 4.9. n vs. $s(n)$.

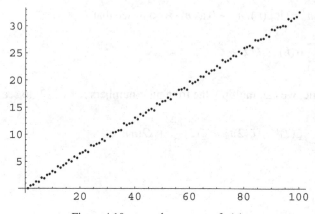

Figure 4.10. n vs. the average of $s(n)$.

implies that

$$\sum_{k=1}^{n} k = \frac{n^2}{2} + O(n \log(n))?$$

Write this out in terms of actual inequalities.

Exercise 4.4.3. Combine the inequalities of Exercises 4.4.1 and 4.4.2 to get an inequality that involve $s(k)$ instead of $\sigma(k)$. Now, convert this back to a statement in Big Oh notation.

Exercise 4.4.4. Find a Big Oh estimate similar to that in Exercise 4.4.2 for

$$\sum_{k=1}^{n} (\zeta(2) - 1)k.$$

Exercise 4.4.5. Using Exercises 4.4.3 and 4.4.4, give the average order of $s(n)$. Because $\zeta(2) = 1.6449340668482264365\ldots$, is an "average" integer abundant or deficient? By how much?

Figure 4.9 shows 100 data points of the form $(n, s(n))$. Compare this with Figure 4.10, which shows points where the second coordinate is

$$\frac{1}{n} \sum_{k \leq n} s(k)$$

for $n \leq 100$. As expected, they lie on a line with slope equal to one half of $\zeta(2) - 1$.

Interlude 1

Calculus

The techniques discussed in the previous chapters can be pushed a little further, at the cost of a lot of work. To make real progress, however, we need to study the prime numbers themselves. How are the primes distributed among the integers? Is there any pattern? This is a very deep question, which was alluded to at the beginning of Chapter 3. This Interlude makes a detour away from number theory to explain the ideas from calculus that we will need. It covers things I wish you had learned but, based on my experience, I expect you did not. I can't force you to read it, but if you skip it, please refer back to it later.

I1.1. Linear Approximations

Although you might not notice, all of differential calculus is about a single idea: Complicated functions can often be approximated, on a small scale anyway, by straight lines. What good is such an approximation? Many textbooks will have a (rather unconvincing) application, something like "approximate the square root of 1.037." In fact, almost everything that happens in calculus is an application of this idea.

For example, one learns that the graph of function $y = f(x)$ increases at point $x = a$ if the derivative $f'(a)$ is positive. Why is this true? It's because of the linear approximation idea: The graph increases if the straight line approximating the graph increases. For a line, it's easy to see that it is increasing if the slope is positive. That slope is $f'(a)$.

There are many different ways to specify a line in the plane using an equation. For us, the most useful will be the "point–slope" form: The line through point (a, b) with slope m has the equation $y - b = m(x - a)$, or $y = b + m(x - a)$. If $y = f(x)$ is some function, the line through point $(a, f(a))$ with slope $f'(a)$ is $y = f(a) + f'(a)(x - a)$. So, if x is close to a,

$$f(x) \approx f(a) + f'(a)(x - a), \tag{I1.1}$$

where $\approx$ means approximately equal in some sense not yet specified.

Exercise I1.1.1. Of course, we don't have to call the variable x. Find the equation of the line tangent to $\log(1 - t)$ at $t = 0$. Be sure your answer really is the equation of a line; if not, your answer is wrong. Plug $t = 1/17$ into the equation for the line, and compare it to the actual value of $\log(16/17)$. This approximation to $\log(1 - t)$ will be crucial in the study of the distribution of prime numbers in the next chapter.

Exercise I1.1.2. Find the linear approximation to $f(x) = (1/4 + x^2)^{1/2}$ at $x = 0$.

I1.2. More on Landau's Big Oh

In this section, we care about small values of a variable, not big ones. We will introduce an analogous way to compare functions, saying

$$f(x) \ll h(x) \quad \text{as} \quad x \to a$$

if there is some constant C and some interval around a such that

$$|f(x)| \le C|h(x)| \quad \text{when } x \text{ is in the interval.}$$

For example,

$$x^3 + x^2 \ll x^2 \quad \text{as} \quad x \to 0,$$

because $x^3 + x^2 = x^2(x + 1)$, and $C = 2$ will work:

$$|x^2(x + 1)| \le 2|x^2| \quad \text{exactly when} \quad |x + 1| \le 2.$$

The inequality on the right holds if $|x| < 1$. A geometric interpretation is given in the top of Figure I1.1. There is a scaling factor C such that the function $x^3 + x^2$ is trapped between the two parabolas $-Cx^2$ and Cx^2, at least in some interval around 0.

As before, we might be sloppy and let the $x \to 0$ be implicit from the context. This can cause confusion; is $x^3 \ll x$? The answer is no if we mean $x \to \infty$, but yes if we mean $x \to 0$. (You should check this.) So we'll try to be explicit.

And there is an analogous Big Oh relation for small variables. That is,

$$f(x) = g(x) + O(h(x)) \quad \text{if } f(x) - g(x) \ll h(x) \quad \text{as } x \to a. \quad \text{(I1.2)}$$

As an example, we will show

$$\exp(x) = 1 + O(x) \quad \text{as } x \to 0.$$

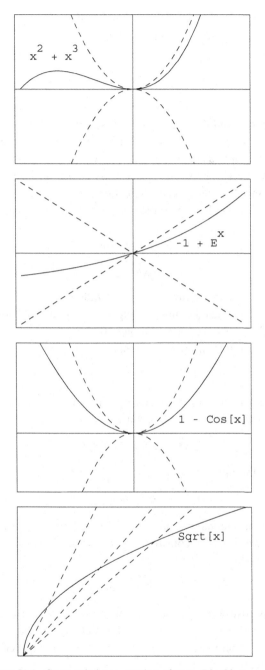

Figure I1.1. Geometric interpretation of some Big Oh examples.

From the definition, we need to show that for some C, $|\exp(x) - 1| \leq C|x|$ on some interval around $x = 0$. From the definition of absolute values, this is equivalent to

$$-C|x| \leq \exp(x) - 1 \leq C|x|.$$

It turns out that $C = 2$ works again. First, consider the case of $x \geq 0$. Because $\exp(x)$ is increasing and $e^0 = 1$, $\exp(x) - 1$ is certainly ≥ 0 for $x \geq 0$, the first inequality, $-2x \leq \exp(x) - 1$, is trivial. The other inequality is $\exp(x) - 1 \leq 2x$, which is the same as $\exp(x) \leq 2x + 1$, which must now be proved. At $x = 0$, it is true that $1 = e^0 \leq 2 \cdot 0 + 1 = 1$. By taking the derivative of $\exp(x)$ at $x = 0$, we find that the slope of the tangent line is 1, less that that of the line $2x + 1$. So, $\exp(x)$ lies under the line $2x + 1$, at least for a little way.

Exercise I1.2.1. Show that for $x \leq 0$ (where $|x| = -x$),

$$2x \leq \exp(x) - 1 \leq -2x.$$

The geometric interpretation is in the middle of Figure I1.1. There is a scaling factor C such that the function $\exp(x) - 1$ is trapped between the two lines $-Cx$ and Cx in some interval around 0.

For another example, we will show that

$$\cos(x) = 1 + O(x^2) \quad \text{as } x \to 0.$$

This is saying that for some C, $|\cos(x) - 1| \leq C|x^2|$, or because $x^2 \geq 0$ always, we can write x^2 instead. We must show that

$$-Cx^2 \leq \cos(x) - 1 \leq Cx^2$$

or, multiplying through by -1,

$$-Cx^2 \leq 1 - \cos(x) \leq Cx^2$$

for some C in an interval around $x = 0$. Because everything in sight is an even function, we need only consider $x \geq 0$. Because $1 - \cos(x)$ is never less than 0, the inequality

$$-Cx^2 \leq 1 - \cos(x)$$

is trivial for any positive C. The other can be shown with, for example, $C = 1$. At $x = 0$, the inequality reduces to $0 \leq 0$, which is true. We are done if we can show that x^2 increases faster than $1 - \cos(x)$. Taking derivatives, this reduces to showing

$$\sin(x) \leq 2x \quad \text{for } x \geq 0.$$

Exercise I1.2.2. Show this inequality. The preceding example of $\exp(x) - 1 \leq 2x$ might be instructive.

This is a lot to digest all at once. Let's consider the much simpler case of $g(x)$ as a constant function equal to some number L and $h(x)$ as the function $x - a$. What does it mean to say that

$$f(x) = L + O(x - a) \quad \text{as} \quad x \to a? \quad \text{(I1.3)}$$

It means that there is some number C and an interval around a such that

$$|f(x) - L| < C|x - a|$$

for every value of x in the interval. This means that we can get $f(x)$ to be as close to L as we need by taking values of x sufficiently close to a. No matter how small the error, ϵ, we are willing to tolerate, if we take $\delta = \epsilon/C$, then whenever $|x - a| \leq \delta$,

$$|f(x) - L| < C|x - a| \leq C\delta = \epsilon.$$

This may sound vaguely familiar to you; it implies that the limit of $f(x)$ is L as x approaches a. If L is the actual value of the function at a, that is, $L = f(a)$, then

$$f(x) = f(a) + O(x - a) \quad \text{as} \quad x \to a \quad \text{(I1.4)}$$

implies that $f(x)$ is continuous at $x = a$.

This is not the same as continuity; the Big Oh statement has more information because it specifies how fast the function is tending to the limit. For example, $\sqrt{x} \to 0$ as $x \to 0$, but it is not true that

$$\sqrt{x} = 0 + O(x) \quad \text{as} \quad x \to 0.$$

Here's why. For a linear error $O(x)$, we can interpret the unknown constant C as a slope, just as in the example with $\exp(x) - 1$. Then, as the bottom of Figure I1.1 indicates, no matter what slope C we pick, eventually the graph of $y = \sqrt{x}$ is above the line $y = Cx$. But examples like this are pathological: $\sqrt{x}$ has no derivative at $x = 0$. For the nice functions we are interested in, it is convenient to do everything with Big Oh notation.

The beauty of this is that we can use it for derivatives, too, to make sense of what $\approx$ means in (I1.1). If there is some number (which we denote $f'(a)$) such that

$$f(x) = f(a) + f'(a)(x - a) + O((x - a)^2) \quad \text{as} \quad x \to a, \quad \text{(I1.5)}$$

then $f(x)$ is differentiable at $x = a$. Why is this consistent with the "difference quotient" definition you learned in calculus? Subtract the $f(a)$ and divide by $x - a$ in (I1.5) to see that this is exactly the same as saying

$$\frac{f(x) - f(a)}{x - a} = f'(a) + O(x - a) \quad \text{as} \quad x \to a. \qquad (\text{I1.6})$$

By (I1.3), this just means that the limit as x approaches a of $(f(x) - f(a))/(x - a)$ is the number $f'(a)$.

It is worth pointing out that this use of Big Oh for small values of $x - a$ is similar to what we did in Chapter 3. We can use it to replace complicated expressions $f(x)$ with simpler ones, such as equations of lines, as long as we are willing to tolerate small errors. When $|x - a|$ is less than 1, powers like $(x - a)^2$ are even smaller. The closer x gets to a, the smaller the error is.

Mostly, in calculus, you look at very nice functions, those that do have derivatives at every point a. So, the rule that assigns to each point a the number $f'(a)$ defines a new function, which we denote $f'(x)$.

I1.3. Fundamental Theorem of Calculus

Integral calculus as well as differential calculus is really all about a single idea. When you took the course, you got a lot of practice with "antiderivatives," that is, undoing the operation of derivative. For example, you write

$$\int x^2 dx = \frac{x^3}{3} + C$$

to mean that a function whose derivative is x^2 must be of the form $x^3/3$ plus some constant C. This is an indefinite integral; it is a collection of *functions*. You also learned about the definite integral; this is a *number* that measures the area under a curve between two points. For example,

$$\int_0^1 x^2 dx$$

is the area under the parabola $y = x^2$ between $x = 0$ and $x = 1$. The symbols

$$\int x^2 dx \quad \text{and} \quad \int_0^1 x^2 dx$$

mean two very different things, even though they look very similar. But why are these two things connected? What does the operation of undoing derivatives have to do with area? After all, the geometric interpretation of derivative is the slope of the tangent line, which has nothing to do with area.

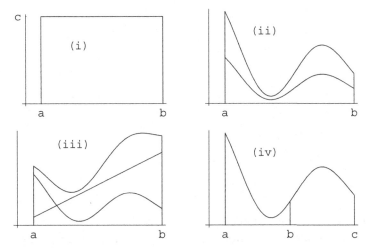

Figure I1.2. Properties of area.

To explain this, we need some basic facts about area, geometric properties that have nothing to do with calculus. In each of these, keep in mind that $\int_a^b f(t)\,dt$ just denotes the area under $y = f(t)$ between a and b, nothing more.

(i) First of all, the area of a rectangle is the height times the width. So, if we have a constant function $f(t) = c$ for all t, then by geometry

$$\int_a^b c\,dt = c \cdot (b - a).$$

(ii) Next, if we scale the function by some constant c to change its height, the area under the graph changes by that same scalar factor. Figure I1.2 shows an example with $c = 2$. So,

$$\int_a^b c \cdot f(t)\,dt = c \cdot \int_a^b f(t)\,dt.$$

(iii) This next one is a little trickier. If we add two functions, $f(t)$ and $g(t)$, together, the area under the new function is the sum of the areas under each one. One way to convince yourself of this is to imagine approximating the area using lots of little rectangles. The height of a rectangle under the graph of $f + g$ is just that of a rectangle under f and another under g. So,

$$\int_a^b f(t) + g(t)\,dt = \int_a^b f(t)\,dt + \int_a^b g(t)\,dt.$$

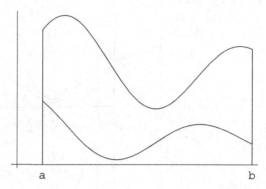

Figure I1.3. Another property of area.

(iv) Finally, if we have three points, a, b, and c, on the t-axis, the area from a to c is just the area from a to b plus the area from b to c. Figure I1.2 is convincing. In equations,

$$\int_a^c f(t)\,dt = \int_a^b f(t)\,dt + \int_b^c f(t)\,dt.$$

That is all we need for now, but two other properties will be useful later on.

(v) It is clear from Figure I1.3 that if $f(t) \le g(t)$, then

$$\int_a^b f(t)dt \le \int_a^b g(t)dt.$$

In fact we were used this property as far back as (3.5). This is the COMPARISON TEST FOR INTEGRALS, analogous to that for infinite series, which we will see in Interlude 2

I lied previously when I said that $\int_a^b f(t)\,dt$ just denotes the area under $y = f(t)$ between a and b. As you remember from calculus, if any portion of the graph dips below the horizontal axis, that area is counted negatively by the definite integral. From Figure I1.4, it is clear that

$$\int_a^b f(t)dt \le \int_a^b |f(t)|dt;$$

they are equal exactly when $f(t)$ is always positive, otherwise, the definite integral of $f(t)$ has a "negative" chunk that the integral of $|f(t)|$ does not.

Exercise I1.3.1. Maybe it isn't clear. Looking at Figure I1.4 again, use property (iv), twice, and property (ii) with $c = -1$, to compare the two integrals.

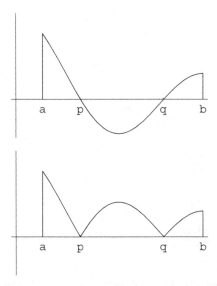

Figure I1.4. Yet another property of area.

Because $-f(t)$ is just another function, it is similarly true that

$$\int_a^b -f(t)dt \leq \int_a^b |-f(t)|dt.$$

The left side is $-\int_a^b f(t)dt$ according to property (ii), while the right side is just $\int_a^b |f(t)|dt$, because $|-f(t)| = |f(t)|$.

(vi) Because both $\int_a^b f(t)dt$ and $-\int_a^b f(t)dt$ are less than or equal to $\int_a^b |f(t)|dt$, we deduce that

$$\left| \int_a^b f(t)dt \right| \leq \int_a^b |f(t)|dt.$$

To prove the Fundamental Theorem of Calculus, we also need an important definition. Suppose $f(x)$ is some function that is nice enough that (I1.4) is true at each point a. We can make a new function $F(x)$, by assigning to each number x the area under f between 0 and x. You should think of this as a definite integral. It can be computed to any degree of accuracy by approximating by rectangles (the so-called Riemann sums) without yet making any reference to antiderivatives. So,

$$F(x) = \int_0^x f(t)dt.$$

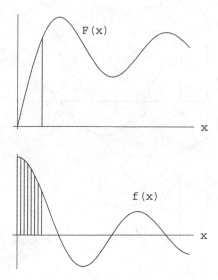

Figure II.5. The Fundamental Theorem of Calculus.

(The variable we rename t, to avoid confusion.) Figure I1.5 shows an example. The height of the line on the upper graph represents the shaded area on the lower graph. As we vary the point x, the amount of area to the left of the point changes, and this is the height on the graph of $F(x)$. When $f(x)$ is negative, the increment of area is negative, so $F(x)$ decreases. It is crucial to recognize the role of the variable x in the function F. It is used to determine a portion of the horizontal axis; the area under f above that portion of the axis is the number $F(x)$. By way of analogy, remember that the harmonic number, H_n, is not $1/n$; rather, it is the sum of the reciprocals of integers up to n.

Approximating area by rectangles is not too hard; Archimedes did it in his *Quadrature of the Parabola*, nearly 2,000 years before Newton.

Exercise I1.3.2. This exercise will indicate why Archimedes was interested in formulas such as (1.7). Suppose you want to approximate the area under $y = x^2$ between $x = 0$ and $x = 1$ with, say, n rectangles of width $1/n$. The height of rectangle k at $x = k/n$ is $(k/n)^2$. So the area is about

$$\sum_{k=1}^{n} \left(\frac{k}{n}\right)^2 \frac{1}{n}.$$

Find a closed-form expression for the sum, as a function of the number of rectangles n. How does it compare to the exact answer, $1/3$?

Now that we can define a function in terms of area and compute it using Riemann sums, we can state the following.

Theorem (Fundamental Theorem of Calculus, Part I). *Suppose $f(x)$ satisfies (I1.4). The function $F(x)$ defined by*

$$F(x) = \int_0^x f(t)dt$$

is differentiable, and $F'(a) = f(a)$ for every point a.

Proof. We need to show that

$$F(x) = F(a) + f(a)(x - a) + O((x - a)^2).$$

But, by definition of F,

$F(x) - F(a)$

$$= \left(\int_0^x f(t)dt - \int_0^a f(t)dt \right)$$

$$= \int_a^x f(t)dt, \qquad\qquad \text{according to property (iv) above,}$$

$$= \int_a^x (f(a) + O(t - a)) \, dt, \qquad \text{according to (I1.4),}$$

$$= \int_a^x f(a)dt + \int_a^x O(t - a)dt, \quad \text{according to property (iii).}$$

Because $f(a)$ is a constant, according to property (i) we get

$$= f(a)(x - a) + \int_a^x O(t - a)dt$$

$$= f(a)(x - a) + \int_a^x O(x - a)dt, \quad \text{because } t \leq x,$$

$$= f(a)(x - a) + O(x - a)(x - a), \text{ by property (i) again.}$$

Because $O(x - a)$ is constant in t, we get

$$= f(a)(x - a) + O((x - a)^2).$$

If the Big Oh manipulations above seem dubious to you, remember we can think of the Big Oh term as some function, either in t or x, satisfying the given bound. $\qquad\qquad\qquad\qquad\qquad\qquad\qquad\qquad\qquad\qquad\qquad \square$

The theorem says in equations that the rate of change of area is height.

Exercise I1.3.3. Compare the graphs of $f(x)$ and

$$F(x) = \int_0^x f(t)dt$$

in Figure I1.5 and convince yourself that $f(x)$ has the right properties to be the derivative of $F(x)$. Where is $F(x)$ increasing or decreasing? Where are the maximums and minimums? Where are the inflection points?

How can we make use of this theorem? The function $F(x)$ computes the area by Riemann sums, which we prefer not to deal with if possible. The answer is

Theorem (Fundamental Theorem of Calculus, Part II). *If $G(x)$ is any anti-derivative for $f(x)$, that is $G'(x) = f(x)$, then*

$$\int_a^b f(t)dt = G(b) - G(a).$$

Thus, we don't need to compute $F(x)$ as long as we can guess *some* antiderivative.

Proof. Because $G'(x) = f(x) = F'(x)$ according to the theorem, $G'(x) - F'(x) = 0$; so, $G(x) - F(x) = C$, some constant, or $G(x) = F(x) + C$. So,

$$
\begin{aligned}
G(b) - G(a) &= (F(b) + C) - (F(a) + C) \\
&= F(b) - F(a) \\
&= \int_0^b f(t)dt - \int_0^a f(t)dt \\
&= \int_a^b f(t)dt.
\end{aligned}
$$

$\square$

Of course, there is nothing special about the base point $x = 0$. We can start measuring the area relative to any base point $x = a$. We get an antiderivative whose value differs from the preceding one using the constant

$$C = \int_0^a f(t)dt.$$

You might have already asked the following question: If we don't care about area, but only what it represents, then why talk about it at all? If so, congratulations; it is a very good question. One answer is that the mind is inherently geometric; to solve a problem, it always helps to have a diagram or picture to refer to.

A more subtle answer is that area is something that can be computed when all else fails. For example, suppose you need to find a function whose derivative is $f(x) = \exp(-x^2)$. The graph of this particular function is the "bell-shape curve"; it arises in probability and statistics. No method you learn in calculus will find an antiderivative in this case. In fact, there is a theorem that says that no combination of elementary functions (polynomial, trigonometric, exponential, etc.) has a derivative that is $\exp(-x^2)$. But some function exists whose derivative is $\exp(-x^2)$. In fact, it is called the error function, $\mathrm{Erf}(x)$; it is related to the probability that a random variable will take on a value $\leq x$. According to the Fundamental Theorem of Calculus, another way to write this is

$$\mathrm{Erf}(x) = \int_0^x \exp(-t^2) dt.$$

So, $\mathrm{Erf}(x)$ can be computed, to any degree of accuracy we like, by approximating the area under the curve with rectangles. These approximations are Riemann sums. For another example, it is perfectly reasonable to *define* the logarithm of x as

$$\log(x) = \int_1^x \frac{1}{t} dt.$$

The Fundamental Theorem of Calculus says the derivative is $1/x$, positive for $x > 0$. So, this function is always increasing and, thus, has an inverse that we can define to be $\exp(x)$. All the properties you know and love can be derived this way. This is a nice example because of the analogy with the harmonic numbers, which we *defined* using a sum in order that $\Delta(H_n) = n^{-1}$.

Exercise II.3.4. Taking the definition of $\log(x)$ to be

$$\log(x) = \int_1^x \frac{1}{t} dt,$$

show that $\log(xy) = \log(x) + \log(y)$ is still true. (Hint: Use property (iv) and change the variables.)

In summary, the Fundamental Theorem of Calculus tells us that if we know an antiderivative, we can use it to compute area easily. If we don't already know an antiderivative, we can use it to define one by computing the area directly.

Chapter 5

Primes

5.1. A Probability Argument

After this long detour through calculus, we are ready to return to number theory. The goal is to get some idea of how prime numbers are distributed among the integers. That is, if we pick a large integer N, what are the chances that N is a prime? A rigorous answer to this question is hard, so in this section we will only give a heuristic argument. The general idea of an argument based on probability is very old. Not only is it known *not* to be a proof (Hardy, Littlewood, 1922), but the way in which it fails to be a proof is interesting.

Because this will be an argument about probability, some explanation is necessary. If you flip a fair coin twelve times, you expect heads to come up about $6 = 12 \times 1/2$ times. You can think of this 6 as $1/2 + 1/2 + \cdots + 1/2$, twelve additions. If you roll a fair die twelve times, you expect to roll a five about $2 = 12 \times 1/6$ times. The 2 is $1/6$ added twelve times. This tells us what to do when the probability changes from one trial to the next. Imagine an experiment in which, at the kth trial, the chance of success is $1/k$. If you repeat the experiment n times, how many successes do you expect? The answer is $1 + 1/2 + 1/3 + \cdots + 1/n = H_n$. Because we already know that the Harmonic number, H_n, is about $\log(n)$ in size, we expect $\log(n)$ successes after n trials.

In order to talk about probability, we also need to know about independent events. If from a deck of fifty-two I deal a card face down and you guess it is an ace, you expect to have a one in thirteen chance of guessing correctly. If I first tell you that the card is a diamond, this does not change your odds. These are independent events. But if I tell you the card is not a seven, it does change the odds, to one in twelve. Being a seven is not independent of being an ace. If I told you it *is* a seven, it would change the odds even more.

Independent events are easier to combine. You expect the chance of getting an ace to be $1/13$, and the chance of getting a diamond to be $1/4$. The chance of getting the ace of diamonds is $1/52 = 1/13 \times 1/4$. Similarly the chance that it is not an ace is $12/13$. The chance that it is not a diamond is $3/4$. The chance that it is neither an ace nor a diamond is $12/13 \times 3/4 = 9/13 = 36/52$. This is correct; there are $52 - 13 = 39$ cards that are not diamonds, but they include the ace of spades, clubs, and hearts. So there are $39 - 3 = 36$ that are neither an ace nor a diamond.

For our large integer N, we will pretend that the chance that it is divisible by one prime, p, is independent of the chance that it is divisible by another prime, q. Call this hypothesis I, for independence. For example, there is 1 chance in 2 that N is divisible by 2, and $1 - 1/2$ chance that it is not. Similarly, there is 1 chance in 3 that N is a multiple of 3, and $1 - 1/3$ chance that it is not. The odds that N is not divisible by either 2 or 3 should be $(1 - 1/2)(1 - 1/3)$. The chance that N is not divisible by 2, 3, or 5 should be $(1 - 1/2)(1 - 1/3)(1 - 1/5)$.

We know that N is a prime if it is divisible by none of the primes p less than N, so we should compute a product over all the primes p less than N. For this, we will use a notation with the Greek letter Π, analogous to the Sigma notation for sums. It will be convenient to replace N with a real variable x, and so the product over primes less than x will be denoted

$$w(x) = \prod_{\substack{p \text{ prime} \\ p < x}} (1 - 1/p)$$

$$= (1 - 1/2)(1 - 1/3)\ldots(1 - 1/q),$$

where q is the biggest prime that is less than x. As in the Fundamental Theorem of Calculus, the role of the variable x here is to tell us how many terms to take in the product. For example, the primes below $x = 4$ are 2 and 3, so $w(4) = (1 - 1/2)(1 - 1/3) = 1/3$; similarly, $w(6.132) = (1 - 1/2)(1 - 1/3)(1 - 1/5) = 4/15$. You should compute $w(10)$, and also $w(\pi^2)$. A graph of $w(x)$ is shown in Figure 5.1; notice that it is constant between primes. It has a jump down every time x passes over a prime, when another term is added to the product. What does this function tell us? We hope it measures the probability, for an integer N close to x, that N is a prime.

Theorem (Prime Number Theorem). *The probability that a large integer N is a prime is about $1/\log(N)$.*

Not a proof. We want to know the size to the function $w(x)$ for large x. We give an argument attributable to M.R. Schroeder (Schroeder, 1997) that the

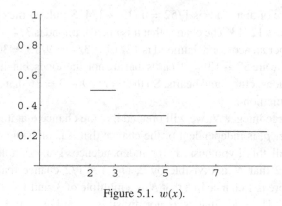

Figure 5.1. $w(x)$.

weaker conclusion, that there should be some constant C such that

$$w(x) \approx \frac{C}{\log(x)},$$

is true, where $\approx$ means "approximately equal" in some vague sense. Because multiplication is harder than addition, we will take the logarithm of $w(x)$ to convert it to a sum. From the definition of $w(x)$, we get that

$$\log(w(x)) \approx \sum_{\substack{p \text{ prime} \\ p < x}} \log(1 - 1/p).$$

But logarithms, too, are complicated. So we will use the approximation idea of Section I1.2. In fact, in Exercise I1.1.1 of that interlude, you computed that for a small number t, $\log(1 - t) = -t + O(t^2)$. For p, a large prime, $t = 1/p$ is small. So,

$$\log(w(x)) \approx \sum_{\substack{p \text{ prime} \\ p < x}} \left\{ -\frac{1}{p} + O(1/p^2) \right\}.$$

Because this is not a proof, we will just throw away the error term,

$$\sum_{\substack{p \text{ prime} \\ p < x}} O(1/p^2),$$

and write

$$\log(w(x)) \approx \sum_{\substack{p \text{ prime} \\ p < x}} -\frac{1}{p}. \tag{5.1}$$

A sum over primes is hard to deal with, because we don't know which integers are prime and which are not. We would like to replace this with the analogous sum over *all* integers $n < x$,

$$\sum_{n<x} -\frac{1}{n},$$

but, of course, this is too large; it has more terms. What we want to do is increment the sum by $-1/n$ if n is a prime number p, and by 0 if n is composite. The way to estimate this is to weight each term in the sum by the probability that n is prime, which is just the unknown function $w(x)$ evaluated at $x = n$. So,

$$\log(w(x)) \approx \sum_{n<x} -\frac{w(n)}{n}.$$

As usual, to estimate a sum, we compare it to an integral of a similar form,

$$\log(w(x)) \approx \int_2^x -\frac{w(t)}{t}dt.$$

For convenience, we will introduce $a(x) = 1/w(x)$. If $w(x)$ is the chance a number near x is a prime, $a(x)$ is the average distance between primes near x. For example, if the chance is 1 in 7 that a number near x is prime, then on average, the primes are about 7 apart near x. So,

$$\log(a(x)) = -\log(w(x)) \approx \int_2^x \frac{1}{t\,a(t)}dt.$$

Observe that the integral on the right depends on x only in the upper limit of integration. If we take derivatives of both sides, according to the Fundamental Theorem of Calculus, we have

$$\frac{a'(x)}{a(x)} \approx \frac{1}{x\,a(x)},$$

or

$$a'(x) \approx \frac{1}{x} \quad \text{so } a(x) \approx \log(x), \quad w(x) = \frac{1}{a(x)} \approx \frac{1}{\log(x)}.$$

$\square$

It looks like we get exactly $1/\log(x)$, not just some some constant $C/\log(x)$. What is the C for? Well, in the preceding linear approximation of $\log(1 - 1/p)$, we neglected an error term. If we exponentiate to get rid of the logarithm and recover $w(x)$, this becomes a multiplicative error.

Exercise 5.1.1. Table 5.1 gives a list of the primes near 10000. Pick a random integer N between, say, 9500 and 10400, and count how many primes there

Table 5.1. *Primes near* 10,000

9497	9643	9769	9883	10039	10159	10273	10427
9511	9649	9781	9887	10061	10163	10289	10429
9521	9661	9787	9901	10067	10169	10301	10433
9533	9677	9791	9907	10069	10177	10303	10453
9539	9679	9803	9923	10079	10181	10313	10457
9547	9689	9811	9929	10091	10193	10321	10459
9551	9697	9817	9931	10093	10211	10331	10463
9587	9719	9829	9941	10099	10223	10333	10477
9601	9721	9833	9949	10103	10243	10337	10487
9613	9733	9839	9967	10111	10247	10343	10499
9619	9739	9851	9973	10133	10253	10357	10501
9623	9743	9857	10007	10139	10259	10369	10513
9629	9749	9859	10009	10141	10267	10391	10529
9631	9767	9871	10037	10151	10271	10399	10531

are between your N and $N + 100$. The prime number theorem says an integer near 10000 has about

$$\frac{1}{\log(10000)} = 0.108574$$

chance of being prime. So, in an interval of length 100 (including N but not $N + 100$), we expect about 10.85 primes. Repeat this experiment a couple of times with different random N, and average the number of primes you find. How close is it to 10.85?

Exercise 5.1.2. All of this so far has been assuming hypothesis I, that divisibility by different primes are independent events. We need to examine this. Consider integers around size $x = 52$. We argued earlier that the chance such an integer is divisible by 11 should be around 1 in 11. Suppose I tell you that an integer that is close to 52 is divisible by 17; now, what do you think the chance is that it is divisible by 11? (Hint: What is the smallest integer divisible by both 11 and 17?) So, are these independent events? Is hypothesis I justified?

Figure 5.2 shows a plot that compares $w(x)$ to $1/\log(x)$. Clearly, it is not a good fit! And your answer to Exercise 5.1.2 seems to indicate why. On the other hand, this seems to contradict your experimental evidence from Exercise 5.1.1. This is all very confusing; it is time to sort it all out.

First of all, the Prime Number Theorem really is true, but we need to state it in more mathematical language. After all, it is not very precise to talk about

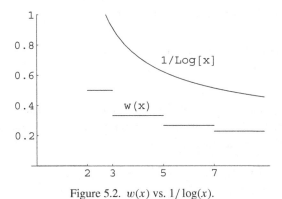

Figure 5.2. $w(x)$ vs. $1/\log(x)$.

the probability that some number N is a prime; an integer is either a prime or it isn't. Rather, we will talk about the density function $\pi(x)$ that counts the number of primes below some parameter x, that is,

$$\pi(x) = \#\{\text{primes } p \mid p < x\}.$$

Here, we use the symbol # to mean the number of elements in a set.

Exercise 5.1.3. For example, there are 1177 primes below 9500, so $\pi(9500) = 1177$. Use this fact and Table 5.1 to compute $\pi(9600)$.

If the chance that an integer N is a prime is $1/\log(N)$, then $\pi(x)$ should be $\sum_{N<x} 1/\log(N)$. This is the way to combine probabilities in successive trials of an experiment, if the probability changes with each trial.

As usual, we prefer to deal with integrals instead of sums whenever possible, so we define a new function:

$$\text{Li}(x) = \int_0^x \frac{dt}{\log(t)}, \quad \text{which should be } \approx \sum_{N<x} \frac{1}{\log(N)}.$$

The function $1/\log(x)$ is another example that does not have an "elementary" antiderivative in terms of polynomial, trigonometric, exponential, or logarithmic functions. So, $\text{Li}(x)$ is a new function, defined in terms of area. $\text{Li}(x)$, pronounced "lie of x," is known as the LOGARITHMIC INTEGRAL FUNCTION. According to the Fundamental Theorem of Calculus, this is an antiderivative of $1/\log(x)$. In the language of probability, $\text{Li}(x)$ is the density function for the probability distribution $1/\log(x)$.

We can now make a precise statement about the distribution of primes.

Theorem (Prime Number Theorem). *As x goes to infinity,*

$$\pi(x) \sim Li(x).$$

Here, of course, $\sim$ means that $\pi(x)$ is asymptotic to $Li(x)$ in the sense used in Chapter 4. There is another way to think of this, to avoid mention of probability. Consider a function

$$\omega(n) = \begin{cases} 1, & \text{if } n \text{ is prime,} \\ 0, & \text{if } n \text{ is composite.} \end{cases}$$

We can smooth out the irregular behavior of $\omega(n)$ by forming averages, just as in Chapter 4. Observe that

$$\sum_{n<x} \omega(n) = \pi(x).$$

From this point of view, the theorem says

Theorem (Prime Number Theorem). *The function $\omega(n)$ has average order* $1/\log(n)$.

Figure 5.3 compares the function $\pi(x)$ to $Li(x)$ for $x \leq 50$, $x \leq 500$, and $x \leq 5000$, respectively. The function $\pi(x)$ is constant between the primes and jumps by 1 every time x passes over a prime. As you can see, for larger values of x, the function $Li(x)$ gives a very close approximation.

Exercise 5.1.4. Table 5.2 gives the numerical values of $\pi(x)$ and $Li(x)$ for various powers of 10. The Prime Number Theorem says that the ratio $Li(x)/\pi(x)$ tends to 1 as x goes to infinity, which means that the error as a percentage tends to 0. Use the data in the table to compute this ratio and fill in the last column. You can also compute the function $x/\log(x)$ for these same values and compare them to the values for $\pi(x)$. Which gives the better approximation, $Li(x)$ or $x/\log(x)$?

The second important point we need to make in concluding this section is to resolve the paradox we created. As Exercise 5.1.2 shows, hypothesis I is *not* valid. And as Figure 5.2 indicates, the function $w(x)$ is *not* asymptotic to $1/\log(x)$. In fact, soon we will be able to prove

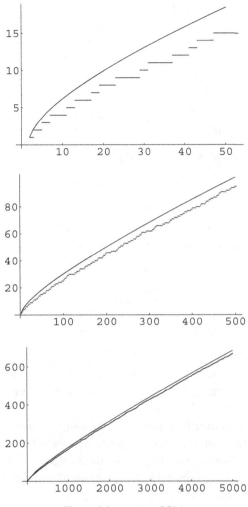

Figure 5.3. $\pi(x)$ vs. Li(x).

Theorem (Mertens' Formula). *As x goes to infinity,*

$$\prod_{\substack{p \text{ prime} \\ p < x}} (1 - 1/p) \sim \frac{\exp(-\gamma)}{\log(x)}, \tag{5.2}$$

where γ is still Euler's constant.

The numerical value of $\exp(-\gamma)$ is about $0.561459\ldots$. In Figure 5.2, the function $w(x)$ is about 56% smaller than $1/\log(x)$. What went wrong

Table 5.2. *Numerical Evidence for the Prime Number Theorem*

x	$\pi(x)$	$\text{Li}(x)$	$\text{Li}(x)/\pi(x)$
10^2	25	30.12614...	
10^3	168	177.60965...	
10^4	1229	1246.13721...	
10^5	9592	9629.80900...	
10^6	78498	78627.54915...	
10^7	664579	664918.40504...	
10^8	5761455	5762209.37544...	
10^9	50847534	50849234.95700...	
10^{10}	455052511	455055614.58662...	

in the nonproof of the Prime Number Theorem? The first mistake was assuming hypothesis I. The second mistake was throwing away the error terms

$$\sum_{\substack{p \text{ prime} \\ p < x}} O(1/p^2)$$

in approximating $\log(1 - 1/p)$ by $-1/p$. Roughly speaking, these two errors tend to cancel each other out. Another way to think about this is that the factor $\exp(-\gamma)$ in some sense quantifies to what extent divisibility by different primes fail to be independent events.

You may have an idea how to fix the probability argument for the Prime Number Theorem. If an integer N is composite, it must be divisible by some prime $p \le \sqrt{N}$. This suggests that we might change the definition of $w(x)$ to include only those primes below $\sqrt{x}$, instead of all primes below x. But Mertens' Formula indicates that

$$\prod_{\substack{p \text{ prime} \\ p < x^{1/2}}} (1 - 1/p) \sim \frac{\exp(-\gamma)}{\log(x^{1/2})} = \frac{2\exp(-\gamma)}{\log(x)}.$$

Because $2\exp(-\gamma) = 1.12292$, this would indicate than an integer near x has about $1.12292/\log(x)$ chance of being prime. But this is too big; it does not agree with the numerical evidence of Exercise 5.1.1. In some sense, this was too easy a fix to the problem caused by the failure of hypothesis I. We will come back to this idea later.

Exercise 5.1.5. Because $1/2$ does not work, what exponent a should you pick so that

$$\prod_{\substack{p \text{ prime} \\ p < x^a}} (1 - 1/p) \sim \frac{1}{\log(x)}?$$

Assume that (5.2) is true and solve for a.

5.2. Chebyshev's Estimates

In Section 10.3, we will show that

$$\mathrm{Li}(x) \sim \frac{x}{\log(x)},$$

so the Prime Number Theorem will also imply the approximation

$$\pi(x) \sim \frac{x}{\log(x)}.$$

This version is not as good, not only because it is less accurate (as you saw in Exercise 5.1.4), but also because it obscures the natural heuristic interpretation as a probability density. Nonetheless, it is useful because we can prove by elementary methods the

Theorem. *As x goes to infinity,*

$$\pi(x) \ll \frac{x}{\log(x)}. \tag{5.3}$$

In fact, we will show that

$$\pi(x) < 2\frac{x}{\log(x)}.$$

An estimate of this kind with the better constant $C = 1.1055$ was first proved by the Russian mathematician Pafnuty Lvovich Chebyshev around 1850. Of course, the closer you get to the optimal constant $C = 1$, the harder you have to work.

Proof. We start with the strange observation that

$$2^{2n} = (1+1)^{2n} = \binom{2n}{0} + \cdots + \binom{2n}{n} + \cdots + \binom{2n}{2n}$$

$$> \binom{2n}{n} = \frac{2n!}{n!n!}$$

$$= \frac{(2n) \cdot (2n-1) \cdot (2n-2) \cdots (n+1)}{n \cdot (n-1) \cdot (n-2) \cdots 1}.$$

Every prime number p between n and $2n$ divides the numerator and cannot appear in the denominator. So, each must divide the quotient, and therefore the product does:

$$\prod_{n < p \le 2n} p \text{ divides } \binom{2n}{n}.$$

The product has exactly $\pi(2n) - \pi(n)$ factors, namely the primes p, and each p is bigger than n. So,

$$n^{\pi(2n)-\pi(n)} < \prod_{n < p \le 2n} p \le \binom{2n}{n} < 2^{2n}.$$

Take logarithms and divide by $\log(n)$ to see that

$$\pi(2n) - \pi(n) < \frac{2n \log(2)}{\log(n)}. \tag{5.4}$$

The proof now goes by induction. To start, we need the following.

Exercise 5.2.1. Using calculus or a graphing calculator, plot $2x/\log(x)$ for $2 \le x \le 100$. By comparing the graph to Figure 5.3, you see that the theorem is easily true in this range.

For the inductive step, we can assume that $\pi(n) < 2n/\log(n)$, and $2 \log(2)$ is less than 1.39, so (5.4) says that

$$\pi(2n) = \frac{2n \log(2)}{\log(n)} + \pi(n) < 3.39 \frac{n}{\log(n)}. \tag{5.5}$$

Exercise 5.2.2. Use calculus and your calculator to show that

$$0 < x - \frac{3.44}{4}(\log(2) + x) \quad \text{for} \quad x \ge 4.6.$$

So,

$$3.44 < \frac{4x}{\log(2) + x} \quad \text{for} \quad x \ge 4.6.$$

Because $\log(100) = 4.60517 > 4.6$, with $x = \log(n)$, this says that

$$3.39 < 3.44 < 4\frac{\log(n)}{\log(2n)} \qquad \text{for} \quad n \geq 100.$$

Multiply through by $n/\log(n)$ to get

$$3.39\frac{n}{\log(n)} < 4\frac{n}{\log(2n)} \qquad \text{for} \quad n \geq 100.$$

So, according to (5.5),

$$\pi(2n) < 4\frac{n}{\log(2n)} = 2\frac{2n}{\log(2n)}.$$

This gets us from n to $2n$, but we also need to consider the odd integers. There is at most one prime between $2n$ and $2n + 1$, so

$$\pi(2n + 1) \leq \pi(2n) + 1 < 3.39\frac{n}{\log(n)} + 1 \qquad (5.6)$$

according to (5.5) again. Because $\log(n)/n$ is decreasing to 0 and $\log(100)/100 = 0.0460517$ is already less than 0.05, the inequalities above imply that

$$3.39 + \frac{\log(n)}{n} < 3.44 < 4\frac{\log(n)}{\log(2n)} \qquad \text{for} \quad n \geq 100.$$

Multiply by $n/\log(n)$ to get

$$3.39\frac{n}{\log(n)} + 1 < 4\frac{n}{\log(2n)} < 2\frac{2n + 1}{\log(2n)} \qquad \text{for} \quad n \geq 100.$$

According to (5.6),

$$\pi(2n + 1) < 2\frac{2n + 1}{\log(2n)} \qquad \text{for} \quad n \geq 100.$$

A fully rigorous proof would try your patience, so we will make an estimate. With $f(x) = 2x/\log(x)$ and $f'(x) = 2(\log(x) - 1)/\log(x)^2 \approx 2/\log(x)$, a linear approximation gives that

$$2\frac{2n + 1}{\log(2n)} = \frac{4n}{\log(2n)} + \frac{2}{\log(2n)}$$
$$\approx f(2n) + f'(2n) \cdot 1$$
$$\approx f(2n + 1) = 2\frac{2n + 1}{\log(2n + 1)}.$$

So,

$$\pi(2n+1) < 2\frac{2n+1}{\log(2n+1)},$$

which finishes the induction.   □

In the other direction, we can also prove

Theorem. *As x goes to infinity,*

$$\frac{x}{\log(x)} \ll \pi(x).$$

In fact, we will show that for $x \geq 15$,

$$\frac{1}{2}\frac{x}{\log(x)} < \pi(x).$$

As with the previous theorem, if we are willing to work harder, we can get a constant closer to the optimal value of $C = 1$. Chebyshev proved you could take $C = 0.921292$.

Lemma. *For a prime number p, let $v_p(n!)$ be the largest power of p that divides $n!$. Then,*

$$v_p(n!) = [n/p] + [n/p^2] + [n/p^3] + \cdots = \sum_{j=1}^{\infty} [n/p^j]. \qquad (5.7)$$

Notice that the sum really is finite; for any given n and p, eventually $[n/p^j] = 0$ when $p^j > n$.

Proof. We get a copy of p dividing $n!$ for each multiple of p below n; there are $[n/p]$ of these. But, then, we get an extra copy of p dividing $n!$ for each multiple of p^2 below n; there are $[n/p^2]$ of these. An example will perhaps clarify. Take $n = 13$ and $p = 2$. Then,

$$13! = 1 \cdot 2 \cdot 3 \cdot 4 \cdot 5 \cdot 6 \cdot 7 \cdot 8 \cdot 9 \cdot 10 \cdot 11 \cdot 12 \cdot 13.$$

We get a 2 dividing 13! from the 2, 4, 6, 8, 10, 12; there are $6 = [13/2]$ of these. But we get another 2 in the 4, 8, 12; there are $3 = [13/4]$ of these. Finally, we get yet another 2 in the multiples of 8 below 13; there is only $1 = [13/8]$. And no multiple of 16 is below 13. So, $v_2(13!) = 10$.   □

Lemma. *For any real numbers a and b,*

$$[a + b] - [a] - [b] \quad \text{is either 0 or 1.}$$

Proof. Write $a = [a] + \delta$, and $b = [b] + \epsilon$, with $0 \leq \delta, \epsilon < 1$. Then,

$$[a + b] = [[a] + \delta + [b] + \epsilon] = [[a] + [b] + (\delta + \epsilon)]. \tag{5.8}$$

From the bounds on δ and ϵ, $0 \leq \delta + \epsilon < 2$. There are two cases. If $0 \leq \delta + \epsilon < 1$, (5.8) says that $[a + b] = [a] + [b]$. But if $1 \leq \delta + \epsilon < 2$, then (5.8) says that $[a + b] = [a] + [b] + 1$. $\qquad\square$

Lemma. *For a prime number p, let $v_p\left(\binom{n}{k}\right)$ be the largest power of p dividing $\binom{n}{k}$. Then,*

$$p^{v_p\left(\binom{n}{k}\right)} \leq n.$$

Proof. Because $\binom{n}{k} = n!/(k!(n-k)!)$, we have

$$v_p\left(\binom{n}{k}\right) = v_p(n!) - v_p(k!) - v_p((n-k)!)$$

$$= \sum_{j=1}^{\infty} \{[n/p^j] - [k/p^j] - [(n-k)/p^j]\}.$$

According to the previous lemma, with $a = k/p^j$ and $b = (n-k)/p^j$, every term in the sum is either a 0 or a 1. And there are at most $\log(n)/\log(p)$ nonzero terms because $[n/p^j] = 0$ when $p^j > n$; that is, $j > \log(n)/\log(p)$. So, $v_p(\binom{n}{k}) < \log(n)/\log(p)$. Multiply both sides by $\log(p)$ and exponentiate to get the lemma. $\qquad\square$

Proof of Theorem. We see that the last lemma gives that for $0 \leq k \leq n$,

$$\binom{n}{k} = \prod_{p \leq n} p^{v_p\left(\binom{n}{k}\right)} \leq n^{\pi(n)}.$$

$\qquad\square$

If we add up these $n + 1$ inequalities for $0 \leq k \leq n$, we get

$$2^n = (1+1)^n = \sum_{k=0}^{n} \binom{n}{k} \leq (n+1)n^{\pi(n)}.$$

Take logarithms and solve for $\pi(n)$ to see that

$$\frac{n \log(2)}{\log(n)} - \frac{\log(n+1)}{\log(n)} \leq \pi(n).$$

Finally, since $\log(n+1)/n$ is decreasing to 0 and $1/2 < \log(2) - \log(16)/15 = 0.50830$, we know that

$$\frac{1}{2} < \log(2) - \frac{\log(n+1)}{n} \quad \text{for } n \geq 15.$$

Multiply by $n/\log(n)$ to get

$$\frac{1}{2}\frac{n}{\log(n)} < \pi(n) \quad \text{for } n \geq 15.$$

Interlude 2

Series

In the previous chapter, we used the idea of a linear approximation on the function $\log(1 - t)$ in the discussion of the Prime Number Theorem. Approximations that are better than linear will be even more useful, so in this interlude we develop higher order approximations. Almost every single exercise here is referred to later on, so if you don't do them now, you will need to do them eventually.

I2.1. Taylor Polynomials

Suppose we want an approximation to $y = f(x)$ at $x = a$ that is a little better than (I1.1). We might look for a quadratic polynomial. If we want it to be at least as good as the linear approximation, it should touch the graph (i.e., pass through the point $(a, f(a))$) and also be tangent to the graph, so the first derivative should be $f'(a)$. But we also want the second derivatives to match up; the second derivative of the quadratic should be $f''(a)$. Here's the formula, which a little thought will show does what it is supposed to:

$$f(x) \approx f(a) + f'(a)(x - a) + \frac{1}{2} f''(a)(x - a)^2. \qquad (I2.1)$$

Notice that $f(a)$, $f'(a)$, and $f''(a)$ are all numbers, so the right side really is a quadratic polynomial, whereas the left side can be a more complicated function. Compare this formula to the one for the linear approximation (I1.1). All we have done is add an extra term, the $\frac{1}{2} f''(a)(x - a)^2$. This extra term has the property that it is zero at $x = a$, and the derivative of this term, $f''(a)(x - a)$, also is zero at $x = a$. So, adding it on doesn't ruin the linear approximation. On the other hand, taking two derivatives of the linear approximation gives zero (check this), whereas two derivatives of our extra term is exactly what we want: $f''(a)$. The $1/2$ is there to compensate for the fact that taking derivatives will introduce a factor of 2 in that term.

111

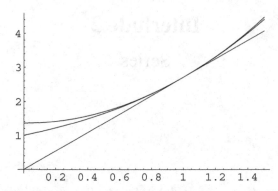

Figure I2.1. Graph of e^x, with linear and quadratic approximations.

Here's an example. Take $f(x) = \exp(x)$ and $a = 1$. Because $f(1)$, $f'(1)$, and $f''(1)$ are all equal to e, we get that

$$e^x \approx e + e(x - 1) + \frac{e}{2}(x - 1)^2,$$

for x close to 1. Notice that the right side is a quadratic polynomial with the property that its value at $x = 1$ and its first and second derivatives at $x = 1$ are all equal to e. Figure I2.1 shows the graph of $y = \exp(x)$, together with the linear approximation $y = e + e(x - 1)$ (on the bottom) and the quadratic approximation $y = e + e(x - 1) + e/2(x - 1)^2$ (on the top). Notice that the quadratic is much closer to $\exp(x)$ near $x = 1$ but curves away eventually.

Next, take $f(x) = \sin(x)$ and $a = \pi/2$. We have $\sin(\pi/2) = 1, \cos(\pi/2) = 0$, and $-\sin(\pi/2) = -1$. So,

$$\sin(x) \approx 1 - \frac{1}{2}(x - \pi/2)^2$$

for x near $\pi/2$.

Figure I2.2 shows the graph of $y = \sin(x)$, together with the linear approximation $y = 1 = 1 + 0(x - \pi/2)$, and the quadratic approximation $1 - 1/2(x - \pi/2)^2$.

The most common mistake is to forget to evaluate at $x = a$ at the appropriate time. This gives a result that is not a quadratic polynomial. For example, if you don't plug in $a = \pi/2$, you would write something like $\sin(x) + \cos(x)(x - \pi/2) - \frac{1}{2}\sin(x)(x - \pi/2)^2$. You get an ugly mess, not a nice simple polynomial. Don't do this.

Exercise I2.1.1. Find a linear approximation and a quadratic approximation to $\exp(x)$ at $x = 0$. Plot them together with $\exp(x)$ on your calculator.

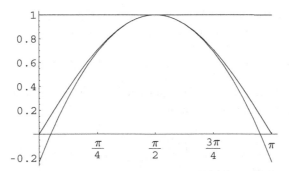

Figure I2.2. Graph of sin(x), with linear and quadratic approximations.

Compare your answer to the example above; the point here is that the polynomial that is the best approximation will depend on the base point a.

Exercise I2.1.2. Find a quadratic approximation to $\log(x)$ at $x = 1$. Plot the quadratic together with $\log(x)$ on your calculator. Do the same for $1/(1 - x)$ at $x = 0$.

Exercise I2.1.3. Find a quadratic approximation to $\sin(x)$ at $x = 0$. Sometimes the linear approximation *is* the best quadratic approximation.

Exercise I2.1.4. Find a quadratic approximation to $3x^2 + 7x - 4$ at $x = 0$. Draw some conclusion about your answer. Now, find the quadratic approximation at $x = 1$. Notice that your answer is really just $3x^2 + 7x - 4$, written in a funny way.

Having carefully read the explanation of how the quadratic approximation works, you should be able to guess the formula to approximate a function with a degree 3 polynomial. (If not, go back and reread the quadratic part.) What we need to do is keep the quadratic approximation and add on a term that makes the third derivatives match up:

$$f(x) \approx f(a) + f'(a)(x - a) + \frac{1}{2}f''(a)(x - a)^2 + \frac{1}{6}f'''(a)(x - a)^3,$$

when x is close to a. The factor of $1/6$ is there because the third derivative of $(x - a)^3$ is exactly $6 = 2 \cdot 3$. The notation for higher derivatives starts to get confusing, so from now on we will also write $f^{(n)}(a)$ for the nth derivative of $f(x)$ at $x = a$. Remember that $f^{(0)}(a)$ means no derivatives at all; it's just the original function $f(x)$ evaluated at $x = a$.

We can get a polynomial of any degree n to approximate a function $f(x)$, which we will call the nTH DEGREE TAYLOR POLYNOMIAL OF f AT a,

$$f(x) \approx f^{(0)}(a) + f^{(1)}(a)(x - a) + \cdots + \frac{1}{n!} f^{(n)}(a)(x - a)^n, \quad \text{(I2.2)}$$

for x near a. Notice that $1! = 1$, and we just define $0! = 1$; so, every term in the polynomial is of the same form,

$$\frac{1}{k!} f^{(k)}(a)(x - a)^k,$$

where k goes from 0 up to n.

Exercise I2.1.5. Compute the degree three Taylor polynomial of each of the functions for which you previously computed the quadratic approximation. In each case, graph the polynomial with the function being approximated.

Exercise I2.1.6. Try to compute the degree three Taylor polynomial at $x = 0$ for the function $x^2 \cos(3x)$. It is very painful exercise. The next section will show you some useful shortcuts.

I2.2. Taylor Series

So far, we've discussed how to get various approximations to a function $f(x)$ at a base point a. For example, in Exercises I2.1.1 and I2.1.5, with $f(x) = \exp(x)$, you computed that

$$\exp(x) \approx 1 + x$$

$$\approx 1 + x + \frac{x^2}{2}$$

$$\approx 1 + x + \frac{x^2}{2} + \frac{x^3}{6}$$

for x near 0. But how near is "near," and how close is the polynomial to the function? The two questions are clearly related. In fact, for linear approximations, we made this explicit in terms of Big Oh notation in Section I1.2 of Interlude 1. A similar statement is true about the higher order approximations:

$$\exp(x) = 1 + x + O(x^2)$$

$$= 1 + x + \frac{x^2}{2} + O(x^3)$$

$$= 1 + x + \frac{x^2}{2} + \frac{x^3}{6} + O(x^4).$$

In fact, the functions we will consider all will be differentiable, as many times as we want (unlike the example $\sqrt{x}$ at $x = 0$.) In this case, the Big Oh notation can be used. That is, the precise meaning of $\approx$ in (I2.2) will turn out to be that

$$f(x) = f^{(0)}(a) + f^{(1)}(a)(x - a) + \cdots$$
$$+ \frac{1}{n!} f^{(n)}(a)(x - a)^n + O((x - a)^{n+1}). \quad \text{(I2.3)}$$

We will come back to this later; for now, we will simply define (I2.2) to mean that the right side is the nth Taylor polynomial of the left side.

In this section, we will define the TAYLOR SERIES. Think of the Taylor series as a polynomial of infinite degree, which goes on forever. The coefficients are determined by formula (I2.2). For example, every derivative of $\exp(x)$ is $\exp(x)$, and $e^0 = 1$. So, the Taylor series for $\exp(x)$ at $x = 0$ is

$$1 + x + \frac{x^2}{2} + \frac{x^3}{6} + \cdots + \frac{x^n}{n!} + \cdots.$$

This is just a way to keep track of the fact that Taylor approximations of all possible degrees exist for this function, even though in any given computation we would use one of finite degree. Analogously, the decimal digits of $\pi = 3.141592653589793\ldots$ go on forever, even though we can only use finitely many of them in a given calculation.

Another way to write the Taylor series more compactly is with the Sigma notation. The Taylor series for $\exp(x)$ at $x = 0$ is

$$1 + x + \frac{x^2}{2} + \frac{x^3}{6} + \cdots + \frac{x^n}{n!} + \cdots = \sum_{n=0}^{\infty} \frac{x^n}{n!}. \quad \text{(I2.4)}$$

The Sigma notation just means you add up all terms of the form $x^n/n!$, for all n going from 0 to infinity.

The functions $\sin(x)$ and $\cos(x)$ have a nice pattern in their Taylor polynomials, because the derivatives repeat after the fourth one: $\sin(x)$, $\cos(x)$, $-\sin(x)$, $-\cos(x)$, $\sin(x)$, $\cos(x)$, etc. When we plug in 0, we get a sequence $0, 1, 0, -1, 0, 1, 0, -1, \ldots$. So according to formula (I2.2), the Taylor series for $\sin(x)$ at $x = 0$ is

$$x - \frac{x^3}{3!} + \frac{x^5}{5!} - \cdots + (-1)^n \frac{x^{2n+1}}{(2n+1)!} + \cdots = \sum_{n=0}^{\infty} (-1)^n \frac{x^{2n+1}}{(2n+1)!},$$
$$\text{(I2.5)}$$

and the Taylor series for $\cos(x)$ at $x = 0$ is

$$1 - \frac{x^2}{2!} + \frac{x^4}{4!} - \cdots + (-1)^n \frac{x^{2n}}{(2n)!} + \cdots = \sum_{n=0}^{\infty} (-1)^n \frac{x^{2n}}{(2n)!}.$$

(I2.6)

Another basic function that is very useful is $(1 - x)^{-1}$. Convince yourself that the nth derivative of this function is $n!(1 - x)^{-n-1}$. So, the Taylor series for $(1 - x)^{-1}$ is just the (infinite) Geometric series

$$1 + x + x^2 + \cdots + x^n + \cdots = \sum_{n=0}^{\infty} x^n.$$

(I2.7)

We saw a finite version of this in Exercise 1.2.10. The Taylor series (I2.4), (I2.5), (I2.6), and (I2.7) come up so often that you will certainly need to know them.

Why are we doing this? It seems a little abstract. Here's the reason: In the previous sections, when computing Taylor polynomials, we had to take a lot of derivatives. This can get very tedious, and it is hard to do accurately. (Recall Exercise I2.1.6.) The Taylor series provide a convenient notation for learning the shortcuts to these computations.

We can add Taylor series together, multiply them, or make substitutions. This will give the Taylor series of new functions from old ones. For example, to get the Taylor series of $\sin(2x)$, we just take (I2.5) and substitute $2x$ for x. Thus, the series for $\sin(2x)$ is

$$2x - \frac{8x^3}{3!} + \frac{32x^5}{5!} - \cdots + (-1)^n \frac{2^{2n+1} x^{2n+1}}{(2n+1)!} + \cdots = \sum_{n=0}^{\infty} (-1)^n \frac{2^{2n+1} x^{2n+1}}{(2n+1)!}.$$

Similarly, if we want the series for $\exp(x^3)$, we take (I2.4) and substitute x^3 for x. Thus, the series for $\exp(x^3)$ is

$$1 + x^3 + \frac{x^6}{2} + \frac{x^9}{6} + \cdots + \frac{x^{3n}}{n!} + \cdots = \sum_{n=0}^{\infty} \frac{x^{3n}}{n!}.$$

We can also add or subtract series. The series for $\exp(x) - 1 - x$ is

$$\frac{x^2}{2} + \frac{x^3}{6} + \cdots + \frac{x^n}{n!} + \cdots = \sum_{n=2}^{\infty} \frac{x^n}{n!},$$

because $1 + x$ is its own Taylor series (see Exercise I2.1.4). To get the series for $x^2 \exp(x)$, just multiply every term in the series for $\exp(x)$ by x^2, to get

$$x^2 + x^3 + \frac{x^4}{2} + \frac{x^5}{6} + \cdots + \frac{x^{n+2}}{n!} + \cdots = \sum_{n=0}^{\infty} \frac{x^{n+2}}{n!}.$$

Moral: It is almost always easier to start with a known series and use these shortcuts than it is to take lots of derivatives and use (I2.2).

Exercise I2.2.1. Use (I2.7) to compute the Taylor series at $x = 0$ for each of these functions: $(1 + x)^{-1}, (1 - x^2)^{-1}, (1 + x^2)^{-1}$.

Exercise I2.2.2. Use the methods of this section to compute the Taylor series at $x = 0$ of $x^2 \cos(3x)$. Compare this to what you did in Exercise I2.1.6. Reread the moral above.

Exercise I2.2.3. Compute the Taylor series at $x = 0$ of $\sin(x)/x$ and $(\exp(x) - 1)/x$.

Earlier, we saw how to add, subtract, and make substitutions. Next, we will take derivatives and integrate, multiply, and divide.

Working with derivatives will be easy, as you already know the rules for derivatives of polynomials. Thus, because the derivative of $(1 - x)^{-1}$ is $(1 - x)^{-2}$, we get the Taylor series for $(1 - x)^{-2}$ by taking the derivative of every term in (I2.7):

$$\frac{d}{dx}(1 + x + x^2 + x^3 + \cdots + x^n + \cdots) =$$

$$0 + 1 + 2x + 3x^2 + \cdots + nx^{n-1} + \cdots =$$

$$\sum_{n=1}^{\infty} nx^{n-1} = \sum_{n=0}^{\infty} (n + 1)x^n.$$

Notice that there are two ways to write the series in Sigma notation: by keeping track of the power of x, or by keeping track of the coefficient. Because the derivative of $\exp(x)$ is $\exp(x)$, we should have that the Taylor series (I2.4) is its own derivative, and it is:

$$\frac{d}{dx}\left(1 + x + \frac{x^2}{2} + \frac{x^3}{6} + \cdots + \frac{x^n}{n!} + \cdots\right) =$$

$$0 + 1 + x + \frac{x^2}{2} + \cdots + \frac{nx^{n-1}}{n!} + \cdots.$$

Every term shifts down by one. Notice that $n/n!$ is just $1/(n-1)!$.

We can also integrate Taylor series term by term. For example, because the antiderivative of $(1-x)^{-1}$ is $-\log(1-x)$, we get the Taylor series for $-\log(1-x)$ by computing the antiderivative of each term in (I2.7):

$$x + \frac{x^2}{2} + \frac{x^3}{3} + \cdots + \frac{x^n}{n} \cdots = \sum_{n=1}^{\infty} \frac{x^n}{n}. \qquad (I2.8)$$

If we want instead $\log(1-x)$, of course we have to multiply every term by -1. There is a subtle point here; the antiderivative of a function is determined only up to a constant (the $+C$ term). In this example, $-\log(1-x)$ is the unique choice of antiderivative that is zero at $x = 0$. That value of the function determines the constant term of the series expansion. So, in this case, the constant term is 0.

This idea lets us get a handle on functions that do not have a simple antiderivative. For example in Section I1.3, we said that the function $\exp(-x^2)$ has no simple antiderivative. If we take a Taylor series for $\exp(-x^2)$,

$$1 - x^2 + \frac{x^4}{2} - \frac{x^6}{6} + \cdots + \frac{(-1)^n x^{2n}}{n!} + \cdots = \sum_{n=0}^{\infty} \frac{(-1)^n x^{2n}}{n!},$$

and integrate term by term, we get

$$x - \frac{x^3}{3} + \frac{x^5}{10} - \frac{x^7}{42} + \cdots + \frac{(-1)^n x^{2n+1}}{(2n+1)n!} + \cdots = \sum_{n=0}^{\infty} \frac{(-1)^n x^{2n+1}}{(2n+1)n!}.$$

This is the Taylor series for the antiderivative of $\exp(-x^2)$, which is 0 at $x = 0$. The function that has this as its Taylor series is the "error function,"

$$\mathrm{Erf}(x) = \int_0^x \exp(-t^2)\,dt.$$

We can use this to approximate $\mathrm{Erf}(1)$; take the first six terms of the series and plug in $x = 1$ to get

$$\int_0^1 \exp(-t^2)\,dt \approx 1 - \frac{1}{3} + \frac{1}{10} - \frac{1}{42} + \frac{1}{216} - \frac{1}{1320} = 0.746729\ldots,$$

which compares well with the exact answer, $0.746824\ldots$.

Figure I2.3 shows the graph of $y = \exp(-x^2)$, (the bell curve) together with the graph of $y = \mathrm{Erf}(x)$ discussed earlier. Convince yourself by comparing the graphs that the bell curve has the right properties to be the derivative of the Error function.

Figure I2.4 shows the degree 1, 5, and 9 Taylor approximation to $\mathrm{Erf}(x)$ together with the function itself. Because any polynomial eventually tends to

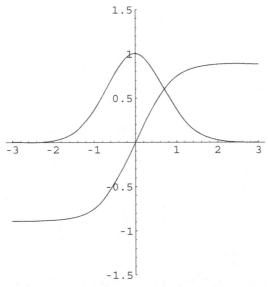

Figure I2.3. Graph of the "bell curve" e^{-x^2} and its antiderivative Erf (x).

infinity, the approximations all eventually curve away from Erf(x). But the higher degree ones "last longer." The function Erf(x) itself tends to ± 1 as x tends to $\pm\infty$.

Another way to think about this is that we have used a Taylor series to solve the differential equation

$$y' = \exp(-x^2), \quad y(0) = 0.$$

Here are the last tricks of this section: Instead of multiplying a Taylor series by a number, or a power of x, we can multiply two different series together. In general, the coefficients of the product are complicated, so we will not use the Sigma notation to write the general, coefficient; we will just write the first few. For example, the series for $\sin(x)\cos(x)$ comes from multiplying

$$\left(x - \frac{x^3}{3!} + \frac{x^5}{5!} - \cdots\right)\left(1 - \frac{x^2}{2!} + \frac{x^4}{4!} - \cdots\right)$$

$$= x - \left(\frac{1}{2!} + \frac{1}{3!}\right)x^3 + \left(\frac{1}{5!} + \frac{1}{2!3!} + \frac{1}{4!}\right)x^5 + \cdots =$$

$$x - \frac{2}{3}x^3 + \frac{2}{15}x^5 + \cdots.$$

Notice that there are two ways of getting an x^3 term in the series for $\sin(x)\cos(x)$: one from the product of the x term in $\sin(x)$ and the x^2 term in

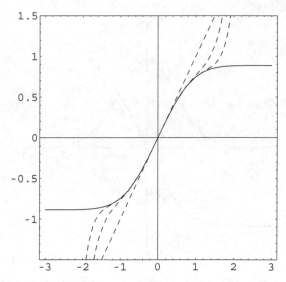

Figure I2.4. Graph of Erf (x) and some Taylor polynomial approximations.

$\cos(x)$, and another from the product of the constant term in $\cos(x)$ and the x^3 term in $\sin(x)$. There are three ways of getting an x^5 term, and so forth.

One can also divide one series by another, just like long division. It is easiest to name the coefficients of the answer $a_0 + a_1 x + a_2 x^2 + a_3 x^3 + \cdots$, multiply both sides by the denominator, and equate coefficients. For example, if we want a series for $1/(2 - x + x^2)$, this leads to

$$1 + 0 \cdot x + 0 \cdot x^2 + 0 \cdot x^3 + \cdots = 1 =$$
$$(2 - x + x^2 + 0 \cdot x^3 + 0 \cdot x^4 + \cdots)(a_0 + a_1 x + a_2 x^2 + a_3 x^3 + \cdots).$$

We get the equations

$$
\begin{array}{ll}
1 = 2 \cdot a_0, & \text{coefficient of } x^0, \\
0 = 2 \cdot a_1 - 1 \cdot a_0, & \text{coefficient of } x^1, \\
0 = 2 \cdot a_2 - 1 \cdot a_1 + 1 \cdot a_0, & \text{coefficient of } x^2, \\
0 = 2 \cdot a_3 - 1 \cdot a_2 + 1 \cdot a_1 + 0 \cdot a_0, & \text{coefficient of } x^3, \\
\quad \vdots & \quad \vdots
\end{array}
$$

We can solve each equation in turn to see that $a_0 = 1/2, a_1 = 1/4, a_2 = -1/8$, $a_3 = -3/16, \ldots$.

Exercise I2.2.4. In Exercise I2.2.1, you computed the Taylor series for $1/(1 + x^2)$. Use this to compute the Taylor series for $\arctan(x)$.

Exercise I2.2.5. What function is $x\frac{d}{dx}(1 + x)^{-1}$? Compute its Taylor series at $x = 0$.

Exercise I2.2.6. Use the answer to Exercise I2.2.1 to compute the Taylor series at $x = 0$ for $-x + \log(1 + x)$.

Exercise I2.2.7. Compute the Taylor series for $-\log(1 - x)/x$ at $x = 0$.

Exercise I2.2.8. In Exercise I2.2.3, you computed the Taylor series for $(\exp(x) - 1)/x$. Use the techniques of this section to find the Taylor series for the "Exponential integral" function, $\text{Ei}(x)$, defined by

$$\text{Ei}(x) = \int_0^x \frac{\exp(t) - 1}{t} dt.$$

In other words, $\text{Ei}(x)$ is the antiderivative of $(\exp(x) - 1)/x$, which is 0 at $x = 0$. This is another function that comes up in physics and engineering.

Exercise I2.2.9. Euler's dilogarithm $L(x)$ is defined by

$$L(x) = \int_0^x \frac{-\log(1 - t)}{t} dt;$$

in other words, it is the antiderivative of $-\log(1 - x)/x$, which is 0 at $x = 0$. We will see applications to number theory. Compute the Taylor series expansion for $L(x)$ at $x = 0$.

Exercise I2.2.10. Find the first few terms for the series expansion of $\sec(x) = 1/\cos(x)$ and of $x/(\exp(x) - 1)$.

Exercise I2.2.11. The point of these sections was to develop shortcuts for computing Taylor series, because it is hard to compute many of derivatives. We can turn this around and use a known Taylor series at $x = a$ to tell what the derivatives at $x = a$ are: The nth coefficient is the nth derivative at a, divided by $n!$. Use this idea and your answer to Exercise I2.2.2 to compute the tenth derivative of $x^2 \cos(3x)$ at $x = 0$.

I2.3. Laurent Series

Suppose we want to study the function $f(x) = (3x^2 + 7x - 4)/x$ near $x = 0$.
The function is not even defined for $x = 0$, so it certainly has no derivatives and
no Taylor series for $x = 0$. Nonetheless, the simpler function $3x^2 + 7x - 4$
is nice at $x = 0$; in Exercise I2.1.4 you showed it was equal to its own Taylor
series. Our methods used in previous sections indicate that we should divide
out an x, and what we get certainly is an algebraic identity:

$$f(x) = \frac{3x^2 + 7x - 4}{x} = \frac{-4}{x} + 7 + 3x.$$

That is,

$$f(x) - \frac{-4}{x} = 7 + 3x$$

is defined at $x = 0$ and has the Taylor series given by the right side of the
equation. We can often do this. For example, $\cos(x)/x^2$ is again not defined
at $x = 0$, but because

$$\cos(x) = 1 - \frac{x^2}{2!} + \frac{x^4}{4!} - \cdots + (-1)^n \frac{x^{2n}}{(2n)!} + \cdots = \sum_{n=0}^{\infty} (-1)^n \frac{x^{2n}}{(2n)!},$$

at least formally

$$\frac{\cos(x)}{x^2} = \frac{1}{x^2} - \frac{1}{2!} + \frac{x^2}{4!} - \cdots + (-1)^n \frac{x^{2n-2}}{(2n)!} + \cdots = \sum_{n=0}^{\infty} (-1)^n \frac{x^{2n-2}}{(2n)!}.$$

We interpret this as meaning that $\cos(x)/x^2 - 1/x^2$ has a Taylor expansion:

$$\frac{\cos(x)}{x^2} - \frac{1}{x^2} = -\frac{1}{2!} + \frac{x^2}{4!} - \cdots + (-1)^n \frac{x^{2n-2}}{(2n)!}.$$

Is this really legal? Let $l(x) = \cos(x)/x^2 - 1/x^2$ be the function on the left,
and let $r(x)$ be the function given by the Taylor series on the right. Then,
certainly $x^2 l(x) = x^2 r(x)$ for all values of x; it is just the Taylor expansion of
$\cos(x) - 1$. So, certainly, $l(x) = r(x)$ for all x different from 0. Because $r(x)$
is defined at 0 ($r(0)$ is the constant term $-1/2$), we simply use that value to
define $l(0) = -1/2$. Furthermore, two applications of L'Hopital's rule show
that

$$\lim_{x \to 0} \frac{\cos(x) - 1}{x^2} = \lim_{x \to 0} \frac{-\sin(x)}{2x} = \frac{-\cos(0)}{2} = -\frac{1}{2};$$

so, this makes $l(x)$ continuous at $x = 0$.

   The point of this discussion is that we can extend our techniques for Taylor
series, even to points where the function is not defined, if we allow finitely

many terms with negative exponent in the series expansion. We use the name LAURENT SERIES for such a thing and allow ourselves to write expressions such as

$$f(x) = \frac{c_{-N}}{(x-a)^N} + \cdots + \frac{c_{-1}}{x-a} + c_0 + c_1(x-a) + \cdots$$

$$= \sum_{n=-N}^{\infty} c_n(x-a)^n.$$

As above, the meaning of this is that if we subtract all the terms with negative exponent over to the left side, we get a function that has a Taylor series at the point $x = a$. The terms with negative exponent

$$\frac{c_{-N}}{(x-a)^N} + \cdots + \frac{c_{-1}}{x-a}$$

are what is called the SINGULAR PART of the function, and the coefficient c_{-1} of $1/(x-a)$ is called the RESIDUE at $x = a$. Of course, there need not be a singular part or a residue; a Taylor series is a perfectly good Laurent series, too. Also, as usual, the expansion depends on the base point chosen. For example, the function

$$\frac{1}{1-x} = \sum_{n=0}^{\infty} x^n$$

has the Geometric series as its Taylor series at $x = 0$. But, at $x = 1$,

$$\frac{1}{1-x} = \frac{-1}{x-1};$$

the function has a Laurent series consisting of a single term. If there is a singular part in the expansion at $x = a$, then a is called a POLE. If the singular part consists of a single term $c_{-1}/(x-a)$, we say a is a SIMPLE POLE.

Here's another example. The function $2/(x^2 - 1)$ is not defined at $x = 1$. But

$$\frac{2}{x^2-1} = \frac{1}{x-1} + \frac{-1}{x+1};$$

by partial fractions. (Put the right side over a common denominator.) And $-1/(x+1)$ is defined at $x = 1$; we get its Taylor series by fooling with the Geometric series:

$$\frac{-1}{x+1} = \frac{-1}{2+(x-1)} = \frac{-1}{2}\frac{1}{1+(x-1)/2} = \frac{-1}{2}\sum_{n=0}^{\infty}\left(\frac{-1}{2}\right)^n(x-1)^n.$$

So, we get the Laurent series for $2/(x^2 - 1)$:

$$\frac{2}{x^2 - 1} = \frac{1}{x - 1} + \sum_{n=0}^{\infty} \left(\frac{-1}{2}\right)^{n+1} (x - 1)^n.$$

We have a simple pole at $x = 1$, and the residue is 1.

We will need to use residues to evaluate functions that we otherwise could not. An example will make this clear. The function $\csc(x) = 1/\sin(x)$ is not defined at $x = 0$. Meanwhile, $\exp(x) - 1$ has a simple zero at $x = 0$. What value should we assign to $(\exp(x) - 1)\csc(x)$ at $x = 0$? Well,

$$\exp(x) - 1 = x + O(x^2) \quad \text{and}$$
$$\sin(x) = x + O(x^3), \quad \text{so}$$
$$\frac{1}{\sin(x)} = \frac{1}{x} + O(x),$$

by long division, as we used in Section I2.2. Thus,

$$\frac{\exp(x) - 1}{\sin(x)} = (x + O(x^2)) \left(\frac{1}{x} + O(x)\right) = 1 + O(x).$$

The constant term 1 in the Taylor expansion is the value of the function at $x = 0$.

Here's a more abstract example. Suppose that a function $f(z)$ has a simple zero at $z = a$. If

$$f(z) = a_1(z - a) + O((z - a)^2), \quad \text{then}$$
$$\frac{1}{f(z)} = \frac{1}{a_1} \frac{1}{z - a} + O(1). \quad \text{Meanwhile,}$$
$$f'(z) = a_1 + O(z - a), \quad \text{so}$$
$$\frac{f'(z)}{f(z)} = \frac{1}{z - a} + O(1).$$

In other words, if $f(z)$ has a simple zero at $z = a$, then the function $f'(z)/f(z)$ has a simple pole at $z = a$, with residue 1. The expression $f'(z)/f(z)$ is called the LOGARITHMIC DERIVATIVE of $f(z)$, because it is the derivative of $\log(f(z))$.

The Laurent series, the poles of a function, and particularly the residues are central objects of study in complex analysis. The reasons why are beyond the scope of this book, but here's an idea. The residue at a pole can be computed by means of an integral. This provides a connection between the methods of analysis and the more formal algebraic techniques we've been developing in this chapter. In the preceding example, we saw that a zero of a function $f(z)$

can be detected by considering the residue of $f'(z)/f(z)$. When residues can be computed by means of integrals, this provides a way of counting the total number of zeros of $f(z)$ in any given region.

Exercise I2.3.1. Compute the Laurent expansion of $2/(x^2 - 1)$ at $x = -1$. Do the same for $x = 0$.

Exercise I2.3.2. What value should you assign to the function $x^2/(\cos(x) - 1)$ at $x = 0$?

Exercise I2.3.3. Suppose a function $f(z)$ has a zero of order N at $z = a$, that is,

$$f(z) = a_N(z - a)^N + O((z - a)^{N+1}), \quad \text{with } a_N \neq 0.$$

Compute the residue of $f'(z)/f(z)$ at $z = a$.

I2.4. Geometric Series

The ancient Greeks created a lot of beautiful mathematics, but they were not very comfortable with the idea of infinity. Zeno's paradox is an example. Zeno of Elea (circa 450 B.C.) is known for the following story about the Tortoise and Achilles. (Achilles was the hero of Homer's *Iliad*.) Achilles and the Tortoise were to have a race, and Achilles gave the Tortoise a head start of one kilometer. The Tortoise argued that Achilles could never catch up, because in the time it took Achilles to cover the first kilometer, the Tortoise would go some further small distance, say one tenth of a kilometer. And in the time it took Achilles to cover *that* distance, the Tortoise would have gone still further, and so forth. Achilles gets closer and closer but is never in the same place (asserts Zeno).

Actually, the mathematicians of the ancient world had the techniques needed to resolve the paradox. Archimedes knew the Geometric series

$$1 + x + x^2 + \cdots + x^n = \frac{1 - x^{n+1}}{1 - x}. \tag{I2.9}$$

Exercise I2.4.1. The formula (I2.9) was derived in Exercise 1.2.10, using finite calculus. But there is a shorter proof, which you should find. Start by multiplying

$$(1 - x)(1 + x + x^2 \cdots + x^n).$$

Because

$$\frac{1 - x^{n+1}}{1 - x} = \frac{1}{1 - x} - x^{n+1}\frac{1}{1 - x},$$

for a *fixed* value of x, as n increases, we can write (I2.9) as

$$\sum_{k=0}^{n} x^k = 1 + x + x^2 + \cdots + x^n = \frac{1}{1 - x} + O(x^{n+1}).$$

Of course, for $x > 1$ the error $O(x^{n+1})$ gets bigger as n increases, but what if $|x| < 1$? Then, x^{n+1} tends to 0 as n tends to infinity; so, according to (I1.3), the limit is

$$\lim_{n \to \infty} \sum_{k=0}^{n} x^k = \frac{1}{1 - x}. \qquad (I2.10)$$

That is, the partial sums $1 + x + x^2 + \cdots + x^n$ get close to $1/(1 - x)$ as n gets big. We now *define* the infinite sum as this limit and write

$$1 + x + x^2 + x^3 + x^4 + \cdots = \sum_{k=0}^{\infty} x^k = \frac{1}{1 - x}.$$

This is the Geometric series with infinitely many terms. It looks just like (I2.7), but the point of view is slightly different. Now, x represents a real number, whereas in (I2.7) the x referred to a variable, and the infinite sum was "formal," just a way to keep track of all the Taylor expansions simultaneously.

Exercise I2.4.2. Apparently, Achilles runs 10 times as fast as the Tortoise. So while Achilles covers the 1/10th of a kilometer, the Tortoise must go 1/100th. Use the Geometric series to find the sum of the infinitely many distances

$$1 + \frac{1}{10} + \frac{1}{100} + \frac{1}{1000} + \frac{1}{10000} + \cdots .$$

If Achilles runs a kilometer in three minutes, how long after the start of the race will it be until they are even? Zeno was wrong.

I2.5. Harmonic Series

Far away across the field
The tolling of the iron bell
Calls the faithful to their knees
To hear the softly spoken magic spells

Pink Floyd

By the fourteenth century, universities organized around the trivium and quadrivium were flourishing. One of the greatest minds of the era was the philosopher, mathematician, and theologian Nicole Oresme.

Nicole Oresme (1320–1382). Oresme lived in interesting times. He held a scholarship at the University of Paris in 1348, the year the Black Death struck Europe. He became Grand Master of his College in 1356. Also in that year, Oresme's patron, the Dauphin of France, future King Charles V, became the Regent when King Jean was taken captive by the English in the battle of Poitiers. Oresme was soon signing documents as "secretary of the king." In 1363, he preached a sermon before Pope Urban V and his cardinals at Avignon on the subject of corruption in the church, and in 1370, he became chaplain to King Charles. Paris was then in revolt against the crown, led by the Provost of Merchants, Étienne Marcel. Marcel eventually allowed English troops to enter Paris and was assassinated.

Despite the turmoil, this was a time of learning and scholarship. At the request of the king, Oresme translated Aristotle's *Metaphysics* from Latin into French and wrote commentaries. This was the first translation of works from the ancient world into the vernacular. Unlike many historical figures mentioned previously, Oresme was a strong opponent of astrology. Oresme's treatise *Proportiones Proportionum* discussed the problem of relating ratios exponentially, and the difference between rational and irrational numbers. This argues against the Pythagorean idea of the great year. If the ratios of the planets' periods are not rational numbers, the planets will never again be in the same relative positions. Oresme also admitted the possibility of the motion and rotation of the earth and argued that the velocity of a falling object was proportional to the elapsed time, not the distance traveled (Clagett, 1968). In this work, he was influenced by the English physicist of the "Merton School," particularly Thomas Bradwardine. In turn, Galileo was influenced by Oresme's thoughts on motion.

In *Questiones super geometriam Euclidis* (Oresme, 1961), Oresme was the first to emphasize the fact that some infinite series converge and others do not. Specifically, he showed that

$$1 + \frac{1}{2} + \frac{1}{3} + \frac{1}{4} + \frac{1}{5} + \cdots = \infty.$$

We already know from Exercise 3.2.1 that the nth Harmonic number, H_n, is bigger than $\log(n)$, so the harmonic numbers can not have a finite limit L, as n goes to infinity. But we will look at Oresme's proof anyway; it is very clever. Oresme observed the inequalities

$$\frac{1}{3} + \frac{1}{4} > \frac{1}{4} + \frac{1}{4},$$

$$\frac{1}{5} + \frac{1}{6} + \frac{1}{7} + \frac{1}{8} > \frac{1}{8} + \frac{1}{8} + \frac{1}{8} + \frac{1}{8},$$

$$\frac{1}{9} + \cdots + \frac{1}{16} > \frac{1}{16} + \cdots + \frac{1}{16}, \cdots$$

So,

$$1 + \frac{1}{2} + \frac{1}{3} + \frac{1}{4} + \frac{1}{5} + \frac{1}{6} + \frac{1}{7} + \frac{1}{8} + \frac{1}{9} + \cdots + \frac{1}{16} + \cdots$$

$$> 1 + \underbrace{\frac{1}{2}}_{} + \underbrace{\frac{1}{4} + \frac{1}{4}}_{} + \underbrace{\frac{1}{8} + \frac{1}{8} + \frac{1}{8} + \frac{1}{8}}_{} + \underbrace{\frac{1}{16} + \cdots + \frac{1}{16}}_{} + \cdots.$$

He grouped the terms as indicated; because each group adds up to $1/2$, this is

$$= 1 + \frac{1}{2} + \frac{1}{2} + \frac{1}{2} + \frac{1}{2} + \cdots = \infty.$$

This same inequality gives a lower bound for Harmonic numbers of the form H_{2^k}; there are k groups, which each add to $1/2$, so

$$H_{2^k} > 1 + \frac{k}{2}. \tag{12.11}$$

This says that the sequence of Harmonic numbers is unbounded, so the sequence cannot have a finite limit:

$$\lim_{n \to \infty} H_n = \infty.$$

Oresme's sum is called the HARMONIC SERIES.

One might say that the reason analysis has applications to number theory is exactly because the Harmonic series diverges. For example, we will see that this fact implies that there are infinitely many primes, and that it even determines their distribution in the Prime Number Theorem. This will become clear in subsequent chapters.

Exercise 12.5.1. Use Oresme's bound (12.11) to find an integer n so that $H_n > 10$. Don't try to compute H_n itself.

Exercise 12.5.2. How does the estimate (12.11) compare to that of Exercise 3.2.1? (Hint: Think about logarithms base 2.)

12.6. Convergence

In our study of the Harmonic numbers, H_n, and their higher order analogs,

$$H_n^{(2)} = \sum_{k=1}^{n} \frac{1}{k^2}, \quad H_n^{(3)} = \sum_{k=1}^{n} \frac{1}{k^3}, \quad H_n^{(4)} = \sum_{k=1}^{n} \frac{1}{k^4}, \ldots,$$

some mysterious constants appeared. We showed in (3.3) that

$$H_n = \log(n) + \gamma + O(1/n). \tag{12.12}$$

And, in (4.6), we showed there was a constant $\zeta(2)$ such that

$$H_n^{(2)} = 0 + \zeta(2) + O(1/n). \tag{12.13}$$

In Exercise 4.3.3, you deduced the existance of a constant $\zeta(3)$ such that

$$H_n^{(3)} = 0 + \zeta(3) + O(1/n^2). \tag{12.14}$$

We have already seen that Euler's constant γ is important in number theory. It appears in the average order of the divisor function (4.5). And it appears in Mertens' Formula in connection with the distribution of primes (5.2).

The constants $\zeta(2), \zeta(3), \ldots$, will be just as significant as γ. If you review the proof of (4.6), you find that $\zeta(2)$ measures the total area of the infinitely many rectangles of height $1/k^2$ for $k = 1, 2, 3, \ldots$. The area is finite because they all lie under $y = 1/x^2$ and we computed that the improper integral $\int_1^\infty 1/x^2 dx$ is finite. You made a similar argument in Exercise 4.3.3.

Another way to think about this is to use (11.3), which says that the limit as n tends to infinity of $H_n^{(2)}$ exists; it is the number $\zeta(2)$. Actually, (11.3) was for small values of $|x - a|$, not large values of n, but the idea is the same. Similarly, $\zeta(3) = \lim_{n \to \infty} H_n^{(3)}$.

Basically, what we just did is add infinitely many numbers, the so-called infinite series. This is similar to what we did in Sections 12.4 and 12.5. The analysis in Section 12.4 showed that the Taylor polynomials get close to the number $1/(1 - x)$ if x is a number with $|x| < 1$, but not if $|x| \geq 1$. We can interpret the result in Section 12.5 as saying that the Taylor polynomials of

$$-\log(1 - x) = x + \frac{x^2}{2} + \frac{x^3}{3} + \cdots + \frac{x^n}{n} + O(x^{n+1})$$

at $x = 1$ do not get close to any number as n increases. Notice that $-\log(1 - x)$ is not defined at $x = 1$ either. These examples are in strong contrast to Sections 12.2 and 12.3, where we dealt with Taylor series as formal objects, as way to keep track of all the different Taylor polynomial approximations of all different degrees.

But does it make sense to add up infinitely many numbers? As Oresme found out, the answer is sometimes. We wish to do this kind of analysis generally, so we need to introduce some terminology. Suppose that $\{c_n\}$ is any infinite list of numbers (a SEQUENCE). When we write $\sum_{n=0}^{\infty} c_n$ (an INFINITE SERIES or just a SERIES), we mean a *new* list of numbers $\{s_n\}$ formed by taking partial sums:

$$s_0 = c_0,$$
$$s_1 = c_0 + c_1,$$
$$s_2 = c_0 + c_1 + c_2,$$
$$\vdots$$

For example, take the preceding Geometric series, with $x = 1/2$. So, c_n will be $(1/2)^n$. By $\sum_{n=0}^{\infty}(1/2)^n$, we mean the sequence of numbers

$$s_0 = (1/2)^0 = 1,$$
$$s_1 = (1/2)^0 + (1/2)^1 = 3/2,$$
$$s_2 = (1/2)^0 + (1/2)^1 + (1/2)^2 = 7/4,$$
$$s_3 = (1/2)^0 + (1/2)^1 + (1/2)^2 + (1/2)^3 = 15/8,$$
$$\vdots$$

Exercise I2.6.1. Just for practice, write down the first five partial sums for the series $\sum_{n=1}^{\infty}(-1)^{n-1}/n$.

We will say that, in general, a series CONVERGES if the sequence of partial sums s_n approaches some limiting value, and we'll say that the series DIVERGES otherwise. According to the discussion in Section I2.4,

$$s_n = \frac{1 - (1/2)^{n+1}}{1 - 1/2} = \text{ by some algebra } 2 - (1/2)^n.$$

So the partial sums s_n converge to 2 as n goes to infinity. The Harmonic series $\sum_{k=1}^{\infty} 1/k$ diverges because (I2.12) says that the partial sums grow like $\log(n)$, and, thus, have no finite limit. On the other hand, (I2.13) and (I2.14) say that

$$\sum_{k=1}^{\infty} \frac{1}{k^2} = \zeta(2) \quad \text{and} \quad \sum_{k=1}^{\infty} \frac{1}{k^3} = \zeta(3);$$

the partial sums $H_n^{(2)}$ and $H_n^{(3)}$ differ from the constants $\zeta(2)$ and $\zeta(3)$ by an error that tends to zero.

For another example, take the Taylor series (12.4) for exp(x) at 0 and plug in $x = 1$. According to $\sum_{n=0}^{\infty} 1/n!$, we mean the sequence

$s_0 = 1/0! = 1,$
$s_1 = 1/0! + 1/1! = 2,$
$s_2 = 1/0! + 1/1! + 1/2! = 5/2 = 2.5,$
$s_2 = 1/0! + 1/1! + 1/2! + 1/3! = 8/3 = 2.6666\ldots,$
$\vdots$

Later, we'll see that this sequence really does converge to

$$\exp(1) = e = 2.71828182846\ldots.$$

Try not to confuse Taylor series (which are formal objects with a variable x) and infinite series (which are actually sequences of numbers, the partial sums). This may seem pedantic, but it is a useful point of view because it lets us talk about all infinite series, convergent or not, in a uniform way.

Exercise 12.6.2. In Chapter 1, we studied the triangular numbers $t_n = n(n+1)/2$, and the tetrahedral numbers $T_n = n(n+1)(n+2)/6$. In Exercise 1.2.12, what you computed was exactly the partial sums of the infinite series

$$\sum_{n=1}^{\infty} \frac{1}{t_n} \quad \text{and} \quad \sum_{n=1}^{\infty} \frac{1}{T_n}.$$

What do these infinite series converge to?

Both these sums were known to Pietro Mengoli in the seventeenth century. Mengoli was the parish priest of Santa Maria Maddalena in Bologna. He was the first to sum infinite series other than the Geometric series. He wrote in an obscure style of Latin, which did not help his reputation, but his work on series, limits, and definite integrals seems quite modern, even though the work precedes Newton's by thirty years.

Exercise 12.6.3. In this exercise you will show in three steps that

$$\sum_{n=1}^{\infty} \frac{(-1)^{n-1}}{n} = \log(2).$$

1. Show by induction on N that

$$1 - \frac{1}{2} + \frac{1}{3} - \cdots + \frac{1}{2N-1} - \frac{1}{2N} =$$

$$\frac{1}{N+1} + \frac{1}{N+2} + \cdots + \frac{1}{2N-1} + \frac{1}{2N}.$$

Observe that the left side is the $2N$th partial sum S_{2N} of the infinite series.

2. Write the right side as an algebraic expression in terms of the Harmonic numbers.

3. Use the identity you just proved and (3.3) to show that the sequence $\{S_{2N}\}$ converges to $\log(2)$.

Exercise I2.6.4. In this exercise you will show in four steps that

$$\sum_{n=0}^{\infty} \frac{(-1)^n}{2n+1} = \frac{\pi}{4}.$$

1. Imitate what you did in Exercise I2.9 to show that

$$\sum_{n=0}^{2N} (-1)^n t^{2n} = \frac{1}{1+t^2} + \frac{t^{4N+2}}{1+t^2}$$

2. Integrate both sides from $t = 0$ to $t = 1$. Observe that the sum on the left is a partial sum for the infinite series we want. The first integral on the right should give you a number.

3. We want to get an upper bound on the remainder

$$\int_0^1 \frac{t^{4N+2}}{1+t^2} dt.$$

How does this integral compare to

$$\int_0^1 t^{4N+2} dt?$$

(Hint: Use property v of integrals from Interlude 1.

4. Compute this latter integral and observe that it tends to 0 as N goes to ∞.

We will also need a more delicate notion of convergence. It sometimes happens that an infinite series will converge, but only because positive and negative terms cause many cancellations. For technical reasons, this is not optimal. So, we'll also consider a new series formed by taking absolute values and say that the series $\sum_n c_n$ CONVERGES ABSOLUTELY if the series of absolute values $\sum_n |c_n|$ converges, that is, if the sequence

$$s_0 = |c_0|,$$
$$s_1 = |c_0| + |c_1|,$$
$$s_2 = |c_0| + |c_1| + |c_2|,$$
$$\vdots$$

converges. The point here is that the numbers being added are all positive, so the partial sums can only increase. There is never any cancellation. Absolute convergence is the very best possible kind of convergence. In particular, if a series converges absolutely, then it converges. But more is true as well. In Section I2.2 we integrated series term by term; this is justified in a course on analysis, under the hypothesis that the series converges absolutely. This is a very powerful tool that we will use frequently later on.

Another application of the idea of absolute convergence is the COMPARISON TEST. It uses information about one known series to get information about another. Here's how it works. Suppose that $\sum_n c_n$ is some series that we know converges absolutely. Given another series, $\sum_n b_n$, if $|b_n| \leq |c_n|$ for all n, then the comparison test says that the series $\sum_n b_n$ also converges absolutely. The reason this works is that if $\sum |c_n|$ converges, its partial sums don't increase to infinity. The partial sums for $\sum_n |b_n|$ are all less than the partial sums for $\sum |c_n|$.

Conversely, if the series with the smaller terms, $\sum b_n$ does not converge absolutely, then neither does the series with the larger terms, $\sum c_n$. Oresme's proof that the Harmonic series diverges is an application of the comparison test.

In fact, all we need is $\ll$ instead of strictly $\leq$ to compare coefficients. A constant multiple C will not change whether a series converges, nor will a finite number of terms where an inequality fails to hold.

The known series can converge or diverge, and the inequality can be $\geq$ or $\leq$, so there are four possibilities in all. Two give no information. That is, saying that your series is less than a divergent series does not tell you your series diverges. If your series is greater than a convergent series, again you can't deduce anything.

Exercise I2.6.5. Use the comparison test and what we know about $\zeta(2)$ to show that for any integer $k \geq 2$, the series $\zeta(k) = \sum_{n=1}^{\infty} 1/n^k$ converges. The notation here is slightly different.

Exercise I2.6.6. Use the comparison test and the Harmonic series to show that $\sum_n 1/\sqrt{n}$ diverges.

What is amazing is that we can use the comparison test and our knowledge of the Geometric series to get information about any series at all. This is important enough to state in fancy language, but the ideas are simple enough for you to follow.

Theorem. *If* $\sum_{n=0}^{\infty} c_n$ *is any series, let* L *be the limit*

$$L = \lim_{n \to \infty} \frac{|c_{n+1}|}{|c_n|}.$$

Then, the series converges absolutely if $L < 1$, *and it diverges if* $L > 1$. *There is no information if* $L = 1$; *anything can happen.*

Proof. Suppose that L, as in the theorem, satisfies $L < 1$; that is,

$$L = \lim_{n \to \infty} \frac{|c_{n+1}|}{|c_n|} < 1.$$

Because L is less than 1, it is a property of real numbers that there is another number r between L and 1. (For example, $r = (L + 1)/2$ works.)

Because r is bigger than the limit L, it is true that

$$\frac{|c_{n+1}|}{|c_n|} < r$$

for n that is big enough, say bigger than some N. (N depends on r, but it is then fixed independent of everything else.) This says, for example, that

$$\frac{|c_{N+1}|}{|c_N|} < r \quad \text{or} \quad |c_{N+1}| < r|c_N|.$$

Similarly,

$$\frac{|c_{N+2}|}{|c_{N+1}|} \frac{|c_{N+1}|}{|c_N|} < r^2 \quad \text{or} \quad |c_{N+2}| < r^2|c_N|.$$

In general, we find that for any k,

$$|c_{N+k}| < r^k|c_N|.$$

Because $r < 1$, the geometric series $|c_N| \sum_{k=0}^{\infty} r^k$ converges absolutely. According to the comparison test, $\sum_{k=0}^{\infty} |c_{N+k}| = \sum_{n=N}^{\infty} |c_n|$ converges; that is, the original series converges absolutely.

In the case where $L > 1$, one can show similarly that the individual terms $|c_n|$ are increasing, so the series diverges. □

We typically use this with $c_n = a_n(x - a)^n$ coming from some Taylor series $\sum_n a_n(x - a)^n$ centered at $x = a$. The RADIUS OF CONVERGENCE R is the "largest" value of x that makes the limit L equal 1. It is the radius of a one-dimensional circle, the interval $(a - R, a + R)$. The ratio test then can be interpreted as saying that the series converges absolutely when $|x - a| < R$. We get no information when $|x - a| = R$. (The term radius makes more sense when this is generalized to functions of a complex variable; then, the series converges inside an actual circle in the complex plane.) The ratio test is extremely useful because so many Taylor series involve $n!$, and the ratios cancel out so nicely.

Here's an example. In the series for $\exp(x)$, we have

$$\frac{|a_{n+1}x^{n+1}|}{|a_n x^n|} = \frac{|x|^{n+1}/(n+1)!}{|x|^n/n!} = \frac{|x|^{n+1}}{|x|^n}\frac{n!}{(n+1)!} = \frac{|x|}{n+1}.$$

So, the limit $L = 0$ for any x and therefore the series always converges absolutely, for any x. We say that the radius $R = \infty$.

For the geometric series, $a_n = 1$ for every n; so,

$$\frac{|a_{n+1}x^{n+1}|}{|a_n x^n|} = \frac{|x^{n+1}|}{|x^n|} = |x|.$$

Thus, the the series converges absolutely if $|x| < 1$, as we already know. The radius of convergence is 1.

The series for $\log(1 - x)$ is $\sum_{n=1}^{\infty} -x^n/n$; so,

$$\frac{|a_{n+1}x^{n+1}|}{|a_n x^n|} = \frac{|x|^{n+1}/(n+1)}{|x|^n/n} = |x|\frac{n}{(n+1)} = |x|\frac{1}{1 + 1/n}.$$

The last step above comes from multiplying numerator and denominator by $1/n$, which is useful for seeing that the limit is $|x| \cdot 1 = |x|$. Thus, the series converges absolutely for $|x| < 1$, or $R = 1$.

Exercise I2.6.7. Use the ratio test to find the radius of convergence for $-x + \log(1 + x)$. (See Exercise I2.2.6.)

Exercise I2.6.8. Use the ratio test to show that $R = 1$ for the series $L(x)$ that you found in Exercise I2.2.9. Notice that the series $L(1) = \zeta(2)$ does converge, even though this is on the boundary, where the ratio test gives no information. The Geometric series also has radius $R = 1$, but it certainly does not converge when $x = 1$. (What is the Nth partial sum for the Geometric series with $x = 1$?) This shows that anything can happen on the boundary.

Exercise I2.6.9. Show that the radius of convergence of $\sum_{n=0}^{\infty} n! x^n$ is 0. This shows that the worst case can happen.

We need to clear up the question of how well the Taylor polynomials approximate the original function $f(x)$. Another way of looking at this is the following: Suppose that the Taylor series converges at some numerical value x. Does that mean that the limit of the partial sums is actually equal to $f(x)$? To attack this problem, we will write the error, that is, the difference between $f(x)$ and its degree n Taylor approximation, in a clever way. From the Fundamental Theorem of Calculus, we know that

$$f(x) = f(0) + \int_0^x f'(t)\, dt.$$

Now, integrate by parts, choosing $u = f'(t)$ and $dv = dt$. Remember that t is the variable in the integral; we think of x as fixed here. So, $du = f''(t)\, dt$, and we can choose $v = t - x$. This is an unusual choice for v but legal, as $dv = dt$. This gives

$$f(x) = f(0) + f'(t)(t-x)\Big|_{t=0}^{t=x} - \int_0^x (t-x) f''(t)\, dt$$

$$= f(0) + f'(0)x + \int_0^x (x-t) f''(t)\, dt.$$

Exercise I2.6.10. Integrate by parts again to show that

$$f(x) = f(0) + f'(0)x + \frac{f''(0)}{2}x^2 + \frac{1}{2}\int_0^x (x-t)^2 f^{(3)}(t)\, dt.$$

This was stated with a Taylor expansion based at 0 for convenience, but obviously we can do this at any base point a. Repeated integration by parts and proof by induction shows that

Theorem. *If $f(x)$ has $n + 1$ derivatives, then*

$$f(x) = f^{(0)}(a) + f^{(1)}(a)(x-a) + \cdots$$

$$+ \frac{1}{n!} f^{(n)}(a)(x-a)^n + \frac{1}{n!}\int_a^x (x-t)^n f^{(n+1)}(t)\, dt.$$

The theorem gives the error, the difference between $f(x)$ and its degree n Taylor polynomial, explicitly as an integral:

$$E_n(x) = \frac{1}{n!} \int_a^x (x-t)^n f^{(n+1)}(t)\, dt. \qquad (12.15)$$

This is good because we can use what we know about integrals to estimate the error. Suppose that the $n + 1$st derivative $f^{(n+1)}(t)$ is bounded in absolute value by some constant M_{n+1} on the interval $a \le t \le x \le b$. According to property (vi) of integrals from Interlude 1, we see that

$$|E_n(x)| \le \frac{1}{n!} \int_a^x |x-t|^n |f^{(n+1)}(t)|\, dt$$

$$\le \frac{1}{n!} \int_a^x M_{n+1}(x-a)^n\, dt.$$

According to hypothesis and property (v), because $|x - t|^n \le (x - a)^n$ for $a \le t \le x$, the integral is

$$\le \frac{M_{n+1}(x-a)^{n+1}}{n!}$$

according to property (i), because the integrand is constant in t. We can finally answer the question of whether the Taylor polynomials really *do* approximate the original function. The answer is yes; the use of Big Oh notation in (12.3) is consistent with the definition (I1.2), as long as there is some bound M_{n+1} on the $n + 1$st derivative. The constant C, which will depend on n, can be taken to be $M_{n+1}/n!$.

Exercise I2.6.11. This exercise sketches another proof that

$$\sum_{n=1}^{\infty} \frac{(-1)^{n-1}}{n} = \log(2).$$

In Exercise I2.2.1, you computed that

$$\frac{1}{1+x} = \sum_{n=0}^{\infty} (-1)^n x^n, \qquad \text{for } |x| < 1, \text{ so}$$

$$\log(1+x) = \sum_{n=0}^{\infty} (-1)^n \frac{x^{n+1}}{n+1}, \qquad \text{for } |x| < 1,$$

$$= \sum_{n=1}^{\infty} (-1)^{n-1} \frac{x^n}{n}.$$

Show that the $n + 1$st derivative of $\log(1 + x)$ is $(-1)^n n!/(1 + x)^{n+1}$. Use the integral form (I2.15) for the difference between $\log(1 + x)$ and its degree n Taylor approximation. Set $x = 1$; use the fact that

$$1/(1 + t)^{n+1} \leq 1$$

and property (v) of integrals of Interlude 1. This gives

$$|E_n(1)| \leq \frac{1}{n + 1}.$$

This says that the difference between $\log(1 + 1)$ and the nth partial sum of the series is bounded by $1/(n + 1)$, which tends to 0. But it does not tend to 0 particularly rapidly, which you observed in Exercise I2.6.1.

For *finite* sums, you are used to the fact that the order of the summands is irrelevant. It is an apparent paradox that this is not necessarily the case for infinite series. Previously, you computed that

$$\frac{1}{1} - \frac{1}{2} + \frac{1}{3} - \frac{1}{4} + \frac{1}{5} - \frac{1}{6} + \frac{1}{7} - \cdots = \log(2). \qquad (I2.16)$$

Suppose that you now rearrange the terms so that each positive term is followed by *two* negative terms:

$$\frac{1}{1} - \frac{1}{2} - \frac{1}{4} + \frac{1}{3} - \frac{1}{6} - \frac{1}{8} + \frac{1}{5} - \frac{1}{10} - \frac{1}{12} + \cdots . \qquad (I2.17)$$

We will group pairs of positive and negative terms as follows:

$$\underbrace{\frac{1}{1} - \frac{1}{2}} - \frac{1}{4} + \underbrace{\frac{1}{3} - \frac{1}{6}} - \frac{1}{8} + \underbrace{\frac{1}{5} - \frac{1}{10}} - \frac{1}{12} + \cdots .$$

The terms in braces clearly simplify to give

$$\frac{1}{2} - \frac{1}{4} + \frac{1}{6} - \frac{1}{8} + \frac{1}{10} - \frac{1}{12} + \cdots . \qquad (I2.18)$$

We can factor out a $1/2$ to get

$$\frac{1}{2} \left(\frac{1}{1} - \frac{1}{2} + \frac{1}{3} - \frac{1}{4} + \frac{1}{5} - \frac{1}{6} + \cdots \right) = \frac{\log(2)}{2}.$$

After rearranging the terms, the series converges to $1/2$ of its original value. The paradox is resolved if we remember the definition of what an infinite series really is: the sequence of partial sums. The partial sums of (I2.16) are

$$\left\{ 1, \frac{1}{2}, \frac{5}{6}, \frac{7}{12}, \frac{47}{60}, \frac{37}{60}, \frac{319}{420}, \cdots \right\},$$

a sequence which happens to converge to log(2). Meanwhile, the partial sums of (I2.17) are

$$\left\{1, \frac{1}{2}, \frac{1}{4}, \frac{7}{12}, \frac{5}{12}, \frac{7}{24}, \frac{59}{120}, \cdots\right\},$$

which is a different sequence of numbers and which has a different limit. The partial sums of (I2.18) are yet another sequence,

$$\left\{\frac{1}{2}, \frac{1}{4}, \frac{5}{12}, \frac{7}{24}, \frac{47}{120}, \frac{37}{120}, \frac{319}{840}, \cdots\right\},$$

which happens to converge to log(2)/2. This is disturbing, but even the appearance of paradox is avoided when we are dealing with series that are absolutely convergent. In that case, rearranging the term still gives a new sequence of partial sums, but there is a theorem that says it will have the same limit; in other words, the sum of the infinite series is independent of the order of the terms.

Exercise I2.6.12. In earlier chapters, we made much use of comparing sums to integrals. Because you already know calculus, this is a very powerful technique. In this exercise, we show that the series $\sum_{n=2}^{\infty} 1/(n\log(n))$ diverges. View each term $1/(n\log(n))$ being summed as the area of a rectangle whose base is the interval $(n, n+1)$ of width 1 on the x axis, and height $1/(n\log(n))$. Draw a rough diagram of the first few rectangles. On this same graph, sketch the function $y = 1/(x\log(x))$. It should lie *under* all the rectangles. Thus, the area under $y = 1/(x\log(x))$, from 2 up to some big integer N, is less than the sum of the areas of the rectangles, which is just s_N, the Nth partial sum of the series. Use calculus to compute the area under $y = 1/(x\log(x))$ from $x = 2$ up to $x = N$. Use this to show that as $N \to \infty$, $s_N \to \infty$.

This trick can often be made to work. It is called the INTEGRAL TEST for convergence. We were doing this as far back as (4.6).

Exercise I2.6.13. The integral test works equally well to show convergence, if we shift each rectangle to the left so that the graph lies *over* the rectangles. In fact, we can not only show convergence, we can also get an estimate for how big the sum is. Use this to show that for $k \geq 2$,

$$\zeta(k) - 1 = \sum_{n=2}^{\infty} \frac{1}{n^k} < \int_1^{\infty} \frac{1}{t^k} dt = \left.\frac{1}{(1-k)t^{k-1}}\right|_1^{\infty} = \frac{1}{k-1}.$$

See (4.6) for the case of $k = 2$.

Basically, the integral test says that an infinite series and corresponding improper integral will either both converge or both diverge.

I2.7. Abel's Theorem

In Exercise I2.6.3, you showed that

$$\sum_{n=1}^{\infty} \frac{(-1)^{n-1}}{n} = \log(2).$$

Here is a argument that is easier but, unfortunately, not legal. In Exercise I2.2.1, you computed that

$$\frac{1}{1+x} = \sum_{n=0}^{\infty}(-1)^n x^n, \quad \text{for } |x| < 1, \text{ so}$$

$$\log(1+x) = \sum_{n=0}^{\infty}(-1)^n \frac{x^{n+1}}{n+1}, \quad \text{for } |x| < 1,$$

$$= \sum_{n=1}^{\infty}(-1)^{n-1}\frac{x^n}{n}.$$

To compute the series we want, it looks like we should just plug $x = 1$ into the function $\log(1 + x)$:

$$\sum_{n=1}^{\infty}(-1)^{n-1}\frac{1}{n} = \log(1 + 1) = \log(2).$$

Well, not exactly. The Geometric series only converges for $|x| < 1$, not $x = 1$. It is *not* true that

$$\frac{1}{1+1} = \sum_{n=0}^{\infty}(-1)^n = 1 - 1 + 1 - 1 \ldots.$$

The partial sums s_n are $1, 0, 1, 0 \ldots$; they have no limit.

But all is not lost. There are two nice results (Abel's Theorem, Parts I and II, below) about series of the form $f(x) = \sum_n a_n x^n$, which fix this problem. For completeness, we will include proofs here, but they are hard and you may skip them. Both depend on the Summation by Parts formula in Chapter 1. They will have many applications in Chapter 11.

Theorem (Abel's Theorem, Part I). *Suppose that an infinite series converges, say* $\sum_{k=0}^{\infty} a_k = L$. *For* $0 \le x \le 1$, *form the power series*

$f(x) = \sum_{k=0}^{\infty} a_k x^k$. *Then,*

$$\sum_{k=0}^{\infty} a_k = \lim_{x \to 1} f(x). \tag{12.19}$$

To prove this we need a lemma.

Lemma. *Let*

$$f_m(x) = \sum_{k=0}^{m} a_k x^k$$

be the mth partial sum of the power series for $f(x)$, a polynomial. Then, we can choose m such that the error $|f(x) - f_m(x)|$ is bounded independent *of the variable x.*

In analysis language, this says that the polynomials $f_m(x)$ converge to $f(x)$ uniformly.

Proof. Suppose that an allowable error ϵ is specified. The hypothesis that the original series $\sum_{k=0}^{\infty} a_k$ converges means that we can find an m such that the error

$$\left| \sum_{k=0}^{m-1} a_k - L \right| = \left| \sum_{k=m}^{\infty} a_k \right|$$

is as small as we like, in particular less than $\epsilon/2$. Of course, m depends on ϵ. So, we can assume that

$$\left| \sum_{m \le k < n} a_k \right| < \epsilon/2 \quad \text{for all } n > m.$$

We will use Summation by Parts on the sum $\sum_{m \le k < n} a_k x^k$. Let $u(n) = x^n$ and $\Delta v(n) = a_n$. Because m is now fixed, we can choose the constant $+C$ such that $v(m) = 0$. This just means that $v(n) = \sum_{m \le k < n} a_k$ (see the Fundamental Theorem of Finite Calculus, Part II, in Chapter 1). So,

$$\Sigma(a_n x^n) = x^n v(n) - \Sigma\, Ev(n)(x^{n+1} - x^n).$$

Applying the Fundamental Theorem, Part I, we see that

$$\sum_{m \le k < n} a_k x^k = (x^n v(n) - x^m v(m)) - \sum_{m \le k < n} v(k+1)(x^{k+1} - x^k).$$

12. Series

Use $v(m) = 0$ and factor out $a-1$ to write this as

$$\sum_{m \leq k < n} a_k x^k = x^n v(n) + \sum_{m \leq k < n} v(k+1)(x^k - x^{k+1}).$$

We want to estimate the absolute value of the sum on the left side; so, we replace it with the sum on the right. According to the triangle inequality, this is $\leq$ the sum of the absolute values of all terms on the right side. But $|x^n| \leq 1$, because $0 \leq x \leq 1$. For the same reason, each $x^{k+1} \leq x^k$; so, $|x^k - x^{k+1}| = x^k - x^{k+1}$. Furthermore,

$$|v(k)| = \left| \sum_{m \leq j < k} a_j \right| < \epsilon/2$$

according to our hypothesis on m. So, combining these estimates, we get that

$$\left| \sum_{m \leq k < n} a_k x^k \right| < \frac{\epsilon}{2} \cdot 1 + \sum_{m \leq k < n} \frac{\epsilon}{2}(x^k - x^{k+1})$$

$$< \frac{\epsilon}{2} \cdot 1 + \frac{\epsilon}{2} \sum_{m \leq k < n} (x^k - x^{k+1})$$

$$< \frac{\epsilon}{2} \cdot 1 + \frac{\epsilon}{2}(x^m - x^n).$$

Again, because $0 \leq x \leq 1$ and $m \leq n$, we have $0 \leq x^m - x^n \leq 1$; so,

$$\left| \sum_{m \leq k < n} a_k x^k \right| < \frac{\epsilon}{2} \cdot 1 + \frac{\epsilon}{2} \cdot 1 = \epsilon.$$

Because this is true for *every* n, we can take the limit as $n \to \infty$ to see that

$$\left| \sum_{m \leq k < \infty} a_k x^k \right| \leq \epsilon$$

and that the error ϵ does not depend on x, as the lemma requires.   □

This also proves that the power series actually does converge for $0 \leq x \leq 1$, and so actually defines a function $f(x)$.

Proof of Theorem. By definition of L and $f(x)$, we know that $L = f(1)$. Suppose that an allowable error ϵ is specified. To prove the theorem, we must specify some range, δ, such that

$$|f(x) - f(1)| < \epsilon \quad \text{whenever} \quad |x - 1| < \delta.$$

According to the lemma, we can choose an integer m such that $|f(x) - f_m(x)| < \epsilon/3$ for all x. Because $f_m(x)$ is just a polynomial, it is continuous. So, for some $\delta > 0$, it is true that $|f_m(x) - f_m(1)| < \epsilon/3$ whenever $|x - 1| < \delta$. So, then, when $|x - 1| < \delta$, we see that

$$|f(x) - f(1)| = |f(x) - f_m(x) + f_m(x) - f_m(1) + f_m(1) - f(1)|$$
$$< |f(x) - f_m(x)| + |f_m(x) - f_m(1)| + |f_m(1) - f(1)|$$
$$< \frac{\epsilon}{3} + \frac{\epsilon}{3} + \frac{\epsilon}{3} = \epsilon.$$

$\square$

This is the standard proof of the fact that if a sequence of continuous functions $f_m(x)$ converges uniformly to a function $f(x)$, then $f(x)$ is continuous.

Theorem (Abel's Theorem, Part II). *Suppose that in an infinite series $\sum_{k=1}^{\infty} a_k$, not necessarily convergent, the partial sums $\sum_{k=1}^{n} a_k$ are all bounded in absolute value. Then, the series*

$$\sum_{k=1}^{\infty} \frac{a_k}{k^{\sigma}} \quad \text{converges for all } \sigma > 0. \tag{12.20}$$

Proof. According to hypothesis, there is a bound B such that

$$\left| \sum_{m \le k < n} a_k \right| < B \quad \text{for all } m \le n.$$

Suppose that an allowable error ϵ is specified. We must specify an m such that $|\sum_{k=m}^{\infty} a_k/k^{\sigma}| \le \epsilon$. It will turn out that the right choice is to pick m such that

$$m > \exp\left(\frac{\log(\epsilon/(2B))}{-\sigma}\right);$$

then, any $k \ge m$ also satisfies this bound, and a little algebra shows that

$$k^{-\sigma} < \frac{\epsilon}{2B} \quad \text{for all } k \ge m.$$

We now use Summation by Parts on $\sum_{m \le k < n} a_k k^{-\sigma}$. Pick $u(n) = n^{-\sigma}$ and $\Delta v(n) = a_n$. We can choose the constant $+C$ such that $v(m) = 0$; that is, $v(n) = \sum_{m \le k < n} a_k$. Then,

$$\Sigma\, a_n n^{-\sigma} = n^{-\sigma} v(n) - \Sigma\, E v(n)((n+1)^{-\sigma} - n^{-\sigma})$$
$$= n^{-\sigma} v(n) + \Sigma\, E v(n)(n^{-\sigma} - (n+1)^{-\sigma}).$$

This implies that

$$\sum_{m \le k < n} a_k k^{-\sigma} = v(n)n^{-\sigma} + \sum_{m \le k < n} v(k+1)(k^{-\sigma} - (k+1)^{-\sigma}),$$

where we have made use of the fact that $v(m) = 0$.

As in the lemma above, we want to estimate the absolute value of the sum on the left. We replace it with the sum on the right and use the triangle inequality. According to our hypothesis, $|v(n)| \le B$ and $n^{-\sigma} < \epsilon/(2B)$ because $n \ge m$. Also according to the hypothesis, each $|v(k+1)| < B$ and $k^{-\sigma} - (k+1)^{-\sigma} > 0$ because $\sigma > 0$. So, we can say that

$$\left| \sum_{m \le k < n} a_k k^{-\sigma} \right| < B \cdot \frac{\epsilon}{2B} + \sum_{m \le k < n} B \cdot (k^{-\sigma} - (k+1)^{-\sigma})$$

$$= \frac{\epsilon}{2} + B \cdot \sum_{m \le k < n} (k^{-\sigma} - (k+1)^{-\sigma})$$

$$= \frac{\epsilon}{2} + B \cdot (m^{-\sigma} - n^{-\sigma}).$$

But because $0 < n^{-\sigma} < m^{-\sigma} < \epsilon/(2B)$, their difference is also bounded:

$$< \frac{\epsilon}{2} + B \cdot \frac{\epsilon}{2B} = \epsilon.$$

This is true for every $n > m$, so

$$\left| \sum_{k=m}^{\infty} a_k k^{-\sigma} \right| \le \epsilon$$

as desired.   □

As an application of this, take $a_k = (-1)^{k-1}$. The series $\sum_k a_k$ does not converge, but the partial sums are $1, 0, 1, 0, \dots$, so they are certainly bounded. Then Abel's Theorem, Part II, proves that $\sum_{k=1}^{\infty} (-1)^{k-1}/k^s$ converges for $s > 0$, particularly, for $s = 1$. Once we have this, Abel's Theorem, Part I, justifies the argument at the beginning of the section, that

$$\sum_{k=1}^{\infty} \frac{(-1)^{k-1}}{k} = \lim_{x \to 1} \sum_{k=1}^{\infty} \frac{(-1)^{k-1}}{k} x^k$$

$$= \lim_{x \to 1} \log(1 + x) = \log(2).$$

As another application of this, consider Gregory's series for $\pi/4$, which you computed in Exercise I2.6.4. We define a sequence by $a_k = 0$ for k even, whereas for k odd $a_k = (-1)^{(k-1)/2}$. The sequence looks like $1, 0, -1, 0,$

1, ... The partial sums are all 1 or 0, so they are bounded. Abel's Theorem, Part II, says that $\sum_{k=1}^{\infty} a_k k^{-s}$ converges for $s > 0$, particularly for $s = 1$. So,

$$1 - \frac{1}{3} + \frac{1}{5} - \frac{1}{7} + \frac{1}{9} - \cdots$$

converges. Because the series for $\arctan(x)$ is

$$\arctan(x) = \sum_{n=0}^{\infty} (-1)^{n+1} \frac{x^{2n+1}}{2n+1} = \sum_{k=1}^{\infty} a_k x^k,$$

Abel's Theorem, Part I, gives a second proof that

$$\frac{\pi}{4} = \arctan(1) = 1 - \frac{1}{3} + \frac{1}{5} - \frac{1}{7} + \frac{1}{9} - \cdots.$$

Chapter 6

Basel Problem

It was discovered by Pietro Mengoli in 1650, as shown in his book *Novae Quadraturae Arithmetica*, that the sum of the reciprocals of the triangular numbers was equal to 2:

$$\sum_{n=1}^{\infty} \frac{2}{n(n+1)} = 1 + \frac{1}{3} + \frac{1}{6} + \frac{1}{10} + \cdots = 2.$$

The proof is Exercise I2.6.2. But no such simple formula was known to Mengoli for the sum of the reciprocals of the squares, $\sum_n 1/n^2$. It seems like a similar problem: The squares and the triangular numbers are the simplest polygonal numbers. Other mathematicians, including John Wallis and Gottfried Leibnitz, tried to find a formula and failed. The Swiss mathematician Jacob Bernoulli discussed the problem in his *Theory of Series* in 1704. It became known as the Basel problem, named after the city.

Everyone was surprised when Euler, then still relatively young, solved the problem. Even more surprising was the nature of the answer Euler found:

$$\sum_{n=1}^{\infty} \frac{1}{n^2} = 1 + \frac{1}{4} + \frac{1}{9} + \frac{1}{16} + \cdots = \frac{\pi^2}{6}.$$

We already know that at least the series converges to a finite number $\zeta(2)$, as does the series $\zeta(k) = \sum_n 1/n^k$, for any integer $k > 1$. On the other hand the fact that the Harmonic series diverges means that $\zeta(1) = \infty$. We can make a function, the RIEMANN ZETA FUNCTION, by thinking about the exponent of n as a variable s:

$$\zeta(s) = \sum_{n=1}^{\infty} \frac{1}{n^s}.$$

One can show using the integral test of Exercise I2.6.12 that the series converges if $s > 1$. And by comparison to the Harmonic series, $\zeta(s)$ diverges if $s < 1$.

6.1. Euler's First Proof

In this section, we'll see one of Euler's several proofs that $\zeta(2) = \pi^2/6$. Euler's first proof is not completely rigorous, but it is simple and interesting.

Euler's idea is to write the function $\sin(\pi x)/\pi x$ as a product over its zeros, analogous to factoring a polynomial in terms of its roots. For example, if a quadratic polynomial $f(x) = a_2 x^2 + a_1 x + a_0$ has roots α, β different from 0, we can write

$$f(x) = a_0 \left(1 - \frac{x}{\alpha}\right)\left(1 - \frac{x}{\beta}\right),$$

because both sides are quadratic polynomials with the same roots and the same constant term. (Notice also that $a_0 = f(0)$.) On the other hand,

$$\sin(\pi x) = 0 \quad \text{when } x = 0, \pm 1, \pm 2, \pm 3, \dots.$$

Because

$$\frac{\sin(\pi x)}{\pi x} = 1 - \frac{\pi^2 x^2}{6} + O(x^4), \tag{6.1}$$

according to techniques discussed in Interlude 2, we see that $\sin(\pi x)/\pi x$ at $x = 0$ is 1 not 0. So,

$$\frac{\sin(\pi x)}{\pi x} = 0 \quad \text{when } x = \pm 1, \pm 2, \pm 3, \dots.$$

Euler guessed that $\sin(\pi x)/\pi x$ had a factorization as a product:

$$\frac{\sin(\pi x)}{\pi x} = \left(1 - \frac{x}{1}\right)\left(1 + \frac{x}{1}\right)\left(1 - \frac{x}{2}\right)\left(1 + \frac{x}{2}\right)\cdots$$

$$= \left(1 - \frac{x^2}{1}\right)\left(1 - \frac{x^2}{4}\right)\left(1 - \frac{x^2}{9}\right)\left(1 - \frac{x^2}{16}\right)\cdots \tag{6.2}$$

When we multiply out these (infinitely many) terms, we get

$$= 1 - \left(\frac{1}{1} + \frac{1}{4} + \frac{1}{9} + \frac{1}{16} + \cdots\right)x^2 + O(x^4). \tag{6.3}$$

Comparing coefficients of x^2 in (6.1) and (6.3) gives the result. If you are dubious about this, multiply the first two terms of (6.2) and look at the coefficient of x^2. Next multiply the first three terms of (6.2) and look at the coefficient of x^2 and so on.

This is not yet a rigorous proof by modern standards, because $\sin(\pi x)/\pi x$ is not a polynomial. Nonetheless, the idea leads to a valid proof; it can be shown that $\sin(\pi x)/\pi x$ does have such a factorization.

The product formula also gives a new proof of an identity in John Wallis'
Arithmetica Infinitorum, written in 1655:

$$\frac{\pi}{2} = \frac{2 \cdot 2}{1 \cdot 3} \cdot \frac{4 \cdot 4}{3 \cdot 5} \cdot \frac{6 \cdot 6}{5 \cdot 7} \cdot \frac{8 \cdot 8}{7 \cdot 9} \cdots = \prod_{n=1}^{\infty} \frac{4n^2}{4n^2 - 1}.$$

Because $\sin(\pi/2) = 1$, simply plug $x = 1/2$ into (6.2) and invert. The fact
that the product formula reproduces a known result at $x = 1/2$ doesn't
prove that the product formula is true, but it indicates that the idea is worth
investigating.

John Wallis (1616–1703). Wallis was appointed to the Savilian Chair of Geometry at Oxford in 1649, because of his services as a cryptographer during
the English Civil War. In spite of this, he went on to become one of the greatest English mathematicians before Newton. He corresponded with Fermat,
worked on Pell's equation, and tried to solve Fermat's last theorem in the
$n = 3$ case. Wallis was the first to try to compute arc length on an ellipse; the
elliptic integrals that arise lead to the elliptic curves discussed in Chapter 12.
His product formula for π influenced Euler's work on the Gamma function
$\Gamma(s)$, as we will see in Section 8.3. Wallis gets a brief mention in the famous
Diary of Samuel Pepys:

Here was also Dr. Wallis, the famous scholar and mathematician; but he promises little.

Despite a poor start, they later became close, corresponding about, among
other things, the Pythagorean theory of music. One letter in Pepys' *Private
Correspondence* (Samuel Pepys, 1925) comments about Wallis, saying that
at age 83 he was still doing cryptographic work, and

...I believe Death will no more surprise him than a proposition in Mathematicks....

Euler was also able to show that

$$\zeta(4) = \sum_{n=1}^{\infty} \frac{1}{n^4} = \frac{\pi^4}{90}.$$

Exercise 6.1.1. This exercise shows how. First, determine the coefficient of x^4
in the series expansion (6.1). This is not quite $\zeta(4)$. The x^4 term in (6.3) comes
from multiplying all possible pairs of terms $-x^2/m^2$ and $-x^2/n^2$ and adding
in (6.2). Notice that because all the terms in the product are different, n has
to be different from m, and we may as well name the larger n, the smaller m.

So, the coefficient of x^4 is also equal to

$$\sum_{m=1}^{\infty} \sum_{n=m+1}^{\infty} \frac{1}{m^2 n^2} = \frac{1}{2} \sum_{m=1}^{\infty} \sum_{n \neq m}^{\infty} \frac{1}{m^2 n^2}$$

by symmetry, as this new sum includes all the $n < m$ as well. This is

$$= \frac{1}{2} \left(\sum_{m=1}^{\infty} \sum_{n=1}^{\infty} \frac{1}{m^2 n^2} - \sum_{n=1}^{\infty} \frac{1}{n^4} \right)$$

$$= \frac{1}{2} \left(\sum_{m=1}^{\infty} \frac{1}{m^2} \sum_{n=1}^{\infty} \frac{1}{n^2} - \sum_{n=1}^{\infty} \frac{1}{n^4} \right).$$

You can now determine $\zeta(4)$.

Euler continued on in this way to compute $\zeta(6)$, $\zeta(8)$, $\zeta(10)$, ..., but this gets messy. He later developed more-careful proofs that actually gave the factorization formula for $\sin(\pi x)/\pi x$. Euler struggled with the case of $s = 3$, but he was never able to get a satisfactory formula for $\sum_n 1/n^3$.

These results might seem at first to be isolated curiosities of limited interest. But, in fact, as we will see in Chapters 11, 12, and 13, values of the Riemann zeta function and other functions like it are of central importance in modern number theory. Another example we've already seen, in Exercise I2.6.4, is

$$\frac{\pi}{4} = 1 - \frac{1}{3} + \frac{1}{5} - \frac{1}{7} + \frac{1}{9} - \cdots.$$

This identity was known to James Gregory in 1671. As with Euler's formula for $\zeta(2)$, it is surprising, in that one side involves only arithmetic and the other side only geometry.

6.2. Bernoulli Numbers

Sometimes it happens that we are more interested in the coefficients of the Taylor series for some function than in the function itself. Identities relating functions can supply information about the coefficients. In this case, the function is called the generating function for the coefficients. For example, the BERNOULLI NUMBERS are the coefficients in the series expansion for $x/(\exp(x) - 1)$:

$$\frac{x}{\exp(x) - 1} = \sum_{k=0}^{\infty} \frac{B_k x^k}{k!}.$$

What are the B_k? Well, multiply both sides by $\exp(x) - 1$ and write

$$1 \cdot x + 0 \cdot x^2 + 0 \cdot x^3 + \cdots = x = (\exp(x) - 1) \sum_{k=0}^{\infty} \frac{B_k x^k}{k!} =$$

$$\left(x + \frac{x^2}{2} + \frac{x^3}{6} + \cdots \right) \left(B_0 + B_1 x + \frac{B_2 x^2}{2} + \frac{B_3 x^3}{6} + \cdots \right).$$

In Exercise I2.2.10 you computed $B_0 = 1$, $B_1 = -1/2$, and $B_2 = 1/6$.

The Bernoulli numbers are very important in number theory. Jacob Bernoulli was aware of the very ancient formulas for sums of powers, discussed in Chapter 1:

$$\sum_{k=1}^{n-1} k = \frac{n(n-1)}{2},$$

$$\sum_{k=1}^{n-1} k^2 = \frac{n(n-1)(2n-1)}{6},$$

$$\sum_{k=1}^{n-1} k^3 = \frac{n^2(n-1)^2}{4}.$$

Here, they are stated in a slightly altered form, stopping at a power of $n - 1$ instead of n. Is there a general formula that covers all the cases at once for any exponent $m = 1, 2, 3, 4, \ldots$? What is the pattern of the coefficients on the right side? Bernoulli denoted

$$S_m(n) = \sum_{k=1}^{n-1} k^m;$$

so, the sums above are $S_1(n)$, $S_2(n)$, and $S_3(n)$, respectively. Then, he proved

Theorem. *For any $m \geq 1$, the formula for $S_m(n)$ is*

$$(m+1)S_m(n) = \sum_{k=0}^{m} \binom{m+1}{k} B_k n^{m+1-k}. \tag{6.4}$$

The formula contains the Binomial coefficients

$$\binom{m+1}{k} = \frac{m+1!}{k!(m+1-k)!}$$

and looks intimidating. In the case of $m = 1$, it reduces to

$$2S_1(n) = \binom{2}{0} B_0 n^2 + \binom{2}{1} B_1 n$$

$$= n^2 - n,$$

which we know is correct.

Exercise 6.2.1. Write out Bernoulli's formula in the case of $m = 2$ and see that it gives the correct formula for $S_2(n)$.

Proof. Bernoulli's formula starts with the finite Geometric series

$$1 + x + x^2 + \cdots + x^{n-1} = \frac{1 - x^n}{1 - x} = \frac{x^n - 1}{x - 1}$$

from Exercise 1.2.10. Plug in $x = \exp(t)$ to get that

$$1 + \exp(t) + \exp(2t) + \cdots + \exp((n-1)t)$$

$$= \frac{\exp(nt) - 1}{\exp(t) - 1} = \frac{\exp(nt) - 1}{t} \cdot \frac{t}{\exp(t) - 1}$$

$$= \sum_{j=1}^{\infty} \frac{n^j t^{j-1}}{j!} \sum_{k=0}^{\infty} \frac{B_k t^k}{k!},$$

where the first sum comes from Exercise I2.2.3 and the second sum is just a definition. This is

$$= \sum_{j=1}^{\infty} \sum_{k=0}^{\infty} \frac{n^j B_k t^{k+j-1}}{j! k!}.$$

Bernoulli now groups together all the terms where t has the same exponent $m = k + j - 1$. This means that $j = m + 1 - k$, and k has to be one of $0, 1, 2, \ldots, m$. So, we get

$$= \sum_{m=0}^{\infty} \left\{ \sum_{k=0}^{m} \frac{n^{m+1-k} B_k}{k!(m+1-k)!} \right\} t^m$$

$$= \sum_{m=0}^{\infty} \left\{ \sum_{k=0}^{m} \binom{m+1}{k} n^{m+1-k} B_k \right\} \frac{t^m}{m+1!}$$

after multiplying the numerator and denominator by $m + 1!$. On the other hand, for each $k = 1, 2, \ldots, n-1$, we have

$$\exp(kt) = \sum_{m=0}^{\infty} k^m \frac{t^m}{m!},$$

so

$$1 + \exp(t) + \exp(2t) + \cdots + \exp((n-1)t) = \sum_{m=0}^{\infty} S_m(n) \frac{t^m}{m!}.$$

The two different methods of expanding the function as a Taylor series must give the same answer. So, comparing the coefficients of t^m, we get that

$$\frac{S_m(n)}{m!} = \left\{ \sum_{k=0}^{m} \binom{m+1}{k} n^{m+1-k} B_k \right\} \frac{1}{m+1!}.$$

Multiplying both sides by $m+1!$ gives the theorem. $\square$

Jacob Bernoulli gets the credit now for (6.4), but Bernoulli credited Johann Faulhaber.

Johann Faulhaber (1580–1635). Faulhaber is now mostly forgotten, but he was once known as the Great Arithmetician of Ulm. He computed formulas for $S_m(n)$ up to $m = 17$ and realized that for odd m, it is not just a polynomial in m but a polynomial in the triangular number t_m. We saw this for $m = 3$ in (1.8). His mathematical reputation was diminished by his belief in the mystical properties of numbers, particularly the numbers 2300, 1335, and 1290, which occur in the "Book Of Daniel," and 1260 and 666, which occur in the "Book of Revelations." He recognized them as examples of polygonal numbers; for example, $666 = t_{36} = t_{s_6}$. Faulhaber believed the prophecies in the Bible could be interpreted through polygonal numbers. He predicted the world would end in 1605. This resulted in a brief jail term in 1606.

What we have seen so far is that $x/(\exp(x) - 1)$ is the generating function for the Bernoulli numbers, B_k. Similarly,

$$1 + \exp(t) + \exp(2t) + \cdots + \exp((n-1)t)$$

is the generating function for the numbers $S_m(n)$.

Earlier, you computed that $B_1 = -1/2$, and here is a weird fact. If you subtract the $B_1 x$ term from the function, you get

$$\frac{B_0 x^0}{0!} + \sum_{k=2}^{\infty} \frac{B_k x^k}{k!} = \frac{x}{\exp(x) - 1} + \frac{x}{2} = \frac{x \exp(x) + 1}{2 \exp(x) - 1}$$

when you combine everything over a common denominator and simplify. If you now multiply numerator and denominator by $\exp(-x/2)$, you get

$$= \frac{x}{2} \frac{\exp(x/2) + \exp(-x/2)}{\exp(x/2) - \exp(-x/2)}.$$

But this last expression is an even function of x, invariant when you change x to $-x$. (Check this.) So, it can have only even powers of x in its series expansion. This says that

$$B_k = 0, \qquad \text{if } k \text{ is odd, } k > 1, \tag{6.5}$$

$$\frac{x}{\exp(x) - 1} = 1 - \frac{x}{2} + \sum_{k=1}^{\infty} \frac{B_{2k} x^{2k}}{2k!}. \tag{6.6}$$

Exercise 6.2.2. Because the Bernoulli numbers are so useful, use the identity

$$x = (\exp(x) - 1) \sum_{k=0}^{\infty} \frac{B_k x^k}{k!}$$

to try to prove the following:

$$\sum_{k=0}^{m} \binom{m+1}{k} B_k = \begin{cases} 1, & \text{in the case of } m = 0, \\ 0, & \text{in the case of } m > 0. \end{cases} \tag{6.7}$$

Exercise 6.2.3. The previous exercise implicitly defines a recurrence relation. For example, because you already know B_0, B_1, B_2, and B_3, the equation

$$\binom{5}{0} B_0 + \binom{5}{1} B_1 + \binom{5}{2} B_2 + \binom{5}{3} B_3 + \binom{5}{4} B_4 = 0$$

lets you solve for B_4. Now, use (6.7) with $m = 7$ to compute B_6. Don't forget about (6.5).

Exercise 6.2.4. Use Bernoulli's theorem to find the formulas for $S_4(n)$, $S_5(n)$, and $S_6(n)$.

6.3. Euler's General Proof

The point of this section is to compute all the values $\zeta(2n)$ simultaneously. The proof is elementary in the sense that it does not use fancy machinery. Unfortunately, this means that the proof is longer than those proofs that assume you know more.

Theorem.

$$\zeta(2n) = (-1)^{n-1} \frac{(2\pi)^{2n} B_{2n}}{2(2n)!},$$

where the B_{2n} are the Bernoulli numbers discussed in Section 6.2.

This will take several steps.

Lemma. *The function $z \cot(z)$ has the Taylor series expansion*

$$z \cot(z) = \sum_{n=0}^{\infty} (-4)^n B_{2n} \frac{z^{2n}}{(2n)!}. \tag{6.8}$$

Proof. This is easy but requires requires complex numbers. If you haven't seen them before, you can jump ahead to Interlude 3. In Section 6.2 we showed that

$$1 + \sum_{n=1}^{\infty} \frac{B_{2n} x^{2n}}{(2n)!} = \frac{x}{2} \frac{\exp(x/2) + \exp(-x/2)}{\exp(x/2) - \exp(-x/2)}.$$

Because $(2i)^{2n} = (-4)^n$, we substitute $z = 2ix$ to get

$$1 + \sum_{n=1}^{\infty} (-4)^n \frac{B_{2n} z^{2n}}{(2n)!} = iz \frac{\exp(iz) + \exp(-iz)}{\exp(iz) - \exp(-iz)}.$$

But from Interlude 3,

$$\cos(z) = \frac{\exp(iz) + \exp(-iz)}{2}$$

$$\sin(z) = \frac{\exp(iz) - \exp(-iz)}{2i};$$

so, the right side is just equal to $z \cot(z)$.   □

The rest of the proof consists of finding another way to look at the Taylor expansion of $z \cot(z)$ such that the coefficients of the series are the values of the zeta function. We need a complicated identity for the function $\cot(z)$.

Lemma. *For any integer $n \geq 1$,*

$$\cot(z) = \cot(z/2^n) - \tan(z/2^n) + \sum_{k=1}^{2^{n-1}-1} \cot((z + k\pi)/2^n)$$

$$+ \cot((z - k\pi)/2^n). \tag{6.9}$$

Proof. This is not as bad as it looks; we will see what the sum means as we prove it. First, recall that

$$\cot(z + \pi) = \cot(z), \tag{6.10}$$

because $\sin(z)$ and $\cos(z)$ both change sign when shifted by π. And

$$\cot(z + \pi/2) = -\tan(z), \tag{6.11}$$

because $\sin(z + \pi/2) = \cos(z)$, whereas $\cos(z + \pi/2) = -\sin(z)$. Furthermore, because

$$\cos(2z) = \cos(z)^2 - \sin(z)^2,$$
$$\sin(2z) = 2\cos(z)\sin(z),$$

we get that

$$\cot(z) - \tan(z) = 2\cot(2z).$$

Equivalently, change z to $z/2$ everywhere to see that

$$\cot(z) = \frac{1}{2}\cot(z/2) - \frac{1}{2}\tan(z/2) \tag{6.12}$$

$$= \frac{1}{2}\cot(z/2) + \frac{1}{2}\cot(z/2 + \pi/2), \tag{6.13}$$

according to (6.11). Notice that (6.12) proves the lemma in the case of $n = 1$. Euler's idea was to iterate this identity and prove it by induction. That is, we apply the identity (6.13) to each of the two terms on the right side in (6.13). So,

$$\cot(z) = \frac{1}{2}\cot(z/2) + \frac{1}{2}\cot(z/2 + \pi/2)$$

$$= \frac{1}{4}\cot(z/4) + \frac{1}{4}\cot(z/4 + \pi/4) + \frac{1}{4}\cot(z/4 + 2\pi/4)$$

$$+ \frac{1}{4}\cot(z/4 + 3\pi/4).$$

After n iterations, we see that

$$\cot(z) = \frac{1}{2^n}\sum_{k=0}^{2^n-1}\cot((z + k\pi)/2^n).$$

Now, pick out the first term ($k = 0$) and the middle term ($k = 2^{n-1}$) from the sum. These are $1/2^n$ times

$$\cot(z/2^n) + \cot((z + 2^{n-1}\pi)/2^n) = \cot(z/2^n) + \cot(z/2^n + \pi/2)$$
$$= \cot(z/2^n) - \tan(z/2^n),$$

according to (6.11). In what remains, the terms in the sum with $k = 1, 2, \ldots$ up to $k = 2^{n-1} - 1$ are fine; they are exactly as shown in the lemma. For the terms with $k = 2^{n-1} + 1, \ldots$ up to $k = 2^n - 1$, use (6.10) to say

$$\cot\left(\frac{z + k\pi}{2^n}\right) = \cot\left(\frac{z + k\pi}{2^n} - \frac{2^n\pi}{2^n}\right)$$
$$= \cot\left(\frac{z + (2^n - k)\pi}{2^n}\right)$$

and group them with the previous terms.   □

Lemma.

$$z \cot(z) = 1 - \frac{2z^2}{\pi^2 - z^2} - \frac{2z^2}{4\pi^2 - z^2} - \frac{2z^2}{9\pi^2 - z^2} - \cdots \qquad (6.14)$$
$$= 1 - 2 \sum_{k=1}^{\infty} \frac{z^2}{k^2\pi^2 - z^2}.$$

Proof. We multiply each side of (6.9) by z, and let n go to infinity. Because we know, in general, that

$$\sin(t) = t + O(t^3), \quad \cos(t) = 1 + O(t^2),$$

we get that

$$t \tan(t) = t^2 + O(t^4), \quad \cot(t) = \frac{1}{t} + O(t), \quad t \cot(t) = 1 + O(t^2).$$

So, with $t = z/2^n$, we see that $z/2^n \cdot \tan(z/2^n)$ goes to 0 as n goes to infinity. And $z/2^n \cdot \cot(z/2^n)$ goes to 1 as n goes to infinity. Meanwhile,

$$\cot((z \pm k\pi)/2^n) = \frac{2^n}{z \pm k\pi} + O(2^{-n})$$

blows up as n tends to infinity, but

$$z/2^n \cdot \cot((z \pm k\pi)/2^n) = \frac{z}{z \pm k\pi} + O(4^{-n}).$$

Combining $z/(z + k\pi)$ and $z/(z - k\pi)$ over a common denominator, we see that as n goes to infinity,

$$\frac{z}{2^n}(\cot((z+k\pi)/2^n) + \cot((z-k\pi)/2^n)) \quad \text{tends to} \quad \frac{-2z^2}{k^2\pi^2 - z^2}.$$

$\square$

Lemma. *For* $|z| < \pi$, $z\cot(z)$ *also has the Taylor series expansion*

$$z\cot(z) = 1 - 2\sum_{n=1}^{\infty} \zeta(2n)\frac{z^{2n}}{\pi^{2n}}. \tag{6.15}$$

Proof. Write each term in (6.14) as

$$\frac{z^2}{k^2\pi^2 - z^2} = \frac{(z/k\pi)^2}{1-(z/k\pi)^2},$$

this is a Geometric series (starting at $n = 1$) in $(z/k\pi)^2$. So,

$$z\cot(z) = 1 - 2\sum_{k=1}^{\infty}\sum_{n=1}^{\infty}\left(\frac{z}{k\pi}\right)^{2n}$$

$$= 1 - 2\sum_{n=1}^{\infty}\left(\sum_{k=1}^{\infty}\frac{1}{k^{2n}}\right)\left(\frac{z}{\pi}\right)^{2n}.$$

As long as $|z/k\pi| < 1$ for every k, all the Geometric series are absolutely convergent. $\square$

To finish the proof of the theorem, we compare the coefficient of z^{2n} in the expansions (6.8) to the coefficient of z^{2n} in (6.15):

$$(-4)^n\frac{B_{2n}}{(2n)!} = -2\frac{\zeta(2n)}{\pi^{2n}}.$$

In Exercise I2.3.3, you saw the logarithmic derivative $f'(z)/f(z)$ of a function $f(z)$. This operation converts products into sums, because log does. Another way to see this is through calculus:

$$\frac{\frac{d}{dz}(f(z)g(z))}{f(z)g(z)} = \frac{f'(z)}{f(z)} + \frac{g'(z)}{g(z)}.$$

The logarithmic derivative of a function f determines the original function f up to a multiplicative constant. That is, suppose that f'/f is some function h, and H is any antiderivative for h. Then, integrating gives $\log(f) = H + C$, so $f = e^C \cdot e^H$. Using this idea, we can now prove the product formula for sin.

Exercise 6.3.1. Compute the logarithmic derivative of both sides of the conjectured product formula (6.2):

$$\sin(\pi z) \overset{?}{=} \pi z \prod_{n=1}^{\infty} \left(1 - \frac{z^2}{n^2} \right).$$

Now, compare this to what you get when you replace z with πz everywhere in (6.14), and then multiply by z. This shows that the product formula is correct up to a multiplicative constant. Now, divide the πz over to the other side; (6.1) shows that $\sin(\pi z)/\pi z$ is 1 at $z = 0$, as is the product. So, the missing constant is 1.

Chapter 7

Euler's Product

Euler's good taste in studying the Basel problem was further vindicated by a relationship he found to the prime numbers. He was able to take the function $\zeta(s)$ and factor it into pieces, with a contribution for each prime number p.

Theorem. *For $s > 1$,*

$$\zeta(s) = \sum_{n=1}^{\infty} \frac{1}{n^s} = \prod_{\text{all primes } p} \left(1 - \frac{1}{p^s}\right)^{-1}.$$

The product on the right side is called the Euler product. It is an infinite product (analogous to the infinite sum discussed in Section I2.6), defined as a limit of partial products.

Proof. Consider, first, what happens when we subtract

$$\zeta(s) - \frac{1}{2^s}\zeta(s) = 1 + \frac{1}{2^s} + \frac{1}{3^s} + \frac{1}{4^s} + \frac{1}{5^s} + \cdots$$
$$- \frac{1}{2^s} - \frac{1}{4^s} - \frac{1}{6^s} - \frac{1}{8^s} - \cdots$$
$$= 1 + \frac{1}{3^s} + \frac{1}{5^s} + \frac{1}{7^s} + \frac{1}{9^s} + \cdots.$$

So,

$$\left(1 - \frac{1}{2^s}\right)\zeta(s) = \sum_{\substack{n=1 \\ n \text{ odd}}}^{\infty} \frac{1}{n^s}.$$

We can do this again with the prime 3, first by distributing

$$\left(1 - \frac{1}{3^s}\right)\left(1 - \frac{1}{2^s}\right)\zeta(s) = \left(1 - \frac{1}{2^s}\right)\zeta(s) - \frac{1}{3^s}\left(1 - \frac{1}{2^s}\right)\zeta(s).$$

This subtraction removes anything divisible by 3 to give

$$= 1 + \frac{1}{5^s} + \frac{1}{7^s} + \frac{1}{11^s} + \cdots$$

$$= \sum_{\substack{n=1 \\ (n,6)=1}}^{\infty} \frac{1}{n^s},$$

where $(n, 6) = 1$ means that we omit from the sum all those n divisible by either 2 or 3. Repeating this argument for all the primes $2, 3, 5, 7, \ldots$ that are less than a fixed x, we have

$$\prod_{p<x} \left(1 - \frac{1}{p^s}\right) \zeta(s) = 1 + \frac{1}{q^s} + \cdots,$$

where we omit all n divisible by any prime $\leq x$. After 1, the next term remaining comes from the first prime q after x. In the limit, as x goes to infinity, we have

$$\prod_{\text{all primes } p} \left(1 - \frac{1}{p^s}\right) \zeta(s) = 1,$$

which is equivalent to the theorem. □

The connection between the zeta function and the prime numbers means that it is closely connected with the question about primes. As a very simple example, we present an analytic proof that

Theorem. *There are infinitely many primes.*

Proof. Suppose that there were only finitely many primes, say $2, 3, \ldots, q$. Then, certainly the product

$$\left(1 - \frac{1}{2^s}\right)^{-1} \left(1 - \frac{1}{3^s}\right)^{-1} \cdots \left(1 - \frac{1}{q^s}\right)^{-1}$$

makes sense (is a finite number) at $s = 1$. But if this is the product over *all* the primes, then it is $\zeta(1) = \sum_n 1/n$. This is the Harmonic series, which we know diverges. Therefore, our supposition that there were only finitely many primes must be wrong. □

Of course, we already know something better than this, namely Chebyshev's estimate $x/\log(x) \ll \pi(x)$. Because $x/\log(x)$ is unbounded, this implies that

$\pi(x)$ is, too. The point, though, is that the simple fact that the Harmonic series diverges contains some information about primes.

Here's the Laurent expansion of $\zeta(s)$ at the pole $s = 1$. This is a different, useful way of seeing that the harmonic series diverges at $s = 1$.

Lemma.

$$\zeta(s) = \frac{1}{s-1} + O(1), \quad as\ s \to 1,$$

$$\zeta(s) = 1 + O\left(\frac{1}{s-1}\right), \quad as\ s \to \infty.$$

Proof. For all $s > 1$, we know from calculus that t^{-s} is a decreasing function; so,

$$(n+1)^{-s} \le t^{-s} \le n^{-s} \quad for\ n \le t \le n+1.$$

So, if we integrate all three and notice that the first and last are constant in the variable t, we get

$$(n+1)^{-s} \le \int_n^{n+1} t^{-s}\, dt \le n^{-s},$$

and summing these inequalities over all n, we get that

$$\zeta(s) - 1 \le \int_1^\infty t^{-s} dt = \frac{1}{s-1} \le \zeta(s).$$

After being rearranged, this says that

$$0 \le \zeta(s) - \frac{1}{s-1} \le 1 \quad for\ all\ s > 1,$$

which gives the first part of the lemma. Rearrange again to see that

$$0 \le \zeta(s) - 1 \le \frac{1}{s-1} \quad for\ all\ s > 1,$$

which gives the second part of the lemma. $\qquad\square$

We proved the lemma using the definition of $\zeta(s)$ as a sum, but of course the same is true for the Euler product:

$$\prod_p (1 - p^{-s})^{-1} = \frac{1}{s-1} + O(1), \quad as\ s \to 1, \qquad (7.1)$$

and

$$\prod_{p}(1 - p^{-s})^{-1} = 1 + O\left(\frac{1}{s-1}\right), \qquad \text{as } s \to \infty. \qquad (7.2)$$

In this chapter we will use Euler's product and the techniques discussed in Interlude 2 to obtain more information about how prime numbers are distributed.

7.1. Mertens' Theorems

The theorems in this section have lovely applications to the distribution of abundant numbers, Sophie Germain primes, and Mersenne primes in subsequent sections. The tools needed are exactly the ones we have developed, but we will use them very intensely. For this reason, you may prefer to delay a careful reading of the proofs here.

Theorem. *As x goes to infinity,*

$$\sum_{\substack{p \text{ prime} \\ p<x}} \frac{\log(p)}{p} = \log(x) + O(1). \qquad (7.3)$$

Proof. The idea of the proof is to estimate $\log(n!)$ two different ways, and to compare them. First, in (5.7), we showed that for any prime p, the power of p dividing $n!$ is

$$v_p(n!) = [n/p] + [n/p^2] + [n/p^3] + \cdots = \sum_{j=1}^{\infty}[n/p^j].$$

Because

$$n! = \prod_{p<n} p^{v_p(n!)} = \prod_{p<n} p^{[n/p]+[n/p^2]+[n/p^3]+\cdots},$$

we can use logarithms to see that

$$\log(n!) = \sum_{p<n}[n/p]\log(p) + \sum_{p<n}\log(p)\sum_{j=2}^{\infty}[n/p^j].$$

Notice that we have separated out the term corresponding to $j = 1$; the rest will be treated separately.

Lemma.

$$\sum_{p<n} \log(p) \sum_{j=2}^{\infty} [n/p^j] \ll n. \tag{7.4}$$

Proof. We have

$$\sum_{p<n} \log(p) \sum_{j=2}^{\infty} [n/p^j] < \sum_{p<n} \log(p) \sum_{j=2}^{\infty} n/p^j$$

$$= n \sum_{p<n} \log(p) \sum_{j=2}^{\infty} 1/p^j = n \sum_{p<n} \frac{\log(p)}{p^2} \sum_{j=0}^{\infty} 1/p^j.$$

By summing the Geometric series, we get

$$= n \sum_{p<n} \frac{\log(p)}{p^2} \frac{1}{1 - 1/p} = n \sum_{p<n} \frac{\log(p)}{p(p-1)}.$$

We can make the sum bigger by summing over *all* primes. To get that the sum on the right is $\ll n$, it is enough to show that the series converges, that is, that

$$\sum_{p \text{ prime}}^{\infty} \frac{\log(p)}{p(p-1)} < \infty. \tag{7.5}$$

It suffices to show, using the comparison test, that the even bigger series including *all* integers is finite:

$$\sum_{k=2}^{\infty} \frac{\log(k)}{k(k-1)} < \infty.$$

But this is true according to the comparison test. We know that $\log(k)/(k(k-1))$ is less than $\log(k)/(k-1)^2$. Because $\log(k) \ll \sqrt{k-1}$, we deduce that $\log(k)/(k-1)^2 \ll 1/(k-1)^{3/2}$. And

$$\sum_{k=2}^{\infty} 1/(k-1)^{3/2} = \zeta(3/2) < \infty.$$

$\square$

Now, we will treat the $j = 1$ term.

Lemma.

$$\sum_{p<n} [n/p] \log(p) = n \sum_{p<n} \frac{\log(p)}{p} + O(n).$$

Proof.

$$\sum_{p<n} [n/p] \log(p) = \sum_{p<n} \{n/p + O(1)\} \log(p)$$

$$= n \sum_{p<n} \frac{\log(p)}{p} + \sum_{p<n} O(1) \log(p).$$

In the second sum, each $\log(p) < \log(n)$ and there are $\pi(n)$ terms, which is $\ll n/\log(n)$ according to Chebyshev's estimate (5.3). So,

$$\sum_{p<n} O(1) \log(p) < \log(n) \sum_{p<n} O(1) \ll n.$$

$\square$

The two lemmas combined say that

$$\log(n!) = n \sum_{p<n} \frac{\log(p)}{p} + O(n).$$

So,

$$\frac{\log(n!)}{n} = \sum_{p<n} \frac{\log(p)}{p} + O(1).$$

On the other hand, from (3.4) we already know that

$$\log(n!) = n \log(n) - n + O(\log(n)),$$

which implies that

$$\frac{\log(n!)}{n} = \log(n) + O(1)$$

because the error $O(\log(n)/n)$ is $\ll 1$. Setting these two estimates equal to each other gives the theorem. $\square$

Theorem. *There is a constant a such that as x goes to infinity,*

$$\sum_{\substack{p \text{ prime} \\ p<x}} \frac{1}{p} = \log(\log(x)) + a + O(1/\log(x)). \tag{7.6}$$

Notice that $\log(\log(x))$ goes to infinity, but extremely slowly. For example, if we take x to be the largest currently known prime, $M_{13466917}$, then

$$x = 2^{13466917} - 1 \approx 9.25 \times 10^{4053945}, \qquad \log(\log(x)) \approx 16.$$

Proof. An easy integration by substitution gives the improper integral

$$\int_p^\infty \frac{dt}{t \log(t)^2} = \frac{1}{\log(p)}.$$

So,

$$\sum_{p<x} \frac{1}{p} = \sum_{p<x} \frac{\log(p)}{p} \int_p^\infty \frac{dt}{t \log(t)^2}$$

$$= \sum_{p<x} \frac{\log(p)}{p} \left\{ \int_p^x \frac{dt}{t \log(t)^2} + \int_x^\infty \frac{dt}{t \log(t)^2} \right\}$$

$$= \sum_{p<x} \frac{\log(p)}{p} \int_p^x \frac{dt}{t \log(t)^2} + \sum_{p<x} \frac{\log(p)}{p} \int_x^\infty \frac{dt}{t \log(t)^2}.$$

This looks much more complicated than what we started with. But the last integral is the one above, except with x instead of p. So,

$$\sum_{p<x} \frac{1}{p} = \sum_{p<x} \frac{\log(p)}{p} \int_p^x \frac{dt}{t \log(t)^2} + \frac{1}{\log(x)} \sum_{p<x} \frac{\log(p)}{p}.$$

In the other term, we would like to interchange the sum and the integral. But the limits of integration depend on the variable p in the sum. This is tricky, like interchanging a double integral in calculus. We know that $2 \le p < x$ in the sum and $p \le t \le x$ in the integral, so we know that $2 \le p \le t \le x$ always. The integration variable t must go between 2 and x, and we then sum on $p < t$. So,

$$\sum_{p<x} \frac{\log(p)}{p} \int_p^x \frac{dt}{t \log(t)^2} = \int_2^x \sum_{p<t} \frac{\log(p)}{p} \frac{dt}{t \log(t)^2}.$$

Returning to our original sum, we have

$$\sum_{p<x} \frac{1}{p} = \int_2^x \sum_{p<t} \frac{\log(p)}{p} \frac{dt}{t \log(t)^2} + \frac{1}{\log(x)} \sum_{p<x} \frac{\log(p)}{p}.$$

It still looks like we've only made things worse. The point is that (7.3) gives a very simple formula for sums of terms $\log(p)/p$. To be precise, let $E(x)$ be the error

$$E(x) = \sum_{\substack{p \text{ prime} \\ p < x}} \frac{\log(p)}{p} - \log(x).$$

So,

$$\sum_{p<x} \frac{1}{p} = \int_2^x \{\log(t) + E(t)\} \frac{dt}{t \log(t)^2} + \frac{1}{\log(x)} \{\log(x) + E(x)\}$$

$$= \int_2^x \frac{dt}{t \log(t)} + \int_2^x \frac{E(t)dt}{t \log(t)^2} + 1 + \frac{E(x)}{\log(x)}.$$

The first integral is another easy substitution:

$$\int_2^x \frac{dt}{t \log(t)} = \log(\log(x)) - \log(\log(2)).$$

For the second integral, write

$$\int_2^x \frac{E(t)dt}{t \log(t)^2} = \int_2^\infty \frac{E(t)dt}{t \log(t)^2} - \int_x^\infty \frac{E(t)dt}{t \log(t)^2}.$$

We now let

$$a = \int_2^\infty \frac{E(t)dt}{t \log(t)^2} - \log(\log(2)) + 1;$$

so,

$$\sum_{p<x} \frac{1}{p} = \log(\log(x)) + a - \int_x^\infty \frac{E(t)dt}{t \log(t)^2} + \frac{E(x)}{\log(x)}.$$

Now, (7.3) says that $E(x) \ll 1$, so the last term is $\ll 1/\log(x)$. Similarly,

$$\frac{E(t)}{t \log(t)^2} \ll \frac{1}{t \log(t)^2},$$

so

$$\int_x^\infty \frac{E(t)dt}{t \log(t)^2} \ll \int_x^\infty \frac{dt}{t \log(t)^2} = \frac{1}{\log(x)}$$

(which also justifies the fact that the improper integral appearing in the definition of a is actually finite.) $\qquad\square$

Theorem. *There is a constant C such that as x goes to infinity,*

$$\prod_{p<x}(1 - 1/p) = \frac{C}{\log(x)} + O(1/\log(x)^2). \tag{7.7}$$

Notice that the theorem implies that

$$\prod_{p<x}(1-1/p) \sim \frac{C}{\log(x)}.$$

This theorem looks very different from the previous one. How will we relate sums with $1/p$ to products with $1 - 1/p$? A lemma indicates the relation.

Lemma. *There is a constant b such that as x goes to infinity,*

$$\sum_{p<x}\left\{\log\left(1-\frac{1}{p}\right)+\frac{1}{p}\right\} = b + O(1/x). \tag{7.8}$$

Proof. We first show the infinite series defined by summing over *all* p converges,

$$\sum_{p}\left\{\log\left(1-\frac{1}{p}\right)+\frac{1}{p}\right\} = \sum_{p}\left\{-\sum_{k=1}^{\infty}\frac{1}{kp^k}+\frac{1}{p}\right\},$$

by using the series expansion (I2.8) with $x = 1/p$,

$$= -\sum_{p}\sum_{k=2}^{\infty}\frac{1}{kp^k}.$$

But for each p, because $p \geq 2$, we know that

$$\sum_{k=2}^{\infty}\frac{1}{kp^k} = \frac{1}{p^2}\sum_{k=2}^{\infty}\frac{1}{kp^{k-2}} \leq \frac{1}{p^2}\sum_{k=2}^{\infty}\frac{1}{k2^{k-2}} < \frac{1}{p^2}\sum_{k=2}^{\infty}\frac{1}{2^{k-2}} = \frac{2}{p^2}$$

by summing the Geometric series. So, our double series converges absolutely according to the comparison test:

$$\sum_{p}\sum_{k=2}^{\infty}\frac{1}{kp^k} < \sum_{p}\frac{2}{p^2} < \sum_{n=1}^{\infty}\frac{2}{n^2} = 2\zeta(2).$$

Let b denote the sum of the infinite series; then,

$$\sum_{p<x}\left\{\log\left(1-\frac{1}{p}\right)+\frac{1}{p}\right\} - b = -\sum_{p\geq x}\left\{\log\left(1-\frac{1}{p}\right)+\frac{1}{p}\right\}$$

$$\ll \sum_{p\geq x}\frac{2}{p^2} < \sum_{n\geq x}\frac{2}{n^2} < \int_{x}^{\infty}\frac{2}{t^2}dt = \frac{2}{x},$$

where the $\ll$ is exactly the same argument as was used earlier, and then we use the integral test. $\square$

Proof of Theorem. Observe that if we subtract (7.6) from (7.8), we get

$$\sum_{p<x} \log\left(1 - \frac{1}{p}\right) = -\log(\log(x)) + b - a + O(1/\log(x)), \quad (7.9)$$

because the error $O(1/\log(x))$ is bigger than the error $O(1/x)$.

Exponentiating (7.9), we get

$$\prod_{p<x}\left(1 - \frac{1}{p}\right) = \frac{\exp(b-a)}{\log(x)} \exp(O(1/\log(x))).$$

The constant is $C = \exp(b - a)$. We need to understand the error term $\exp(O(1/\log(x)))$. But we know that

$$\exp(t) = 1 + O(t) \quad \text{as } t \to 0.$$

So, because $t = 1/\log(x) \to 0$ as $x \to \infty$, we have

$$\exp(1/\log(x)) = 1 + O(1/\log(x)) \quad \text{as } x \to \infty,$$

and, similarly for any error bounded by a constant times $1/\log(x)$,

$$\exp(O(1/\log(x))) = 1 + O(1/\log(x)) \quad \text{as } x \to \infty.$$

This shows nicely how useful the different ideas are of what Big Oh means. $\qquad\qquad\square$

The constants a and b are mysterious, but

Theorem (Mertens' Formula). *The constant C of the previous theorem is* $\exp(-\gamma)$, *where γ is Euler's constant. That is, as x goes to infinity,*

$$\prod_{p<x}(1 - 1/p) = \frac{\exp(-\gamma)}{\log(x)} + O(1/\log(x)^2). \quad (7.10)$$

The proofs in most standard texts rely on esoteric identities such as

$$\gamma = -\int_0^\infty \exp(-x)\log(x)\,dx,$$

but the following, adapted from the fiendishly clever proof in Tennenbaum and Mendés France (2000), shows directly the connection to the Harmonic numbers.

Proof. We want to make use of both (7.7) and (7.1), so we have to relate the variables x and s to a third variable, ϵ, close to 0, by setting

$$x = \exp(1/\epsilon), \quad s = 1 + \epsilon$$

therefore

$$\log(x) = \frac{1}{\epsilon} = \frac{1}{s-1}.$$

If we substitute these into (7.7) and (7.1) and multiply them, we get

$$\prod_{p < \exp(1/\epsilon)} (1 - p^{-1}) \prod_{\text{all } p} (1 - p^{-1-\epsilon})^{-1} = \left(C \cdot \epsilon + O(\epsilon^2)\right)(1/\epsilon + O(1))$$

$$= C + O(\epsilon).$$

As ϵ goes to 0, the first product goes to 0 like $C \cdot \epsilon$, whereas the second product goes to infinity like $1/\epsilon$. We now rearrange some of the terms in the products, so that

$$\prod_{p < \exp(1/\epsilon)} \frac{1 - p^{-1}}{1 - p^{-1-\epsilon}} \prod_{p > \exp(1/\epsilon)} \frac{1}{1 - p^{-1-\epsilon}} = C + O(\epsilon).$$

The point is that, now, we will be able to show that each infinite product has a finite limit as ϵ goes to 0. Multiplying those limits must give the constant C. Take logarithms of both sides and multiply by -1 to get that

$$\sum_{p < \exp(1/\epsilon)} \log\left(\frac{1 - p^{-1-\epsilon}}{1 - p^{-1}}\right) + \sum_{p > \exp(1/\epsilon)} \log(1 - p^{-1-\epsilon}) = -\log(C + O(\epsilon)).$$

Because logarithms are messy to deal with, we will make some approximations. Using the Taylor expansion (I2.8), we see that

$$\log(1 + t) = t + O(t^2) \quad \text{as } t \to 0. \tag{7.11}$$

So, for $\epsilon \to 0$, $p \geq \exp(1/\epsilon) \to \infty$ and $t = -p^{-1-\epsilon} \to 0$, which means that

$$\log(1 - p^{-1-\epsilon}) \approx -p^{-1-\epsilon}.$$

Meanwhile,

$$\frac{1 - p^{-1-\epsilon}}{1 - p^{-1}} = \frac{1 - p^{-1} + p^{-1} - p^{-1-\epsilon}}{1 - p^{-1}} = 1 + \frac{1 - p^{-\epsilon}}{p - 1}.$$

Exercise 7.1.1. Show that for fixed p,

$$1 - p^{-\epsilon} \leq \epsilon \log(p)$$

as a function of ϵ. (Hint: Compute the tangent line approximation to $1 - p^{-\epsilon}$ at $\epsilon = 0$.) Using calculus, show that the graph of $1 - p^{-\epsilon}$ is concave down always; what does this tell you about the tangent line?

By dividing by $p - 1$, the exercise says that

$$\frac{1 - p^{-\epsilon}}{p - 1} \leq \frac{\log(p)\epsilon}{p - 1};$$

so, $(1 - p^{-\epsilon})/(p - 1) \to 0$ as $\epsilon \to 0$ and the linear approximation

$$\log\left(\frac{1 - p^{-1-\epsilon}}{1 - p^{-1}}\right) = \log\left(1 + \frac{1 - p^{-\epsilon}}{p - 1}\right) \approx \frac{1 - p^{-\epsilon}}{p - 1}$$

is good.

You may recall that we got into trouble making linear approximations like these in (5.1). The distinction is that earlier, we used the approximation, which gets better as p goes to ∞, for *every* prime p. The errors introduced by the small primes never go away. In the current case, the error tends to 0 for every prime p as the parameter ϵ goes to 0.

To summarize the work so far, we want to estimate

$$\sum_{p < \exp(1/\epsilon)} \frac{1 - p^{-\epsilon}}{p - 1} - \sum_{p > \exp(1/\epsilon)} p^{-1-\epsilon} \overset{\text{definition}}{=} A(\epsilon) - B(\epsilon)$$

as ϵ goes to 0. The limit will be $-\log(C)$; we want to show that it is Euler's constant γ. Consider, first,

$$A(\epsilon) = \sum_{p < \exp(1/\epsilon)} \frac{1 - p^{-\epsilon}}{p - 1}$$

$$= \sum_{p < \exp(1/\epsilon)} \frac{1 - p^{-\epsilon}}{p} + \sum_{p < \exp(1/\epsilon)} \{1 - p^{-\epsilon}\}\left\{\frac{1}{p - 1} - \frac{1}{p}\right\}.$$

Exercise 7.1.2. Show that for $\epsilon \to 0$,

$$\sum_{p < \exp(1/\epsilon)} \{1 - p^{-\epsilon}\}\left\{\frac{1}{p - 1} - \frac{1}{p}\right\} \ll \epsilon.$$

(Hint: Use Exercise 7.1.1 and (7.5).)

So, we know that

$$A(\epsilon) = \sum_{p < \exp(1/\epsilon)} \frac{1 - p^{-\epsilon}}{p} + O(\epsilon).$$

Lemma.

$$A(\epsilon) = \sum_{n < 1/\epsilon} \frac{1 - \exp(-\epsilon n)}{n} + O(\epsilon \log(\epsilon)).$$

Observe that the lemma allows an error term $O(\epsilon \log(\epsilon))$ that is bigger than $O(\epsilon)$ but still tends to 0 as ϵ goes to 0, according to L'Hopital's rule.

Proof. The first step is to use calculus to express $1 - p^{-\epsilon}$ as an integral:

$$\sum_{p < \exp(1/\epsilon)} \frac{1 - p^{-\epsilon}}{p} = \sum_{p < \exp(1/\epsilon)} \frac{\epsilon}{p} \int_1^p \frac{dt}{t^{1+\epsilon}}.$$

We want to interchange the sum and the integral, but again the limits of the integral depend on the variable p in the sum. Because $1 < p < \exp(1/\epsilon)$ in the sum and $1 < t < p$ in the integral, we always have $1 < t < p < \exp(1/\epsilon)$; so, this is

$$= \epsilon \int_1^{\exp(1/\epsilon)} \left\{ \sum_{t < p < \exp(1/\epsilon)} \frac{1}{p} \right\} \frac{dt}{t^{1+\epsilon}}.$$

For the second step, we write the term in braces as

$$\sum_{p < \exp(1/\epsilon)} \frac{1}{p} - \sum_{p < t} \frac{1}{p}$$

$$= (\log(1/\epsilon) + a + O(\epsilon)) - (\log(\log(t)) + a + O(1/\log(t))),$$

according to (7.6). Fortunately, the unknown constant a cancels out, and we know that $t < \exp(1/\epsilon)$, which implies that $\epsilon < 1/\log(t)$; so, the larger error dominates:

$$= \log(1/\epsilon) - \log(\log(t)) + O(1/\log(t)).$$

Use (4.4) once with the variable set to $1/\epsilon$ and once with the variable set to $\log(t)$ to get

$$= \left(H_{1/\epsilon} - \gamma + O(\epsilon) \right) - \left(H_{\log(t)} - \gamma + O(1/\log(t)) \right) + O(1/\log(t))$$

$$= H_{1/\epsilon} - H_{\log(t)} + O(1/\log(t))$$

$$= \sum_{\log(t) < n < 1/\epsilon} \frac{1}{n} + O(1/\log(t)).$$

The point of this second step is to use Mertens' previous theorems to remove any mention of prime numbers from $A(\epsilon)$. We substitute this expression into the braces to see that

$$\sum_{p<\exp(1/\epsilon)} \frac{1-p^{-\epsilon}}{p} = \epsilon \int_1^{\exp(1/\epsilon)} \left\{ \sum_{\log(t)<n<1/\epsilon} \frac{1}{n} + O(1/\log(t)) \right\} \frac{dt}{t^{1+\epsilon}}$$

$$= \epsilon \int_1^{\exp(1/\epsilon)} \left\{ \sum_{\log(t)<n<1/\epsilon} \frac{1}{n} \right\} \frac{dt}{t^{1+\epsilon}}$$

$$+ \epsilon \int_1^{\exp(1/\epsilon)} O(1/\log(t)) \frac{dt}{t^{1+\epsilon}}.$$

The third step is to estimate the integral of the error. We use $1/t^{1+\epsilon} < 1/t$ and

$$\epsilon \int_1^{\exp(1/\epsilon)} \frac{dt}{t\log(t)} = \epsilon \log(\log(t))|_1^{\exp(1/\epsilon)} = \epsilon \log(\epsilon).$$

That is, we know so far that

$$A(\epsilon) = \epsilon \int_1^{\exp(1/\epsilon)} \left\{ \sum_{\log(t)<n<1/\epsilon} \frac{1}{n} \right\} \frac{dt}{t^{1+\epsilon}} + O(\epsilon \log(\epsilon)).$$

In the fourth step, we undo the calculus operations of the first step. We know that $1 < t < \exp(1/\epsilon)$ in the integral and that in the sum $\log(t) < n < 1/\epsilon$, so $t < \exp(n)$. Thus, we get

$$A(\epsilon) = \epsilon \sum_{n<1/\epsilon} \frac{1}{n} \int_1^{\exp(n)} \frac{dt}{t^{1+\epsilon}}$$

$$= \sum_{n<1/\epsilon} \frac{1-\exp(-\epsilon n)}{n} + O(\epsilon \log(\epsilon)).$$

$\square$

We next treat the other term,

$$B(\epsilon) = \sum_{p>\exp(1/\epsilon)} p^{-1-\epsilon}.$$

Lemma.

$$B(\epsilon) = \sum_{n>1/\epsilon} \frac{\exp(-\epsilon n)}{n} + O(\epsilon).$$

Proof in Exercises. This takes four steps, just like the lemma for $A(\epsilon)$.

Exercise 7.1.3. Show that

$$\sum_{p > \exp(1/\epsilon)} p^{-1-\epsilon} = \epsilon \int_{\exp(1/\epsilon)}^{\infty} \left\{ \sum_{\exp(1/\epsilon) < p < t} \frac{1}{p} \right\} \frac{dt}{t^{1+\epsilon}}.$$

Exercise 7.1.4. Verify that

$$\sum_{\exp(1/\epsilon) < p < t} \frac{1}{p} = \sum_{1/\epsilon < n < \log(t)} \frac{1}{n} + O(\epsilon).$$

Notice that the only difference is that now $\exp(1/\epsilon) < t$; so, the error $O(\epsilon)$ dominates.

Exercise 7.1.5. Show that

$$\epsilon^2 \int_{\exp(1/\epsilon)}^{\infty} \frac{dt}{t^{1+\epsilon}} = \epsilon \exp(-1)$$

therefore the integral of the error term is $\ll \epsilon$.

Exercise 7.1.6. Show that

$$\epsilon \int_{\exp(1/\epsilon)}^{\infty} \left\{ \sum_{1/\epsilon < n < \log(t)} \frac{1}{n} \right\} \frac{dt}{t^{1+\epsilon}} = \sum_{n > 1/\epsilon} \frac{\exp(-\epsilon n)}{n}.$$

$\square$

We now consider

$$A(\epsilon) - B(\epsilon) = \sum_{n < 1/\epsilon} \frac{1 - \exp(-\epsilon n)}{n} - \sum_{n > 1/\epsilon} \frac{\exp(-\epsilon n)}{n} + O(\epsilon \log(\epsilon))$$

$$= \sum_{n < 1/\epsilon} \frac{1}{n} - \sum_{\text{all } n} \frac{\exp(-\epsilon n)}{n} + O(\epsilon \log(\epsilon))$$

$$= \{\log(1/\epsilon) + \gamma + O(\epsilon)\} + \log(1 - \exp(-\epsilon)) + O(\epsilon \log(\epsilon)),$$

according to (4.4) and (I2.8). We use the properties of logarithm and some algebra to write this as

$$= \gamma - \log\left(\frac{-\epsilon}{\exp(-\epsilon) - 1}\right) + O(\epsilon \log(\epsilon)).$$

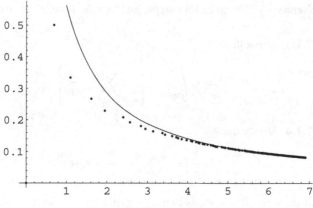

Figure 7.1. Mertens' Formula.

According to (6.6), this is

$$= \gamma - \log(1 + \epsilon/2 + O(\epsilon^2)) + O(\epsilon \log(\epsilon)),$$

which according to (7.11) is

$$= \gamma + O(\epsilon) + O(\epsilon \log(\epsilon)) = \gamma + O(\epsilon \log(\epsilon)).$$

This is what we wanted; it says that $A(\epsilon) - B(\epsilon) \to \gamma$ as $\epsilon \to 0$. ☐

Figure 7.1 shows a plot of the data points $\prod_{p<x}(1 - 1/p)$, for $x = 2, 3, 5, \ldots$ for all primes below 1000, compared to the graph of $\exp(-\gamma)/\log(x)$. The x-axis here is shown on a logarithmic scale. You should compare the fit here to that of Figure 5.2.

7.2. Colossally Abundant Numbers

In Section 3.4, we showed that

$$\sigma(n) \ll n \log(n),$$

whereas in Section 3.6, we showed that

$$\sigma(n) \text{ is not } \ll n.$$

Using Mertens' Formula, we can now determine the true order of magnitude of $\sigma(n)$, and even determine the "best" choice for the constant C.

Theorem. *For every $C > \exp(\gamma)$, there are only finitely many n with*

$$\sigma(n) > C \cdot n \log(\log(n)). \qquad (7.12)$$

For every $C < \exp(\gamma)$, there are infinitely many n with

$$\sigma(n) > C \cdot n \log(\log(n)). \tag{7.13}$$

In more sophisticated language, this says that

$$\overline{\lim} \frac{\sigma(n)}{n \log(\log(n))} = \exp(\gamma) \approx 1.78107.$$

The notation $\overline{\lim}$ denotes lim sup, pronounced "limb soup." The same theorem will be true for the function $s(n)$, because

$$\frac{s(n)}{n \log(\log(n))} = \frac{\sigma(n) - n}{n \log(\log(n))} = \frac{\sigma(n)}{n \log(\log(n))} - \frac{1}{\log(\log(n))}$$

and $1/\log(\log(n)) \to 0$.

We will prove the two statements separately.

Proof of (7.12). We need to factor the typical integer into a product of powers of primes,

$$n = p_1^{a_1} p_2^{a_2} \cdots p_k^{a_k},$$

and we may as well assume that the primes are listed in decreasing order: $p_1 > p_2 > \cdots > p_k$. Next, we need to keep track of which of the primes dividing n happen to be bigger than $\log(n)$, and which are less. Call l the number of primes that are bigger than $\log(n)$. So,

$$n = \underbrace{p_1^{a_1} \cdots p_l^{a_l}} \underbrace{p_{l+1}^{a_{l+1}} \cdots p_k^{a_k}},$$

where in the first group each prime is $> \log(n)$ and in the second group each prime is $\leq \log(n)$. So,

$$\log(n)^l < p_1^{a_1} \cdots p_l^{a_l} < n,$$

which means that when we take logarithms, we get

$$l < \frac{\log(n)}{\log(\log(n))}.$$

We know that

$$\sigma(n) = \prod_{i=1}^{k} \frac{p_i^{a_i+1} - 1}{p_i - 1} = n \prod_{i=1}^{k} \frac{1 - p_i^{-a_i-1}}{1 - p_i^{-1}}$$

after factoring $p_i^{a_i}$ out of each term. But

$$\prod_{i=1}^{k}(1 - p^{-a_i-1}) < 1;$$

so,

$$\sigma(n) < n\prod_{i=1}^{k}\frac{1}{1 - p_i^{-1}}$$

and, thus,

$$\frac{\sigma(n)}{n} < \prod_{i=1}^{k}\frac{1}{1 - p_i^{-1}} = \prod_{i=1}^{l}\frac{1}{1 - p_i^{-1}}\prod_{i=l+1}^{k}\frac{1}{1 - p_i^{-1}}$$

$$= \frac{1}{(1 - 1/\log(n))^l}\prod_{i=l+1}^{k}\frac{1}{1 - p_i^{-1}},$$

because each of those l primes is bigger than $\log(n)$. (This takes a little work to see.) Now, we add extra terms to the product by including *all* the primes $p < \log(n)$ to get

$$\frac{\sigma(n)}{n} < (1 - 1/\log(n))^{-l}\prod_{p<\log(n)}\frac{1}{1 - p^{-1}},$$

and using the inequaltiy $-l > -\log(n)/\log(\log(n))$, we get

$$\frac{\sigma(n)}{n} < (1 - 1/\log(n))^{-\log(n)/\log(\log(n))}\prod_{p<\log(n)}\frac{1}{1 - p^{-1}}.$$

We now divide both sides by $\log(\log(n))$ and claim that the limit of the awful mess on the right side is $\exp(\gamma)$. Here's why.

Lemma.

$$\left(1 - \frac{1}{t}\right)^{-t/\log(t)} = 1 + O\left(\frac{1}{\log(t)}\right) \qquad \textit{as } t \to \infty.$$

Proof. We know that

$$\left(1 - \frac{1}{t}\right)^{-t/\log(t)} = \exp\left(-\log\left(1 - \frac{1}{t}\right)\frac{t}{\log(t)}\right).$$

According to (7.11), this is

$$= \exp\left(\left(\frac{1}{t} + O\left(\frac{1}{t^2}\right)\right)\frac{t}{\log(t)}\right)$$

$$= \exp\left(\frac{1}{\log(t)} + O\left(\frac{1}{\log(t)^2}\right)\right),$$

because $1/(t\log(t)) \ll 1/\log(t)^2$

$$= 1 + O\left(\frac{1}{\log(t)}\right).$$

$\square$

The lemma implies that with $t = \log(n)$,

$$\left(1 - \frac{1}{\log(n)}\right)^{-\log(n)/\log(\log(n))} \sim 1 \quad \text{as } n \to \infty.$$

On the other hand, Mertens' Formula (7.10) with $x = \log(n)$ tells us that

$$\frac{1}{\log(\log(n))} \prod_{p < \log(n)} \frac{1}{1 - p^{-1}} \sim \exp(\gamma) \quad \text{as } n \to \infty,$$

so the product of the two expressions has limit $\exp(\gamma)$. To summarize, we've shown that $\sigma(n)/n\log(\log(n))$ is less than some function that tends to $\exp(\gamma)$ as n goes to infinity. Therefore, if $C > \exp(\gamma)$ is any constant, we can have $\sigma(n)/n\log(\log(n)) > C$ only finitely many times. $\square$

Sadly, the proof of (7.12) is not effective. That is, if you tell me your favorite $C > \exp(\gamma)$, I can't determine exactly which n satisfy $\sigma(n) > C \cdot n\log(\log(n))$.

Exercise 7.2.1. Find at least ten integers n that satisfy

$$\sigma(n) > 2 \cdot n\log(\log(n)).$$

There might be more, but I searched and didn't find any others.

The next proof will have a similar defect. Recall that in Section 3.6, we used factorials to construct very abundant numbers. For any constant C, we produced infinitely many integers n with $\sigma(n) > C \cdot n$. What we will now do is produce an infinite sequence of integers n that are "colossally abundant," that is, having $\sigma(n) > C \cdot n\log(\log(n))$. This will work only for $C < \exp(\gamma)$.

But the extra factor $\log(\log(n))$ goes to infinity; it is eventually bigger than any constant.

Proof of (7.13). We define the jth integer n_j in the sequence by

$$n_j = \prod_{p < \exp(j)} p^j.$$

Because each prime p is less than $\exp(j)$, we know that $p^j < \exp(j^2)$. There are $\pi(\exp(j))$ terms in the product, so we can say that

$$n_j < \exp(j^2)^{\pi(\exp(j))}.$$

So,

$$\log(n_j) < \pi(\exp(j)) \log(\exp(j^2)) = j^2 \pi(\exp(j)).$$

According to Chebyshev's estimate (5.3), we know that $\pi(\exp(j)) < 2\exp(j)/j$, so

$$\log(n_j) < 2 \cdot j \exp(j), \qquad \log(\log(n_j)) < \log(2) + \log(j) + j.$$

Again, we use the product formula

$$\sigma(n_j) = \prod_{p < \exp(j)} \frac{p^{j+1} - 1}{p - 1} = n_j \prod_{p < \exp(j)} \frac{1 - p^{-j-1}}{1 - p^{-1}}$$

and the inequality

$$\prod_{p < \exp(j)} (1 - p^{-j-1}) > \prod_{\text{all } p}(1 - p^{-j-1}) = \frac{1}{\zeta(j+1)}.$$

Combining these, we can say that

$$\frac{\sigma(n_j)}{n_j \log(\log(n_j))} > \frac{1}{(\log(2) + \log(j) + j)\zeta(j+1)} \prod_{p < \exp(j)} (1 - p^{-1})^{-1}.$$

From (7.2) we deduce that

$$\frac{1}{\zeta(j+1)} \sim 1 \quad \text{as } j \to \infty.$$

On the other hand, Mertens' Formula (7.10) with $x = \exp(j)$ implies that

$$\prod_{p < \exp(j)} (1 - p^{-1})^{-1} \sim j \exp(\gamma) \quad \text{as } j \to \infty.$$

Notice that

$$\frac{j \exp(\gamma)}{(\log(2) + \log(j) + j)} \sim \exp(\gamma) \quad \text{as } j \to \infty.$$

What we have shown is that $\sigma(n_j)/(n_j \log(\log(n_j)))$ is bigger than a function that has limit $\exp(\gamma)$. So, for any constant $C < \exp(\gamma)$, all but finitely many terms are bigger than C. $\qquad\square$

Again, this proof has the defect that it is not effective. If you tell me your favorite $C < \exp(\gamma)$, I can't say how big j has to be to make $\sigma(n_j)/(n_j \log(\log(n_j))) > C$. Also, the sequence n_j goes to infinity much faster than the factorials. In fact,

$$n_1 = 2,$$
$$n_2 = 44100,$$
$$n_3 = 912585499096480209000,$$
$$n_4 = 11279571063695793469838343564441 4386145$$
$$00209657126975306730373885091173744 10000.$$

Exercise 7.2.2. The theorem does not make any claim about what happens with $C = \exp(\gamma)$. Are there infinitely many n with

$$\sigma(n) > \exp(\gamma) \cdot n \log(\log(n))?$$

Try to find as many examples as you can. I found twenty-seven, including, of course, the ten from Exercise 7.2.1.

7.3. Sophie Germain Primes

If q is an odd prime number with the property that $2q + 1$ is also a prime, then q is called a SOPHIE GERMAIN PRIME. So, 3 and 5 are examples, but 7 is not. Interest in this class of primes comes from Germain's work on Fermat's Last Theorem, the question of finding integers x, y, and z that satisfy $x^n + y^n = z^n$. It is not hard to see that if there are no solutions for odd prime exponents, then there can be none for composite exponents. What she proved was that if q is a Sophie Germain prime, then the "first case" of Fermat's Theorem is true for exponent q: In any solution

$$x^q + y^q = z^q,$$

one of x, y, or z must be a multiple of q.

Marie-Sophie Germain (1776–1831). Germain's story is so interesting, I can't resist a digression. She was only 13 years old, in Paris, when the French revolution began, so her parents tried to keep her safe in the family library. There she became interested in mathematics after reading the legend of the death of Archimedes. Her parents tried to prevent her from studying by taking away her candles, but she persisted. Women were not allowed to enroll in the Ecole Polytechnique, but she obtained lecture notes and studied on her own. When another student, M. Antoine-August Le Blanc, dropped out, she adopted his identity and turned in solutions to problems under his name. This ruse was uncovered because her work was much better than Le Blanc's had been.

Her theorem lead to the question of whether there are infinitely many Sophie Germain primes. This is still an open problem, and even though Fermat's Last Theorem is now solved for all exponents, the question remains interesting. For example, Euler showed that for odd primes $q \equiv 3$ modulo 4 (that is, with remainder 3 when divided by 4), the number $2q + 1$ is prime if and only if $2q + 1$ divides the Mersenne number $2^q - 1$. Thus, if we knew infinitely many Sophie Germain primes, we would also know infinitely many composite Mersenne numbers. Recently, Agrawal–Kayal–Saxena made a spectacular breakthrough in primality testing; the distribution of Sophie Germain primes influences the estimates for the running time of their algorithm.

In this section, we will make a conjecture about how the Sophie Germain primes are distributed. This is based on a probability argument similar to that in Section 5.1. Take an integer N, large in the sense that it is near some parameter x that we will make go to infinity. What are the chances that both N and $2N + 1$ are primes? They need to be odd, of course, but $2N + 1$ is always an odd number. So with probability $1/2$, we expect that both N and $2N + 1$ are odd.

What about divisibility by odd primes p? We expect that N is divisible by p with probability $1/p$, and not divisible with probability $1 - 1/p$. We might be tempted to assume the same for $2N + 1$ and thus expect the probability that *neither* is divisible by p to be $(1 - 1/p)^2$. This is how it would work if they were independent events. (Think about the chance of not getting a six after two rolls of a die.)

In fact, independence fails here, and we can understand exactly how. No prime p can divide both $2N + 1$ and N; for if it divides N, it certainly divides $2N$. If it also divides $2N + 1$, then it divides the difference $(2N + 1) - 2N = 1$, which is a contradiction. So, p divides neither N nor $2N + 1$

with probability $1 - 2/p$. (This is more analogous to the chance of rolling neither a six nor a three with a single roll.)

We now want to combine the probabilities from different primes p, just as we did in Section 5.1. But before when we took the product over all primes less than x, we got an answer that was too small. When we took a product over all primes below $\sqrt{x}$, the answer was too big. Only in Exercise 5.1.5 of Interlude 1 did you observe that

$$\prod_{\substack{p \text{ prime} \\ p < x^{\exp(-\gamma)}}} (1 - 1/p) \sim \frac{1}{\log(x)}$$

gives an answer consistent with both the evidence of Exercise 5.1.1 and the Prime Number Theorem. In fact, this is what (7.10) says if we substitute $x^{\exp(-\gamma)}$ for x. We will adopt this cutoff point here as well, for purposes of making a conjecture.

Conjecture (First Version). *For integers near x, the chance of being a Sophie Germain prime is*

$$\frac{1}{2} \prod_{\substack{p \text{ odd prime} \\ p < x^{\exp(-\gamma)}}} (1 - 2/p).$$

There is no justification for *why* $x^{\exp(-\gamma)}$ should be the correct choice; this is a conjecture and not a theorem. (For more discussion on the use of heuristic arguments, see Polya (1966).)

The conjecture will be easier to test if we can rewrite it in a form more like our other products over primes. For that purpose, multiply every term by 1 written cleverly as $(1 - 1/p)^{-2} \cdot (1 - 1/p)^2$. After some arithmetic, we find that

$$(1 - 2/p) \cdot (1 - 1/p)^{-2} = 1 - \frac{1}{(p-1)^2}.$$

So, our conjectured formula becomes

$$\frac{1}{2} \prod_{\substack{p \text{ odd prime} \\ p < x^{\exp(-\gamma)}}} \left(1 - \frac{1}{(p-1)^2}\right) \prod_{\substack{p \text{ odd prime} \\ p < x^{\exp(-\gamma)}}} \left(1 - \frac{1}{p}\right)^2.$$

This looks much worse, but it is not. Except for the fact that the contribution from $p = 2$ is missing, the second product looks like two copies of what appears in Mertens' Formula (7.10) with $x^{\exp(-\gamma)}$ substituted for x. So,

$$\prod_{\substack{p \text{ odd prime} \\ p < x^{\exp(-\gamma)}}} \left(1 - \frac{1}{p}\right)^2 = 4 \prod_{p < x^{\exp(-\gamma)}} \left(1 - \frac{1}{p}\right)^2 \sim \frac{4}{\log(x)^2}.$$

And the first product approaches a finite limit as x goes to infinity:

$$\prod_{p \text{ odd prime}} \left(1 - \frac{1}{(p-1)^2}\right) = C_2 \approx 0.6601618.$$

This limit is known as the twin prime constant, because this same analysis applies to the twin primes, pairs of integers N, $N+2$, which are both prime. So, our revised statement of the conjecture is that

Conjecture. *For integers near x, the chance of being a Sophie Germain prime is $2C_2/\log(x)^2$. The corresponding density that counts the number of Sophie Germain primes below x is*

$$\pi_{SG}(x) = \#\{\text{Sophie Germain primes } p \mid p < x\} \sim 2C_2 \int_0^x \frac{dt}{\log(t)^2}.$$

Exercise 7.3.1. In Table 5.1, the primes that are Sophie Germain primes are 9479, 9539, 9629, 9689, 9791, 10061, 10091, 10163, 10253, 10271, 10313, 10331, and 10529. What does the conjecture predict for the chance a number near 10000 is a Sophie Germain prime? The number in an interval of length 100 should be about 100 times as big. Choose a random integer N between 9500 and 10400 count the number of Sophie Germain primes p with $N \le p < N + 100$, and compare to the predicted value.

The three graphs in Figure 7.2 compare $\pi_{SG}(x)$ to $2C_2 \int_0^x 1/\log(t)^2 dt$ for $x \le 10^3$, $x \le 10^5$, and $x \le 10^8$.

7.4. Wagstaff's Conjecture

Another special type of prime number of interest is the Mersenne prime, that is, a prime number of the form $2^p - 1$. In this section, we will look at a conjecture, of Wagstaff (Wagstaff, 1983), on the distribution of Mersenne primes. Again, it is based on Mertens' Formula, but we will use it in a different way this time.

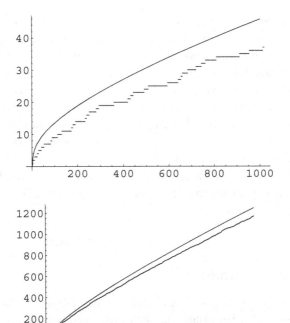

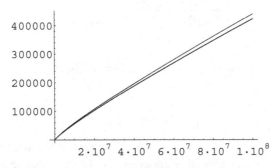

Figure 7.2. $\pi_{SG}(x)$ vs. prediction.

For a given integer p, we ask what the chance is that $M_p = 2^p - 1$ is a prime number.

1. We already know that in order to have M_p be prime, p itself must be a prime (see Exercise 2.1.9). The Prime Number Theorem says this probability is about $1/\log(p)$.
2. A typical number of size $2^p - 1$ is prime with a probability of about $1/\log(2^p - 1)$.

3. But among the numbers of that size, the Mersenne numbers M_p are atypical. Because of the special form of the number, a great many potential prime divisors q are automatically excluded. In fact we show, in Interlude 4 that if a prime q divides $2^p - 1$, then q itself must be of a special form; $q = kp + 1$ for some integer k. In particular, $q > p$. So, no "small" primes can divide a Mersenne number, and that means it has a much better chance of being prime than the Prime Number Theorem's estimate, $1/\log(2^p - 1)$. To quantify this, we will run the probability argument of the last section in reverse. That is, for each prime $q < p$, we don't have to worry about the possibility that q divides M_p, so this increases the chance that M_p is prime by $(1 - 1/q)^{-1}$. The contribution of all the primes below p is

$$\prod_{q<p}(1 - 1/q)^{-1} \sim \exp(\gamma)\log(p),$$

according to Mertens' Formula again.

Combine the three factors for the chance that p is prime, the chance that a number of size $2^p - 1$ is prime, and the compensation for the fact no small primes divide a Mersenne number: For a given integer p, we conjecture that $2^p - 1$ is prime with probability

$$\frac{\exp(\gamma)\log(p)}{\log(p)\log(2^p - 1)} \sim \frac{\exp(\gamma)}{\log(2)p}$$

because

$$\log(2^p - 1) \sim \log(2^p) = p\log(2).$$

If we let

$$\pi_M(x) = \#\left\{\text{Mersenne primes } 2^p - 1 < x\right\},$$

we can estimate this by summing the above probability over the various integers p of a size no more than about $\log(x)$. (Observe that here we are not assuming that p is a prime; the probability that this is so is factored into our estimate.) This gives

$$\sum_{p<\log(x)} \frac{\exp(\gamma)}{\log(2)p} = \frac{\exp(\gamma)}{\log(2)} H_{\log(x)} \sim \frac{\exp(\gamma)}{\log(2)} \log(\log(x)).$$

Conjecture (Wagstaff's Conjecture). *As x goes to infinity,*

$$\pi_M(x) \sim \frac{\exp(\gamma)}{\log(2)} \log(\log(x)).$$

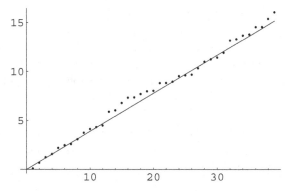

Figure 7.3. log(log(nth Mersenne prime)).

Another way of saying this is that if x is the nth Mersenne prime, so that $n = \pi_M(x)$, then

$$\log(\log(x)) \sim \frac{\log(2)}{\exp(\gamma)} n.$$

Figure 7.3 shows a log log plot of the known thirty-nine Mersenne primes compared to the line $y = n \log(2)/\exp(\gamma)$. A remarkable fit. This conjecture certainly implies, because $\log(\log(x)) \to \infty$, that there are infinitely many Mersenne primes. But the quantity of Mersenne numbers less than x is approximately $\pi(\log(x))$, the quantity of primes less than $\log(x)$. (For the exact number, we should use $\log_2(x)$, but this differs from $\log(x)$ only by a constant $\log(2)$.) Because $\pi(\log(x)) \sim \log(x)/\log(\log(x))$ is much bigger than $\log(\log(x))$, Wagstaff's conjecture implies there are also infinitely many composite Mersenne numbers.

The conjecture also has implications for the distribution of perfect numbers. Assume for the moment that there are no odd perfect numbers (this is almost certainly true.) Then, we know the nth perfect number is of the form $2^{p-1}(2^p - 1)$, where $M_p = 2^p - 1$ is the nth Mersenne prime. So, if y is the nth perfect number, then y is about size $M_p^2/2$, and

$$
\begin{aligned}
\log(\log(y)) &= \log(2\log(M_p) - \log(2)) \\
&\sim \log(2\log(M_p)) \\
&= \log(2) + \log(\log(M_p)) \\
&\sim \frac{\log(2)}{\exp(\gamma)} n.
\end{aligned}
$$

We can finally address the statement by Nicomachus quoted at the beginning of Chapter 3. He claimed that the nth perfect number has n digits. This is equivalent to saying the base 10 logarithm of the nth perfect number is about size n. In fact, he seems to be off by a whole logarithm. He would have been better off conjecturing that the nth perfect number has about 10^n digits. (To be precise, 10 raised to the power $\log_{10}(2)n/\exp(\gamma)$ digits.)

Exercise 7.4.1. Based on Wagstaff's conjecture, how many digits do you expect the fiftieth Mersenne prime to have? Which will be the first Mersenne prime with more than 10^9 digits?

Exercise 7.4.2. Do you think there are infinitely many primes of the form $(3^n - 1)/2$? If $(3^n - 1)/2$ is prime, what can you say about n? You might come back to this after reading Interlude 4.

Interlude 3

Complex Numbers

Unless you've already had a course in complex analysis, you can't skip this interlude; for, surprisingly, it turns out that the study of the distribution of primes requires complex numbers. The French mathematician Jacques Hadamard contributed extensively to this subject, as we will see in Section 10.1. He is quoted as saying,

The shortest path between two truths in the real domain passes through the complex domain.

You know that for a real number x, $x^2 \geq 0$, so no negative number has a real square root. The terminology "real" and "imaginary" is misleading. A complex number $z = (x, y)$ is just a vector in the plane; it is as simple as that. Complex numbers are no more dubious than any other mathematical object you are used to dealing with. As vectors, they add together in the usual way; that is,

$$(x_1, y_1) + (x_2, y_2) = (x_1 + x_2, y_1 + y_2).$$

Any vector (x, y) can be written in polar coordinates; just specify the length R of the vector and the angle θ it makes with the x-axis. Using trigonometry, we get that

$$x = R\cos(\theta) \qquad R = \sqrt{x^2 + y^2}$$
$$\text{or}$$
$$y = R\sin(\theta) \qquad \theta = \arctan(y/x).$$

We will use square brackets, $z = [R, \theta]$, when we mean that a pair of numbers is to be interpreted as polar coordinates of a vector. We can easily define a multiplication of vectors in polar coordinates. The rule is that we multiply the lengths and add the angles. That is, if $z_1 = [R_1, \theta_1]$ and $z_2 = [R_2, \theta_2]$, then

$$z_1 \cdot z_2 = [R_1 R_2, \theta_1 + \theta_2].$$

187

Exercise I3.0.1. Write the vector $(0, 1)$ in polar coordinates. Compute $(0, 1) \cdot (0, 1)$ and convert your answer back to rectangular coordinates.

Exercise I3.0.2. Convert $(1, 0)$ to polar coordinates. What happens if you multiply this vector by any other vector $[R, \theta]$?

This definition is very geometric and very pretty, but it is tedious to convert back and forth between rectangular and polar coordinates. What is the formula for multiplication in rectangular coordinates? We have

$$z_1 = (x_1, y_1) = (R_1 \cos(\theta_1), R_1 \sin(\theta_1)),$$
$$z_2 = (x_2, y_2) = (R_2 \cos(\theta_2), R_2 \sin(\theta_2)),$$

so, by definition,

$$z_1 \cdot z_2 = (R_1 R_2 \cos(\theta_1 + \theta_2), R_1 R_2 \sin(\theta_1 + \theta_2))$$
$$= R_1 R_2 (\cos(\theta_1 + \theta_2), \sin(\theta_1 + \theta_2)).$$

According to some trig identities,

$$\cos(\theta_1 + \theta_2) = \cos(\theta_1) \cos(\theta_2) - \sin(\theta_1) \sin(\theta_2),$$
$$\sin(\theta_1 + \theta_2) = \cos(\theta_1) \sin(\theta_2) + \cos(\theta_2) \sin(\theta_1).$$

Plug this in and regroup the R terms with their matching θs to get that

$$z_1 \cdot z_2 = (R_1 \cos(\theta_1) R_2 \cos(\theta_2) - R_1 \sin(\theta_1) R_2 \sin(\theta_2),$$
$$R_1 \cos(\theta_1) R_2 \sin(\theta_2) + R_2 \cos(\theta_2) R_1 \sin(\theta_1)).$$

But this just says that

$$z_1 \cdot z_2 = (x_1 x_2 - y_1 y_2, x_1 y_2 + x_2 y_1).$$

Earlier, computed that $(0, 1)^2 = (-1, 0)$, and that the vector $(1, 0)$ acts like the number 1. From now on, we will identify a real number x with the vector $(x, 0)$ on the horizontal axis. We use the special symbol i to denote the vector $(0, 1)$. So, $i^2 = -1$ in this notation. A typical complex number can now be written as

$$z = (x, y) = x(1, 0) + y(0, 1) = x \cdot 1 + y \cdot i = x + yi.$$

This is the traditional way to write complex numbers, but you should never forget that they are just vectors in the plane. In this notation,

$$z_1 \cdot z_2 = (x_1 x_2 - y_1 y_2) + (x_1 y_2 + x_2 y_1)i.$$

Here's some terminology that is useful. For $z = x + yi$, we say that x is the REAL PART, $x = \text{Re}(z)$, and y is the IMAGINARY PART, $y = \text{Im}(z)$. This is a little confusing; the imaginary part is still a real number. The COMPLEX CONJUGATE of $z = x + yi$, denoted $\bar{z}$, is just the complex number $x - yi$. Geometrically, this is flipping the vector across the horizontal axis.

Exercise I3.0.3. Compute the real and imaginary parts of z^2 in terms of x and y.

Exercise I3.0.4. Show that

$$z\bar{z} = x^2 + y^2 + 0i = R^2.$$

So, the length of the vector z can be computed by $|z| = R = \sqrt{z\bar{z}}$.

Because we can multiply, we also want to divide. It is enough to compute $1/z$ for any $z \neq 0$, then, z_1/z_2 is just $z_1 \cdot 1/z_2$. But

$$\frac{1}{z} = \frac{1}{x + yi} = \frac{1}{x + yi} \cdot \frac{x - yi}{x - yi} = \frac{x - yi}{x^2 + y^2} = \frac{x}{x^2 + y^2} - \frac{y}{x^2 + y^2}i.$$

Exercise I3.0.5. Compute $1/i$.

Finally, we need to define some of the basic transcendental functions, such as cosine, sine, and the exponential function. Certainly, we want

$$\exp(x + yi) = \exp(x)\exp(yi)$$

to be true, because the analogous identity is true for real numbers. To understand $\exp(yi)$, we use the series expansion

$$\exp(yi) = \sum_{n=0}^{\infty} \frac{(iy)^n}{n!} = \sum_{n=0}^{\infty} i^n \frac{y^n}{n!}.$$

We split the sum up into even $n = 2k$ and odd $n = 2k + 1$ terms,

$$= \sum_{k=0}^{\infty} i^{2k} \frac{y^{2k}}{2k!} + \sum_{k=0}^{\infty} i^{2k+1} \frac{y^{2k+1}}{2k+1!}$$

$$= \sum_{k=0}^{\infty} (-1)^k \frac{y^{2k}}{2k!} + i \sum_{k=0}^{\infty} (-1)^k \frac{y^{2k+1}}{2k+1!} = \cos(y) + \sin(y)i,$$

using the series expansions for sine and cosine from Interlude 2. To summarize,

$$\exp(x + yi) = \exp(x)(\cos(y) + \sin(y)i).$$

Logarithms of complex numbers are tricky. The fact that $|\cos(y) + \sin(y)i| = 1$ implies $|\exp(x + iy)| = \exp(x)$. So, if $x + iy = \log(u + iv)$, then $x = \mathrm{Re}(\log(u + iv)) = \log(|u + iv|)$.

Exercise I3.0.6. What is $e^{\pi i}$?

From the special cases

$$\exp(yi) = \cos(y) + \sin(y)i,$$
$$\exp(-yi) = \cos(y) - \sin(y)i,$$

we can solve algebraically for $\cos(y)$ or $\sin(y)$ to get

$$\cos(y) = \frac{\exp(yi) + \exp(-yi)}{2},$$

$$\sin(y) = \frac{\exp(yi) - \exp(-yi)}{2i}.$$

Because this identity relates cosine and sine to exponentials for real variables, it makes sense to define the complex cosine and sine such that this still holds true. So, we define

$$\cos(z) = \frac{\exp(zi) + \exp(-zi)}{2}, \qquad \sin(z) = \frac{\exp(zi) - \exp(-zi)}{2i}.$$

The concept of graph of a function is more difficult for complex variables. In the real variables case, the graph of a function such as $y = x^2$ is the set of points of the form (x, x^2) in the two-dimensional plane $\mathbb{R}^2$. In contrast, a complex function has two real inputs, x and y, and two real outputs, the real and imaginary parts of $f(z)$. With $f(z) = z^2$, as in the preceding exercise, the real part is $x^2 - y^2$ and the imaginary part is $2xy$. So, the graph of this function is the set of points of the form $(x, y, x^2 - y^2, 2xy)$ inside four-dimensional space $\mathbb{R}^4$. Because we can't see $\mathbb{R}^4$, we can't view the graph directly.

An alternate approach is to think of $u = \mathrm{Re}(f(z))$ and $v = \mathrm{Im}(f(z))$ separately as functions of two variables. Each then defines a surface in $\mathbb{R}^3$. To visualize both surfaces at the same time, we can draw LEVEL CURVES for the surfaces on the x–y plane. These are points in the plane where the surface is at some constant height, as in a topographic map. For example, the function $f(z) = z^2$ has $u = x^2 - y^2$ and $v = 2xy$. The level curves are shown in Figure I3.1. The level curves $x^2 - y^2 = c$ are hyperbolas with asymptote

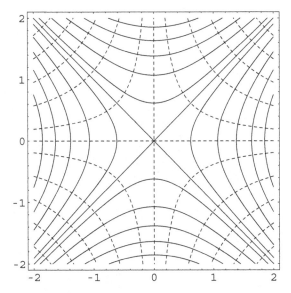

Figure I3.1. Level curves for z^2.

$y = \pm x$. The level curves $2xy = c$ are hyperbolas with the x and y axes as asymptotes; these are shown with dotted lines. Observe that real and imaginary part curves always meet at right angles. This is a beautiful property of analytic functions, that is, those that have a power series expansion at every point.

The function $f(z) = 1/z$ has real part $u = x/(x^2 + y^2)$ and imaginary part $v = -y/(x^2 + y^2)$. The level curves for the real part are of the form $x/(x^2 + y^2) = c$. If $c = 0$, this is $x = 0$, the y axis. Otherwise, this equation is equivalent to

$$x^2 - \frac{x}{c} + y^2 = 0 \quad \text{or}$$

$$\left(x - \frac{1}{c}\right)^2 + y^2 = \frac{1}{4c^2}.$$

These are all circles centered at $(1/c, 0)$ with radius $1/(2c)$.

Exercise I3.0.7. Show that the level curves for the imaginary part are also circles. Where is the center? What is the radius?

The level curves for $1/z$ are shown in Figure I3.2. Again, the imaginary part is shown with dotted lines.

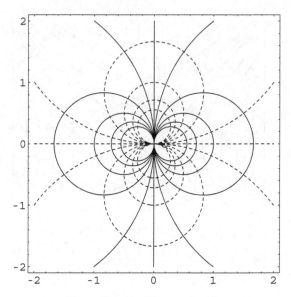

Figure I3.2. Level curves for $1/z$.

Exercise I3.0.8. The level curve plots in Figure I3.1 were created with the *Mathematica* commands

```
re = ContourPlot[x^2 - y^2, {x, -2, 2}, {y, -2, 2},
    PlotPoints -> 100,
    ContourShading -> False,
    DisplayFunction -> Identity];
im = ContourPlot[2x*y, {x, -2, 2}, {y, -2, 2},
    PlotPoints -> 100,
    ContourShading -> False,
    DisplayFunction -> Identity,
    ContourStyle -> Dashing[{0.01}]];
Show[re, im, DisplayFunction -> $DisplayFunction]
```

and Figure I3.2 was created similarly, with the real and imaginary parts of $1/z$. Create similar plots of your own for the functions z^3 and $\exp(z)$. Maple, too, can do contour plots; load the package with the command with(plots);

Chapter 8
The Riemann Zeta Function

Mathematicians have tried in vain to this day to discover some order
in the sequence of prime numbers, and we have reason to believe
that it is a mystery into which the mind will never penetrate.

Leonard Euler

Given the number of times we've mentioned Euler's work, you might think
it is odd that the zeta function is named for Riemann. The next three chapters
will try to counter that impression. In fact, what Riemann showed was that,
despite Euler's quote above, there *is* a certain regularity to the sequence of
prime numbers.

Georg Friedrich Bernhard Riemann (1826–1866). Riemann, just like Euler
before him, began university studies as a student of theology, in accordance
with his father's wishes, before switching to mathematics. Riemann had a very
short career and produced only a few papers before he died of tuberculosis.
In spite of that, he was one of the greatest mathematicians of all time.

For his *Habilitationsvortrag*, the advanced part of his Ph.D. degree,
Riemann prepared three potential topics to lecture on. To his surprise, Gauss
selected "On the Hypotheses that lie at the foundations of geometry." The
first part of the lecture describes the mathematical problem of defining
n-dimensional space with an abstract concept of distance. In the second part,
he discussed the question of the actual geometry of the universe we live in.
According to *Dictionary of Scientific Biography* (1970–1980), "[i]ts reading
on June 10, 1854 was one of the highlights in the history of mathematics:
young, timid, Riemann lecturing to the aged legendary Gauss, who would not
live past the next spring, on consequences of ideas the old man must have
recognized as his own and which he had long secretly cultivated. W. Weber
recounts how perplexed Gauss was, and how with unusual emotion he praised
Riemann's profundity on their way home."

Of the influence of Riemann's geometry on Einstein's theory of general relativity, *Dictionary of Scientific Biography* (1970–1980) says, "In the mathematical apparatus developed from Riemann's paper, Einstein found the frame to fit his physical ideas, his cosmology. The spirit of Riemann's paper was just what physics needed." This is reiterated in Monastyrsky (1987): "Riemannian geometry ... became the mathematical foundation of the general theory of relativity." Thus, Riemann achieved a synthesis between two parts of the ancient quadrivium: geometry and astronomy.

For his election to the Berlin Academy of Sciences, Riemann prepared a short paper called "On the number of primes less than a given magnitude." We will see in the next three chapters how the ideas introduced by Riemann similarly achieved a synthesis between the remaining two subjects of the quadrivium: arithmetic and music. No claim is made that Riemann deliberately set out to do this or had the quadrivium in mind. Nonetheless, the coincidence is striking.

Riemann was the first to realize that $\zeta(s)$ can be defined for complex values of s as well as for real values. If we write $s = \sigma + it$, then by definition

$$
\begin{aligned}
n^{-s} &= \exp(-s \log(n)) = \exp(-\sigma \log(n) - it \log(n)) \\
&= \exp(-\sigma \log(n)) \exp(-it \log(n)) \\
&= n^{-\sigma}(\cos(t \log(n)) - i \sin(t \log(n)),
\end{aligned}
\tag{8.1}
$$

as cosine is an even function and sine is odd. So,

$$
\mathrm{Re}(\zeta(s)) = \sum_{n=1}^{\infty} \frac{\cos(t \log(n))}{n^{\sigma}}, \qquad \mathrm{Im}(\zeta(s)) = -\sum_{n=1}^{\infty} \frac{\sin(t \log(n))}{n^{\sigma}}.
$$

Because $\left|n^{-s}\right| = n^{-\sigma}$, and $\sum_n 1/n^{\sigma}$ converges if $\sigma > 1$, we see that the series for $\zeta(s)$ converges absolutely for s on the right side of the vertical line $\sigma = \mathrm{Re}(s) = 1$.

In Section 8.4 we will see that there is a function, defined for *all* complex $s \neq 1$, that agrees with $\sum_n 1/n^s$ in the half plane $\mathrm{Re}(s) > 1$. We still call this function $\zeta(s)$, even though it is not given by this series if $\mathrm{Re}(s) \leq 1$.

It was Riemann who saw that the crucial thing that controls how prime numbers are distributed is the location of the zeros of $\zeta(s)$. Riemann conjectured that it was possible to factor $\zeta(s)$ into a product over its zeros like a polynomial and, like Euler, assumed existed for the function $\sin(\pi x)/\pi x$. By comparing the product over zeros to Euler's product over primes, Riemann discovered an "explicit formula" for the primes in terms of the zeros. This is the content of Chapter 10.

8.1. The Critical Strip

Euler's product formula for the function $\zeta(s)$ contains information about all prime numbers. Can $\zeta(s)$ have a second factorization formula in terms of zeros? There is a theorem about infinite products that says that a product can only be zero if one of the terms is zero (just as you expect for finite products). Because

$$\frac{1}{1-p^{-s}} = 0 \quad \text{only when} \quad 1 - p^{-s} = \infty,$$

there is no complex number s that makes the product over primes equal to 0. What we will do is find another expression, *not* given by the infinite sum over integers or product over primes, that agrees with the function $\zeta(s)$ for $\mathrm{Re}(s) > 1$ but that also makes sense for other values of s as well. We will then use the new expression to *define* $\zeta(s)$ more generally.

It turns out that there are values of s that make $\zeta(s) = 0$, and the interesting ones are all in the vertical strip $0 < \mathrm{Re}(s) < 1$. For this reason, we call this part of the complex plane the CRITICAL STRIP.

In this section, we will look at a fairly simple idea of Euler's for extending the domain on which $\zeta(s)$ is defined. Write

$$\zeta(s) = 1 + \frac{1}{2^s} + \frac{1}{3^s} + \frac{1}{4^s} + \frac{1}{5^s} \cdots \quad \text{thus}$$

$$\frac{2}{2^s}\zeta(s) = \frac{2}{2^s} + \frac{2}{4^s} + \frac{2}{6^s} + \frac{2}{8^s} + \frac{2}{10^s} \cdots.$$

If we subtract $2/2^s \zeta(s)$ from $\zeta(s)$, we see that the odd n are unchanged and that the even n have a minus sign:

$$\zeta(s) - \frac{2}{2^s}\zeta(s) = 1 - \frac{1}{2^s} + \frac{1}{3^s} - \frac{1}{4^s} + \frac{1}{5^s} \cdots.$$

Call this new series $\phi(s)$:

$$\phi(s) = (1 - 2^{1-s})\zeta(s) = \sum_{n=1}^{\infty} \frac{(-1)^{n-1}}{n^s}.$$

Surprisingly, this new series converges for a larger set, for $\mathrm{Re}(s) > 0$. In effect, the zero of $1 - 2^{1-s}$ at $s = 1$ cancels out the pole of $\zeta(s)$ at $s = 1$, as you will see in Exercise 8.2.1. The term $1 - 2^{1-s}$ is zero *only* at the isolated points $s_n = 1 + 2\pi i n / \log(2)$, for n any integer. So, we can use $\phi(s)$ to define $\zeta(s)$ on this larger set by defining

$$\zeta(s) = (1 - 2^{1-s})^{-1}\phi(s) \quad \text{for } \mathrm{Re}(s) > 0, s \neq s_n.$$

According to what we said earlier, this agrees with the old definition on the set Re(s) > 1, so it is an example of what we call an ANALYTIC CONTINUATION. Of course, the series for $\phi(s)$ does not converge absolutely for Re(s) > 0, because the absolute value of the terms just gives you back the series for ζ(Re(s)):

$$\left|\frac{(-1)^{n-1}}{n^s}\right| = \frac{1}{n^{\text{Re}(s)}},$$

according to (8.1). Nonetheless, $\phi(s)$ does define a function in the critical strip 0 < Re(s) < 1, where it has the same zeros as $\zeta(s)$. The function $\phi(s)$ is zero only when both its real and imaginary parts are simultaneously zero. Writing $s = \sigma + it$ as usual, then according to (8.1) we get

$$\text{Re}(\phi(s)) = \sum_{n=1}^{\infty} \frac{(-1)^{n-1}}{n^\sigma} \cos(t \log(n)),$$

$$\text{Im}(\phi(s)) = \sum_{n=1}^{\infty} \frac{(-1)^n}{n^\sigma} \sin(t \log(n)).$$

The RIEMANN HYPOTHESIS is the conjecture that for 0 < σ < 1 and any t, if $\zeta(s) = 0$ (or equivalently Re($\phi(s)$) = 0 = Im($\phi(s)$)), then $\sigma = 1/2$. Geometrically, this says that all the points s in the critical strip where $\zeta(s) = 0$ lie on a straight line.

Eminent mathematicians have noted (Zagier, 1977) how ironic it is that we can't answer this question. After all, Re($\phi(s)$) and Im($\phi(s)$) are defined in terms of elementary operations, exponentiation and logarithm, cosine and sine. Figure 8.1 show a graph of the two functions for $\sigma = 1/2$ and $0 \le t \le 26$. The t variable is measured on the horizontal axis here. You can see the first three zeros at $t = 14.13473\ldots$, $t = 21.02204\ldots$, and $t = 25.01086\ldots$. These are the places where the two functions cross the horizontal axis together. A close-up view of these three crossings is given in Figure 8.2.

Theorem. *The series*

$$\phi(s) = \sum_{n=1}^{\infty} \frac{(-1)^{n-1}}{n^s}$$

converges if σ = Re(s) > 0.

Proof. What we need to prove is that the partial sums

$$S_N = \sum_{n=1}^{N} \frac{(-1)^{n-1}}{n^s} = \frac{1}{1^s} - \frac{1}{2^s} + \frac{1}{3^s} - \cdots + \frac{(-1)^{N-1}}{N^s}$$

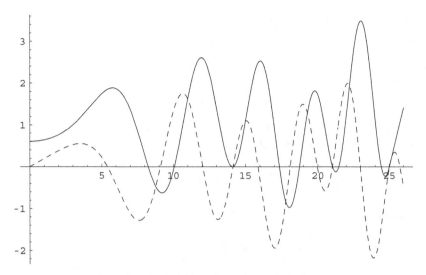

Figure 8.1. Real and imaginary parts of $\phi(1/2 + it)$.

approach some limiting value (depending of course on s) as N gets big. It helps to consider whether N is even $N = 2M$ or odd $N = 2M + 1$. Because

$$S_{2M+1} = S_{2M} + \frac{1}{(2M + 1)^s}$$

and $1/(2M + 1)^s \to 0$ as $M \to \infty$ as long as $\sigma = \text{Re}(s) > 0$, we see that

$$\lim_{M \to \infty} S_{2M+1} = \lim_{M \to \infty} S_{2M} + \lim_{M \to \infty} \frac{1}{(2M + 1)^s}$$
$$= \lim_{M \to \infty} S_{2M} + 0.$$

This says that the even index partial sums, S_{2M}, and the odd indexed sums, S_{2M+1}, will have the same limit. In other words, it suffices to look at only the partial sums indexed by even integers, which we now do.

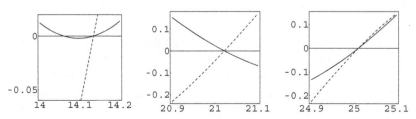

Figure 8.2. Close-up view of three zeros.

We now group the $2M$ terms in the partial sum S_{2M} into M pairs:

$$S_{2M} = \left\{ \frac{1}{1^s} - \frac{1}{2^s} \right\} + \left\{ \frac{1}{3^s} - \frac{1}{4^s} \right\} + \left\{ \frac{1}{5^s} - \frac{1}{6^s} \right\} + \cdots$$

$$+ \left\{ \frac{1}{(2k-1)^s} - \frac{1}{(2k)^s} \right\} + \cdots + \left\{ \frac{1}{(2M-1)^s} - \frac{1}{(2M)^s} \right\}.$$

We can think of this as the Mth partial sum, S'_M, of a new series, and we can write each term as an integral:

$$\frac{1}{(2k-1)^s} - \frac{1}{(2k)^s} = s \int_{2k-1}^{2k} \frac{1}{x^{s+1}} dx.$$

This is just calculus. So,

$$\left| \frac{1}{(2k-1)^s} - \frac{1}{(2k)^s} \right| = |s| \left| \int_{2k-1}^{2k} \frac{1}{x^{s+1}} dx \right|$$

$$\leq |s| \int_{2k-1}^{2k} \left| \frac{1}{x^{s+1}} \right| dx.$$

This is just property (vi) of integrals from Interlude 1. Now, $|1/x^{s+1}| = 1/x^{\sigma+1}$, and this is a decreasing function of x. So, the function is bounded by the value at the left endpoint, $1/x^{\sigma+1} \leq 1/(2k-1)^{\sigma+1}$, and property (v) then says that

$$\left| \frac{1}{(2k-1)^s} - \frac{1}{(2k)^s} \right| \leq |s| \int_{2k-1}^{2k} \frac{1}{(2k-1)^{\sigma+1}} dx = \frac{|s|}{(2k-1)^{\sigma+1}},$$

according to property 1 of integrals.

What does this accomplish? It shows that the new series with partial sums S'_M converges absolutely, by the comparison with the series for $\zeta(\sigma + 1)$:

$$|s| \sum_{k=1}^{\infty} \frac{1}{(2k-1)^{\sigma+1}} < |s| \zeta(\sigma + 1) < \infty,$$

because $\sigma > 0$. Absolute convergence implies convergence; the partial sums S'_M have a limit, and this is equal the limit of the partial sums S_{2M}. We are done.   □

This proof was quite subtle. Even though the series for $\phi(s)$ does not converge absolutely, by grouping the terms in pairs we get a new series that does converge absolutely. One could also adapt the proof of Abel's Theorem, Part II, to complex variables to get a different proof that $\phi(s)$ converges for $\mathrm{Re}(s) > 0$.

Lemma. *The Riemann zeta function $\zeta(s)$ has no zeros on the real axis $0 < s < 1$.*

Proof. It is enough to prove this for $\phi(s)$, as $1 - 2^{1-s}$ is never 0 for $0 < s < 1$. In the partial sums

$$S_{2M} = \left\{ \frac{1}{1^s} - \frac{1}{2^s} \right\} + \left\{ \frac{1}{3^s} - \frac{1}{4^s} \right\} + \left\{ \frac{1}{5^s} - \frac{1}{6^s} \right\} + \cdots ,$$

every term is positive because x^{-s} is a decreasing function of x for fixed $s > 0$. The sum of these positive numbers is positive; $\phi(s) > 0$ for $s > 0$. □

Figure 8.3 shows another viewpoint, this time of the output values $w = \zeta(s)$. In this version, we fix σ such that as t varies, $\sigma + it$ determines a vertical line in the s plane (not shown). Applying the function ζ gives a curve in the w complex plane. As we increase σ, the vertical line in the s plane moves from left to right. We get a sequence of pictures in the w plane, like frames in a movie. In this version, the Riemann Hypothesis says that the curve passes through the origin in one frame only, that for $\sigma = 0.5$.

Exercise 8.1.1. Use *Mathematica*'s `ParametricPlot` feature to graph the real and imaginary parts of $\zeta(\sigma + it)$ as a function of t for some fixed σ. If you make a table of these for various values of σ, *Mathematica* can animate them to make a movie. Maple, too, has the Riemann zeta function built in, as `Zeta`, and can do animation (see Help).

8.2. Summing Divergent Series

Once we know that the series for $\phi(1)$ converges, it is natural to ask what number it converges to. In fact, we computed this in Exercise I2.6.3: $\phi(1) = \log(2)$.

Exercise 8.2.1. Use the techniques of Interlude 2 to get an expansion of $1 - 2^{1-s} = 1 - \exp(-\log(2)(s - 1))$ near $s = 1$. Do the same for $(1 - 2^{1-s})^{-1}$. We know that $\phi(s) = \log(2) + O(s - 1)$; compute the residue of $\zeta(s)$ at $s = 1$. This gives an alternate proof of (7.1).

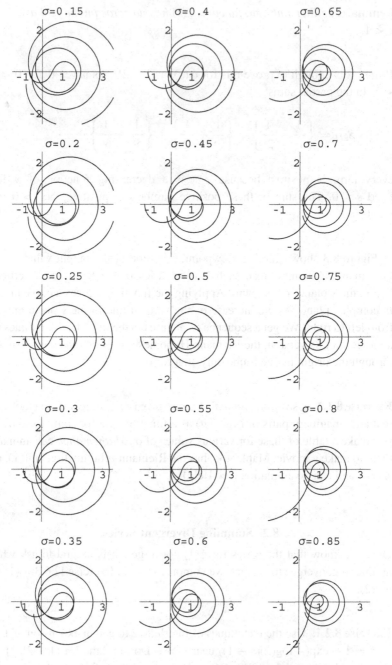

Figure 8.3. Movie of $\zeta(\sigma + it)$.

Euler computed that $\phi(1) = \log(2)$, but he wanted *more*. He wanted to assign values to

$$\phi(0) = \sum_{n=1}^{\infty} (-1)^{n-1} = 1 - 1 + 1 - 1 + \cdots,$$

$$\phi(-1) = \sum_{n=1}^{\infty} (-1)^{n-1} n = 1 - 2 + 3 - 4 + \cdots,$$

$$\phi(-2) = \sum_{n=1}^{\infty} (-1)^{n-1} n^2 = 1 - 4 + 9 - 16 + \cdots.$$

Of course, the the series for $\phi(0)$ makes no sense. The partial sums are alternately 1 and 0; they do not converge. Euler's idea, when confronted with this obstacle, was to make up a new definition of what a sum should be (Ayoub, 1974). Because it is true that

$$\frac{1}{1+x} = \sum_{n=0}^{\infty} (-1)^n x^n = 1 - x + x^2 - x^3 + \cdots \qquad \text{for } |x| < 1,$$

Euler simply defined the value of the series for $\phi(0)$ to be what you get when you plug $x = 1$ into $1/(1+x)$:

$$\phi(0) = 1 - 1 + 1 - 1 + \cdots \stackrel{A}{=} \frac{1}{2}.$$

This idea is now called ABEL SUMMATION, and we will use this name to distinguish it from the more traditional idea of convergence of a series discussed in Interlude 1. The idea is that if a function $f(x)$ has a Taylor series $\sum_n a_n x^n$ for $|x| < 1$, and $\lim_{x \to 1} f(x)$ is some number L, we define $\sum_n a_n$ as equal to L in the sense of Abel summation and write

$$\sum_{n=0}^{\infty} a_n \stackrel{A}{=} L.$$

The symbol $\stackrel{A}{=}$ is meant to indicate that this is a different definition for what an infinite sum means; we reserve the notation

$$\sum_{n=0}^{\infty} a_n = L \qquad \text{to mean} \quad \lim_{n \to \infty} s_n = L.$$

Notice that the definition for Abel summation is the same as the hypothesis for Abel's Theorem, Part I (12.19), except that we don't have the hypothesis that the series converges in the traditional sense.

Euler's idea is not completely unreasonable. After all, the reason the series does not converge in the traditional sense is that the partial sums alternate: $1, 0, 1, 0, \ldots$. The average is $1/2$.

Because $\zeta(s) = (1 - 2^{1-s})^{-1}\phi(s)$ for $\text{Re}(s) > 0$, Euler extended this definition to $s = 0$ as well. So,

$$\zeta(0) = 1 + 1 + 1 + 1 \cdots \overset{A}{=} (1 - 2)^{-1}\phi(0) = -\frac{1}{2}.$$

To compute $\phi(-1)$, start with $1/(1 + x)$, take a derivative, and multiply by x. This operation on a function $f(x)$ is called the Euler operator:

$$\mathcal{E} f(x) = x f'(x).$$

So,

$$\mathcal{E}\left(\frac{1}{1 + x}\right) = \frac{-x}{(1 + x)^2}.$$

Apply the Euler operator to the series to see that for $|x| < 1$,

$$\frac{-x}{(1 + x)^2} = \sum_{n=1}^{\infty}(-1)^n n x^n = -x + 2x^2 - 3x^3 + 4x^4 - \cdots.$$

You computed this series in Exercise I2.2.5. As $-x/(1 + x)^2$ is equal to $-1/4$ when $x = 1$,

$$\sum_{n=1}^{\infty}(-1)^n n = -1 + 2 - 3 + 4 \cdots \overset{A}{=} -\frac{1}{4},$$

$$\phi(-1) = \sum_{n=1}^{\infty}(-1)^{n-1}n = 1 - 2 + 3 - 4 \cdots \overset{A}{=} \frac{1}{4},$$

because $\phi(-1) = 1/4$, $\zeta(-1) = (1 - 2^2)^{-1}\phi(-1)$, or

$$\zeta(-1) = 1 + 2 + 3 + 4 + \cdots \overset{A}{=} -\frac{1}{12}.$$

Exercise 8.2.2. Apply the Euler operator to $-x/(1 + x)^2$ and to its series expansion. Use this to compute $\phi(-2)$ in the sense of Abel summation. From this, deduce that $\zeta(-2) \overset{A}{=} 0$. By repeatedly applying the Euler operator, compute $\zeta(-3)$, $\zeta(-4)$, $\zeta(-5)$, $\zeta(-6)$, $\zeta(-7)$. The series expansions are easy to compute, but to find the functions themselves, you will want a computer algebra package such as Maple or *Mathematica*. In *Mathematica*, you can define the Euler operator by using

```
e[f_] := x*D[f, x] // Together
```

The input $e[1/(1 + x)]$ returns the answer $-x/(1+x)^2$. In Maple, you can analogously define the Euler operator by using

```
e:=f->simplify(x*diff(f,x));
```

The input $e(-x/(1+x)^2)$; returns the answer $x(-1+x)/(1+x)^3$.

Exercise 8.2.3. The zeros of $\zeta(s)$ at $s = -2, -4, -6, \ldots$ are called the TRIVIAL ZEROS. What connection do you see between the numbers

$$\zeta(-1) \quad \text{and} \quad B_2 = \frac{1}{6},$$

$$\zeta(-3) \quad \text{and} \quad B_4 = -\frac{1}{30},$$

$$\zeta(-5) \quad \text{and} \quad B_6 = \frac{1}{42},$$

$$\zeta(-7) \quad \text{and} \quad B_8 = -\frac{1}{30}?$$

Try to make a conjecture.

Exercise 8.2.4. This exercise outlines a proof of what you conjectured above. First, from the starting point

$$\frac{x}{\exp(x) - 1} = \sum_{m=0}^{\infty} B_m \frac{x^m}{m!},$$

develop a series expansions for $1/(\exp(x) - 1) - 2/(\exp(2x) - 1)$. Put this over a common denominator to see that it is $1/(\exp(x) + 1)$. Finally, substitute $x = -y$ to get that

$$\frac{1}{1 + \exp(-y)} = \sum_{m=0}^{\infty} (-1)^{m-1} \frac{B_m(1 - 2^m)}{m} \frac{y^{m-1}}{m - 1!}.$$

You know from the result of Exercise I2.2.11 that the $m - 1$st derivative at $y = 0$ is

$$\frac{d^{m-1}}{dy^{m-1}} \frac{1}{1 + \exp(-y)} \bigg|_{y=0} = (-1)^{m-1} \frac{B_m(1 - 2^m)}{m}. \tag{8.2}$$

Next, for $y > 0$, you should expand $1/(1 + \exp(-y))$ as a Geometric series in powers of $\exp(-y)$. See that the $m - 1$st derivative with respect to y is

$$\frac{d^{m-1}}{dy^{m-1}} \frac{1}{1 + \exp(-y)} = (-1)^m \sum_{n=0}^{\infty} (-1)^{n-1} n^{m-1} \exp(-ny).$$

This is not a Taylor series in y. It is an example of what's called a Lambert series (powers of $\exp(-y)$). It certainly does not converge at $y = 0$. Nonetheless, we can say that they are equal in the sense of Abel summation for $y = 0$, when $\exp(-y) = 1$:

$$\left.\frac{d^{m-1}}{dy^{m-1}}\frac{1}{1+\exp(-y)}\right|_{y=0} \stackrel{A}{=} (-1)^m \sum_{n=0}^{\infty}(-1)^{n-1}n^{m-1}. \qquad (8.3)$$

Compare (8.2) and (8.3) to determine $\phi(1 - m)$ and $\zeta(1 - m)$.

Abel summation is merely a definition, a rule that we have made up. It is part of the beauty of mathematics that we can make any rules we want. It is best, though, if the rules are consistent. For example, Euler's product formula for $\sin(\pi x)/\pi x$ let him rederive Wallis product formula for $\pi/2$; it gave answers that were consistent. Is Abel summation a good rule? We will see later that Euler's intuition was justified; the answers he derived using Abel summation are consistent with other approaches to defining $\zeta(s)$ for $\mathrm{Re}(s) \leq 0$.

8.3. The Gamma Function

Euler was interested in the factorial function $n!$ and the question of how to extend this function beyond just positive integers. Euler succeeded and discovered, for example, that

$$-\frac{1}{2}! = \sqrt{\pi}.$$

It turns out that the answer to this question is connected to the study of prime numbers.

For a real number $s > 0$, Euler defined a complicated-looking function $\Gamma(s)$, the GAMMA FUNCTION, as follows:

$$\Gamma(s) = \int_0^{\infty} \exp(-t)t^{s-1}dt. \qquad (8.4)$$

This is an improper integral, so we need to be a little careful. Write

$$\int_0^{\infty} \exp(-t)t^{s-1}dt = \int_0^1 \exp(-t)t^{s-1}dt + \int_1^{\infty} \exp(-t)t^{s-1}dt$$

and consider the two pieces separately. First, we need to observe that for $t > 0$, every term in the series for $\exp(t)$ is positive. So, because $\exp(t)$ is the sum of all of the terms, it is bigger than each of them: For every positive n,

$$\frac{t^n}{n!} < \exp(t) \quad \text{or} \quad \exp(-t) < \frac{n!}{t^n}.$$

Now, for fixed s, pick any $n > s$; then,

$$\int_1^\infty \exp(-t)t^{s-1}dt < n! \int_1^\infty t^{s-n-1}dt.$$

This is just the comparison test, property v of Interlude 1 again. We can compute this last integral. It is

$$n!\left(\frac{t^{s-n}}{s-n}\right)\Big|_1^\infty = n!\frac{1}{n-s}.$$

The key fact here is that t^{s-n} tends to 0 as t tends to ∞, because we made $s - n < 0$. Notice that this same argument works for $s \le 0$ as well. We only need $s > 0$ for the other piece.

Because we are interested in $t > 0$, $\exp(-t) < 1$; so, $\exp(-t)t^{s-1} < t^{s-1}$ and

$$\int_0^1 \exp(-t)t^{s-1}dt < \int_0^1 t^{s-1}dt.$$

So, it suffices to show that this simpler integral is finite, again using the comparison test. But

$$\int t^{s-1}dt = \frac{t^s}{s}, \quad \text{so} \quad \int_0^1 t^{s-1}dt = \frac{t^s}{s}\Big|_0^1 = \frac{1}{s} - 0$$

is finite *if* $s > 0$. To see this, write $t^s = \exp(s\log(t))$ and notice that as $t \to 0$, $s\log(t) \to -\infty$, so $\exp(s\log(t)) \to 0$. But if $s < 0$, then $s\log(t) \to +\infty$ and $t^s = \exp(s\log(t)) \to +\infty$ as $t \to 0$.

Now that we've done this, we can make the same definition of $\Gamma(s)$ for complex values of s, as long as $\mathrm{Re}(s) > 0$. Remember that $t^{s-1} = \exp((s-1)\log(t))$. Checking that the improper integral is finite works as before, because $|t^s| = t^{\mathrm{Re}(s)}$.

Lemma.

$$\Gamma(s+1) = s\Gamma(s).$$

Proof. We just integrate by parts in $\int_0^\infty \exp(-t)t^s dt$. Let $u = t^s$, $du = st^{s-1}$, $dv = \exp(-t)dt$, $v = -\exp(-t)$. So, as indefinite integrals,

$$\int \exp(-t)t^s dt = -\exp(-t)t^s + s\int \exp(-t)t^{s-1}dt.$$

So, for the definite integral,

$$\int_0^\infty \exp(-t)t^s\,dt = -\exp(-t)t^s\Big|_0^\infty + s\int_0^\infty \exp(-t)t^{s-1}\,dt$$
$$= 0 + s\Gamma(s),$$

because t^s tends to 0 as t tends to 0, and as t tends to ∞, $\exp(-t)$ tends to 0 much faster than t^s increases.   □

It is an easy computation to see that $\Gamma(1) = 1$, and from the lemma, we get that $\Gamma(n+1) = n\Gamma(n)$ for any integer n. By induction,

$$\Gamma(n+1) = n!.$$

For example,

$$\Gamma(4) = 3\Gamma(3) = 3\cdot 2\Gamma(2) = 3\cdot 2\cdot 1\Gamma(1) = 3\cdot 2\cdot 1.$$

Roughly speaking, the Gamma function "interpolates" the factorial; it extends the function from integers to complex numbers in a natural way. The above integral defining the Gamma function is often called Euler's second integral.

Because we will spend a lot of time thinking about this function, a brief digression is in order. What led Euler to think that factorials might be computed using an integral? Euler was fooling around with infinite products and noticed that for a positive integer n,

$$\left(\left(\frac{2}{1}\right)^n \frac{1}{n+1}\right)\left(\left(\frac{3}{2}\right)^n \frac{2}{n+2}\right)\left(\left(\frac{4}{3}\right)^n \frac{3}{n+3}\right)\cdots = n!.$$

Here, the equal sign is just formal; we pay no attention to convergence. Observe that the numerator in each fraction raised to the nth power cancels out the denominator of the next term. And all numerator terms except $1\cdot 2\cdot 3\cdots n$ are cancelled out by the denominator terms. Euler used this same product to compute $(-1/2)!$. He plugged in $n = -1/2$ and cancelled the fractions raised to the $-1/2$ power; as before, he was left with

$$\left(-\frac{1}{2}\right)! = \frac{2}{1}\cdot\frac{4}{3}\cdot\frac{6}{5}\cdot\frac{8}{7}\cdot\frac{10}{9}\cdots.$$

Similarly,

$$\frac{1}{2}! = \frac{2}{3}\cdot\frac{4}{5}\cdot\frac{6}{7}\cdot\frac{8}{9}\cdot\frac{10}{11}\cdots.$$

So,

$$\left(-\frac{1}{2}\right)! \cdot \frac{1}{2}! = \frac{2 \cdot 2}{1 \cdot 3} \cdot \frac{4 \cdot 4}{3 \cdot 5} \cdot \frac{6 \cdot 6}{5 \cdot 7} \cdot \frac{8 \cdot 8}{7 \cdot 9} \cdots = \frac{\pi}{2}$$

from the Wallis product formula. From the identity $s! = s \cdot (s-1)!$, Euler deduced that $(1/2)! = 1/2 \cdot (-1/2)!$, so $(-1/2)! \cdot (-1/2)! = \pi$, and thus,

$$-\frac{1}{2}! = \sqrt{\pi}.$$

A formula involving π led Euler to think about circles and area, and that led to thinking about integrals (Davis, 1959). This is not all just a coincidence; the Gamma function really can be described by an infinite product, similar to the sine function. And there is a connection:

$$\Gamma(s)\Gamma(1-s) = \frac{\pi}{\sin(\pi s)}$$

is true.

Our study of the Gamma functions requires another way of looking at it.

Lemma.

$$\Gamma(s) = \sum_{k=0}^{\infty} \frac{(-1)^k}{k!} \frac{1}{s+k} + \int_1^{\infty} \exp(-t)t^{s-1}dt. \tag{8.5}$$

Proof. This is actually pretty easy. We again break the integral into two pieces,

$$\Gamma(s) = \int_0^{\infty} \exp(-t)t^{s-1}dt = \int_0^1 \exp(-t)t^{s-1}dt + \int_1^{\infty} \exp(-t)t^{s-1}dt,$$

and expand $\exp(-t)$ as a series,

$$\int_0^1 \exp(-t)t^{s-1}dt = \int_0^1 \sum_{k=0}^{\infty} \frac{(-1)^k t^k}{k!} t^{s-1}dt.$$

Now, change the order of the sum and integral,

$$= \sum_{k=0}^{\infty} \frac{(-1)^k}{k!} \int_0^1 t^k t^{s-1}dt,$$

and compute the integral,

$$= \sum_{k=0}^{\infty} \frac{(-1)^k}{k!} \left(\frac{t^{s+k}}{s+k}\right)\Bigg|_{t=0}^{t=1},$$

$$= \sum_{k=0}^{\infty} \frac{(-1)^k}{k!} \frac{1}{s+k}.$$

One detail that we ignored is whether it is legal to interchange the infinite sum and the integral. Both are defined in terms of limits, and a double limit is not necessarily the same if you reverse the order. Here, it is legal, basically, because the integral of the sum of absolute values

$$\int_0^1 \sum_{k=0}^{\infty} \left| \frac{(-1)^k t^k}{k!} t^{s-1} \right| dt = \int_0^1 \exp(t) t^{\mathrm{Re}(s)-1} dt$$

is finite. Absolute convergence saves us.   □

Although this seems like a complicated formula, it has the advantage that it makes sense for *all* s in $\mathbb{C}$ except $s = 0, -1, -2, \ldots$. Here's why: We showed above that $\int_1^{\infty} \exp(-t) t^{s-1} dt$ is finite for *all* values of s. And the series converges, even converges absolutely, for $s \neq 0, -1, -2, \ldots$, in comparison to the series for $\exp(1) = \sum_k 1/k!$. We use (8.5) to *define* $\Gamma(s)$ for all s in $\mathbb{C}$ except $s = 0, -1, -2, \ldots$. According to the lemma, it agrees with the old definition (8.4) if $\mathrm{Re}(s) > 0$; so, this is another example of an analytic continuation.

Notice that the series on the right side of (8.5) is not a Laurent expansion in the variable s. It is something new. If we isolate any individual term, for example the $k = 3$ term, the series is

$$\frac{(-1)^3}{3!} \frac{1}{s+3} + \sum_{k \neq 3} \frac{(-1)^k}{k!} \frac{1}{s+k}.$$

For values of s near -3, the sum over all k that are different from 3 converges, even converges absolutely in the sense of Section I2.6. So, the sum over $k \neq 3$ defines a function of s for s near -3 or even $s = -3$. Thus, $\Gamma(s)$ has a simple pole at $s = -3$ with residue $-1/3!$. We can do this for any integer, not just $s = 3$, of course. So, we see that

Theorem. $\Gamma(s)$ *has a simple pole at* $s = -k$, *for* $k = 0, 1, 2 \ldots$, *with residue* $(-1)^k/k!$. *That is,*

$$\Gamma(s) = \frac{(-1)^k}{k!} \frac{1}{s+k} + O(1) \quad \text{near } s = -k. \tag{8.6}$$

What we have obtained here is a PARTIAL FRACTIONS expansion of the function $\Gamma(s)$, which is analogous to writing $2/(x^2 - 1) = 1/(x - 1) - 1/(x + 1)$. The difference is that now there can be infinitely many poles. The partial fractions expansion is the sum of infinitely many singular parts.

Exercise 8.3.1. Show, by computing the integral in (8.4), that $\Gamma(1) = 1$, as claimed above.

Exercise 8.3.2. Assuming that $s > 0$ is real, change variables by $x = t^s$ in the integral (8.4) to show that

$$\Gamma(s) = \frac{1}{s} \int_0^\infty \exp(-x^{1/s})\, dx \qquad \text{for } s > 0.$$

Now, change s to $1/s$ to prove that

$$\Gamma(1 + 1/s) = \int_0^\infty \exp(-x^s)\, dx \qquad \text{for } s > 0.$$

In Exercise 9.1.2, you will compute

$$\int_0^\infty \exp(-x^2)\, dx = \frac{\sqrt{\pi}}{2};$$

according to the above, this is just $\Gamma(3/2)$. Use the recursion formula to compute $\Gamma(5/2)$, $\Gamma(7/2)$, and $\Gamma(9/2)$. Try to develop a general formula for $\Gamma(n + 1/2)$ for $n \geq 1$. Now, compute $\Gamma(1/2)$ and $\Gamma(-1/2)$. Notice that $\Gamma(1/2)$, which should be $(-1/2)!$, gives the same answer Euler got using infinite products.

Exercise 8.3.3. Show in two stages that the property $\Gamma(s + 1) = s\Gamma(s)$ still holds for (8.5). First, show that

$$\int_1^\infty \exp(-t)t^s\, dt = e^{-1} + s \int_1^\infty \exp(-t)t^{s-1} dt.$$

Next, show that

$$\sum_{k=0}^\infty \frac{(-1)^k}{k!} \frac{1}{s+1+k} = s \sum_{j=0}^\infty \frac{(-1)^j}{j!} \frac{1}{s+j} - e^{-1}.$$

(Hint: Change variables in the sum $j = k + 1$, and write $1/(j-1)!$ as $(s + j - s)/j!$.)

8.4. Analytic Continuation

In the previous section, we introduced the Gamma function for $\text{Re}(s) > 0$ and showed how to extend it to a larger domain. In this section, we will show how $\Gamma(s)$ is connected to the Riemann zeta function $\zeta(s)$ and lets us extend the definition of $\zeta(s)$ beyond $\text{Re}(s) > 0$ as well. Riemann was the first to do this, but this variation of the proof is attributable to Hermite.

Theorem. *For Re(s) > 1,*

$$\Gamma(s)\zeta(s) =$$

$$\frac{1}{s-1} - \frac{1}{2s} + \sum_{k=1}^{\infty} \frac{B_{2k}}{2k!} \frac{1}{s+2k-1} + \int_1^{\infty} \frac{1}{\exp(x)-1} x^{s-1} dx.$$

Proof. First, for an integer n, consider the integral

$$\int_0^{\infty} \exp(-nx) x^{s-1} dx = \frac{1}{n^s} \Gamma(s).$$

This identity is easy to prove. Just change the variables by $nx = t$, so $ndx = dt$ and $x^{s-1} = t^{s-1}/n^{s-1}$. If we sum both sides over all $n \geq 1$, we get

$$\Gamma(s)\zeta(s) = \sum_{n=1}^{\infty} \int_0^{\infty} \exp(-nx) x^{s-1} dx$$

Now, change the sum and integral to get

$$= \int_0^{\infty} \left\{ \sum_{n=1}^{\infty} \exp(-nx) \right\} x^{s-1} dx.$$

Because $\exp(-nx) = \exp(-x)^n$, we see a Geometric series in the variable $\exp(-x)$, but starting with $n = 1$. So, we have

$$= \int_0^{\infty} \frac{\exp(-x)}{1-\exp(-x)} x^{s-1} dx,$$

$$= \int_0^{\infty} \frac{1}{\exp(x)-1} x^{s-1} dx$$

after multiplying numerator and denominator by $\exp(x)$.

Now we can use the same trick we used on the Gamma function. Break the integral into two pieces:

$$\Gamma(s)\zeta(s) = \int_0^1 \frac{1}{\exp(x)-1} x^{s-1} dx + \int_1^{\infty} \frac{1}{\exp(x)-1} x^{s-1} dx.$$

We need to examine the first piece. We can write $1/(\exp(x)-1)$ as a series with the Bernoulli numbers; we just need to factor an x out of (6.6):

$$\int_0^1 \frac{1}{\exp(x)-1} x^{s-1} dx = \int_0^1 \left\{ \frac{1}{x} - \frac{1}{2} + \sum_{k=1}^{\infty} \frac{B_{2k} x^{2k-1}}{2k!} \right\} x^{s-1} dx.$$

Now, multiply in the x^{s-1} term to get

$$= \int_0^1 \left\{ x^{s-2} - \frac{1}{2} x^{s-1} + \sum_{k=1}^{\infty} \frac{B_{2k} x^{2k+s-2}}{2k!} \right\} dx.$$

We next integrate each term separately; that is, we change the sum and integral. They are all of the same form:

$$\int_0^1 x^{s+n} dx = \left(\frac{x^{s+n+1}}{s+n+1} \right) \Big|_0^1 = \frac{1}{s+n+1}, \quad \text{where } n = -2, -1,$$
$$\text{or } 2k - 2.$$

So,

$$\int_0^1 \frac{1}{\exp(x) - 1} x^{s-1} dx = \frac{1}{s-1} - \frac{1}{2s} + \sum_{k=1}^{\infty} \frac{B_{2k}}{2k!} \frac{1}{s+2k-1}.$$

$\square$

This theorem gives the analytic continuation of the function $\Gamma(s)\zeta(s)$, which is analogous to that of $\Gamma(s)$ from the previous section. Here's how. Let

$$F(s) = \frac{1}{s-1} - \frac{1}{2s} + \sum_{k=1}^{\infty} \frac{B_{2k}}{2k!} \frac{1}{s+2k-1}$$

be the sum of all the singular parts, and let

$$G(s) = \int_1^{\infty} \frac{1}{\exp(x) - 1} x^{s-1} dx.$$

So, the theorem says $\Gamma(s)\zeta(s) = F(s) + G(s)$. As in our discussion of the Gamma function, the integral defining $G(s)$ is convergent for any value of s, basically because the exponential decay near infinity of $1/(\exp(x) - 1)$ dominates the polynomial growth of x^{s-1}, for any s. Meanwhile, for $s \neq 1, 0$, or any negative odd integer, the series defining $F(s)$ converges and, thus, defines a function of s.

In other words, $F(s) + G(s)$ defines a function for s different from $1, 0$, or any negative odd integer, which agrees with the previous definition of $\Gamma(s)\zeta(s)$, which worked only if $\operatorname{Re}(s) > 1$. And, as before, it is a partial fractions expansion. It says that $\Gamma(s)\zeta(s)$ has simple poles at 1, at 0, and at all the negative odd integers. For $k \geq 1$,

$$\Gamma(s)\zeta(s) = \frac{B_{2k}}{2k!} \frac{1}{(s+2k-1)} + O(1) \quad \text{near } s = 2k - 1. \tag{8.7}$$

What about $\zeta(s)$ by itself? We can divide out the $\Gamma(s)$ term:

$$\zeta(s) = \frac{F(s) + G(s)}{\Gamma(s)}.$$

Theorem. *The function* $\zeta(s)$ *extends to all values of* s *except* 1. $\zeta(s)$ *has a simple pole at* $s = 1$, *with residue* 1. $\zeta(0) = -1/2$. *Furthermore, for* $n = 1, 2, 3 \ldots$,

$$\zeta(-2n) = 0,$$

$$\zeta(-2n+1) = \frac{-B_{2n}}{2n},$$

where the B_{2n} *are again the Bernoulli numbers.*

Proof. We need one fact about $\Gamma(s)$ that we can't prove without using more sophisticated mathematics: that $\Gamma(s)$ is never equal to 0. Given this fact, it is okay to divide. Furthermore, because $\Gamma(1) = 1$, near $s = 1$ we have

$$\Gamma(s)^{-1} = 1 + O(s - 1),$$

$$F(s) + G(s) = \frac{1}{s-1} + O(1), \quad \text{and thus}$$

$$\zeta(s) = \frac{1}{s-1} + O(1). \tag{8.8}$$

Near $s = 0$ we have, according to the partial fractions expansions of $\Gamma(s)$ and $F(s) + G(s)$,

$$\Gamma(s) = \frac{1}{s} + O(1), \quad \text{and thus}$$

$$\Gamma(s)^{-1} = s + O(s^2);$$

$$F(s) + G(s) = -\frac{1}{2s} + O(1), \quad \text{and thus}$$

$$\zeta(s) = (s + O(s^2)) \cdot \left(-\frac{1}{2s} + O(1) \right)$$

$$= -\frac{1}{2} + O(s).$$

For s near $-2n + 1$, we calculate just as with $s = 0$.

Exercise 8.4.1. What are the singular parts for the Laurent expansions of $\Gamma(s)$ and $F(s) + G(s)$ at $s = -2n + 1$? What is the lead term of the Taylor expansion of $\Gamma(s)^{-1}$ at $s = -2n + 1$? What is $\zeta(-2n + 1)$?

For the last case, notice that $s = -2n$ is not a pole of $F(s) + G(s)$; so, all we can say is that

$$F(s) + G(s) = F(-2n) + G(-2n) + O(s + 2n), \quad \text{whereas}$$

$$\Gamma(s) = \frac{1}{2n!} \frac{1}{s + 2n} + O(1) \quad \text{according to (8.6); so,}$$

$$\Gamma(s)^{-1} = (2n)!(s + 2n) + O(s + 2n)^2,$$

$$\zeta(s) = (F(-2n) + G(-2n))(2n)!(s + 2n) + O(s + 2n)^2.$$

In particular, $\zeta(-2n) = 0$. The constant $F(-2n) + G(-2n)$ is mysterious. □

This theorem leads to a couple of interesting observations. First, it justifies Euler's use of Abel summation to say that

$$1 + 1 + 1 + 1 \cdots \overset{A}{=} -\frac{1}{2},$$

$$1 + 2 + 3 + \cdots + n + \cdots \overset{A}{=} -\frac{1}{12},$$

$$1 + 4 + 9 + \cdots + n^2 + \cdots \overset{A}{=} 0.$$

These are the same values for the zeta function at $s = 0, -1, -2 \ldots$ that we have just computed.

Second, we see (as Euler did) that there is a remarkable symmetry between $\zeta(2n)$ and $\zeta(1 - 2n)$:

$$\zeta(2n) = (-1)^{n+1} \frac{(2\pi)^{2n}}{2} \frac{B_{2n}}{2n!}, \quad \text{whereas} \quad \zeta(1 - 2n) = -\frac{B_{2n}}{2n}.$$

They both involve the mysterious Bernoulli number B_{2n}, as well as some boring powers of 2π and factorials. Our goal in Chapter 9 is to make the powers of 2π and the factorials look symmetric, and to extend this symmetry to all complex values of s, not just even integers.

Figure 8.4 shows the level curves for the real and imaginary parts of $\zeta(s)$. For comparison, there are analogous pictures in Interlude 3 for functions of a variable z instead of s: Figure I3.1 for the function z^2, and Figure I3.2 for the function $1/z$.

Near the point $s = 1$, we know that $\zeta(s)$ should look like $1/(s - 1)$; that is what (8.8) says. You can actually see this; compare Figure 8.4 near the point $s = 1$ to Figure I3.2 near $z = 0$.

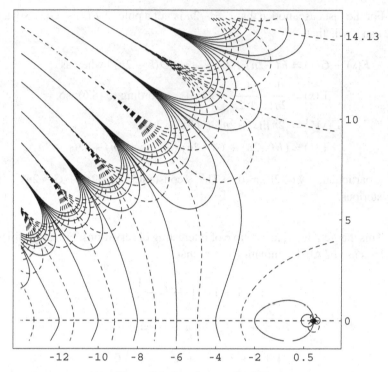

Figure 8.4. Level curves for $\zeta(s)$.

Since $\zeta(\sigma)$ is real for real $\sigma > 1$ from the series definition, it turns out it must be real for all real σ. So, $\mathrm{Im}(\zeta)$ is zero on the real axis. And we see this level curve in the horizontal dotted line in Figure 8.4. The solid curves crossing this one are level curves for $\mathrm{Re}(\zeta) = 0$. This means that we can also see the trivial zeros of $\zeta(s)$ at the negative even integers; any point where $\mathrm{Re}(\zeta) = 0$ and $\mathrm{Im}(\zeta) = 0$ means that $\zeta(s) = 0$.

Finally, in the upper right corner we can see the first nontrivial zero of $\zeta(s)$ at $s = 1/2 + i14.13473\ldots$, or the point $(1/2, 14.13473\ldots)$ in the plane. The other places where the level curves cross represent values other than 0 for either $\mathrm{Re}(\zeta)$ or $\mathrm{Im}(\zeta)$.

Exercise 8.4.2. My proof that the integral that defines the function $G(s)$ is actually finite was a little sketchy. Give a more careful argument, analogous to the one given for the Gamma function. (Hint: Show that for any integer n, $1 + x^n/n! < \exp(x)$.) What can you deduce from this?

Exercise 8.4.3. Show that the infinite series

$$\sum_{k=1}^{\infty} \frac{B_{2k}}{2k!} \frac{1}{s + 2k - 1},$$

which appears in the definition of $F(s)$, converges absolutely for s not a negative odd integer. Hint: According to the comparison test of Section I2.6, $\zeta(2k) < \zeta(2)$.) Now, use the value of $\zeta(2k)$ and the comparison test.

Exercise 8.4.4. Use the same facts as in the previous problem to show that the series

$$\frac{z}{\exp(z) - 1} = 1 - \frac{z}{2} + \sum_{k=1}^{\infty} \frac{B_{2k}}{2k!} z^{2k}$$

converges absolutely for $|z| < 2\pi$. (We certainly wouldn't expect any larger radius of convergence; the function has a pole at $z = 2\pi i$.) This exercise justifies our use of the series for $1/(\exp(x) - 1)$ in the integral between 0 and 1 defining the function $F(s)$.

Chapter 9
Symmetry

When the stars threw down their spears,
And water'd heaven with their tears,
Did he smile his work to see?
Did he who made the Lamb make thee?
Tyger! Tyger! burning bright
In the forests of the night,
What immortal hand or eye,
Dare frame thy fearful symmetry?

William Blake

Much of mathematics is about the search for symmetries. We like them because they tend to propagate themselves. That is, symmetries tend to lead to other symmetries. A good example, which you may have already seen, happens in linear algebra. The eigenvalues of a complex matrix A are complex numbers λ for which

$$A \overset{\leftarrow}{v} = \lambda \overset{\leftarrow}{v}$$

has a nonzero solution $\overset{\leftarrow}{v}$. A class of matrices that are interesting are those that are self-adjoint, that is, equal to their transpose conjugate. This kind of symmetry leads to another one for the eigenvalues, because the eigenvalues of a self-adjoint matrix are actually real numbers. So,

$$A = \overline{A}^t \quad \text{implies that} \quad \lambda = \overline{\lambda}.$$

At the end of the Chapter 8, we rediscovered Euler's observation that there is apparent symmetry between $\zeta(2n)$ and $\zeta(1 - 2n)$ for positive integers n. This chapter will extend the symmetry to all complex numbers s. In the first section, we develop a tool, Stirling's formula, which approximates $n!$ more accurately than we did in Exercise 3.3.2. The next section proves a symmetry property for the classical Jacobi theta function, which arises in the physics of heat conduction, among other places. The last section of this chapter uses this symmetry to prove the functional equation for the Riemann zeta function.

9.1. Stirling's Formula

In Section 8.3, we introduced the function

$$\Gamma(s) = \int_0^\infty \exp(-t)t^{s-1}dt;$$

if we take s to be a positive integer $n + 1$, $\Gamma(n + 1) = n!$. So, if we want to know how fast the factorial function grows as a function of n, that is the same as asking how big is the integral

$$\Gamma(s + 1) = \int_0^\infty \exp(-t)t^s dt$$

as a function of the parameter s. We can use Taylor series to approximate the integrand by using a simpler one, and we can use this approximation to get an approximation for the integral. We will assume that s is real and positive, for simplicity, although this can be done more generally.

First, we have to change the variables, introducing a new variable of integration x, where $t = s(1 + x)$, therefore $dt = sdx$ and the integral goes from -1 to ∞. Convince yourself that after this change, the integral is

$$\Gamma(s + 1) = \exp(-s)s^{s+1}\int_{-1}^\infty (\exp(-x)(1 + x))^s \, dx.$$

The term $\exp(-s)s^{s+1}$ is fine; it won't get any simpler. We next want to simplify the integrand, writing

$$\exp(-x)(1 + x) = \exp(h(x))$$

for some function $h(x)$. Clearly,

$$\begin{aligned}
h(x) &= \log(\exp(-x)(1 + x)) \\
&= \log(\exp(-x)) + \log(1 + x) = -x + \log(1 + x) \\
&= -\frac{1}{2}x^2 + \frac{1}{3}x^3 - \frac{1}{4}x^4 + \cdots,
\end{aligned}$$

where we have used the result of Exercise I2.2.6.

Here's where the approximation idea comes in. Because $h(x) \approx -\frac{1}{2}x^2$, we should have

$$\int_{-1}^\infty (\exp(-x)(1 + x))^s \, dx = \int_{-1}^\infty \exp(sh(x))dx \approx \int_{-1}^\infty \exp\left(-\frac{s}{2}x^2\right) dx.$$

Of course, it is a little more complicated than that; the approximation $h(x) \approx -\frac{1}{2}x^2$ is valid for x close to zero. In fact, in Exercise I2.6.7, you showed that the radius of convergence of the series for $h(x)$ is 1. But we are

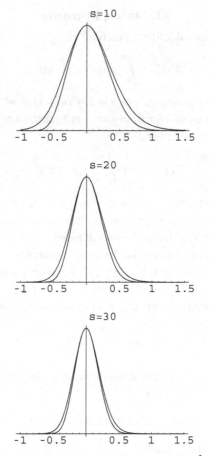

Figure 9.1. Three approximations to $\exp(-sx^2/2)$.

interested in all values of x between -1 and ∞. One can show by some calculus (see Exercise 9.1.1) that both the functions $h(x) = -x + \log(1 + x)$ and $-x^2/2$ are negative and decreasing for $x > 0$. So, if s is very large, both $\exp(sh(x))$ and $\exp(-\frac{s}{2}x^2)$ are eventually close to zero. Figure 9.1 compares the two functions for $s = 10$, 20, and 30.

Now, to compute the value of the approximation, we claim that

$$\int_{-1}^{\infty} \exp\left(-\frac{s}{2}x^2\right) dx \approx \int_{-\infty}^{\infty} \exp\left(-\frac{s}{2}x^2\right) dx,$$

because the function being integrated is very close to zero if $x < -1$, at least if s is large. So, the extra part of the integral we've added, from $-\infty$ to -1, is very small. Change variables again, with $y = x\sqrt{s/2}$, to get

$$\int_{-\infty}^{\infty} \exp\left(-\frac{s}{2}x^2\right) dx = \sqrt{2/s} \int_{-\infty}^{\infty} \exp(-y^2)\, dy.$$

This last integral is well known; it is $\sqrt{\pi}$ (see Exercise 9.1.2).

Stirling's formula says that our rough analysis is correct, that

$$\Gamma(s+1) \sim \exp(-s)s^{s+1}\sqrt{2\pi/s} = \left(\frac{s}{e}\right)^s \sqrt{2\pi s}.$$

Because for a positive integer n, $\Gamma(n+1) = n!$, we have

Theorem. (Stirling's Formula). *For large n,*

$$n! \sim \left(\frac{n}{e}\right)^n \sqrt{2\pi n}. \tag{9.1}$$

For example,

$50! =$
$30414093201713378043612608166064768844377641568960$
$512000000000000.$

On the other hand,

$$\left(\frac{50}{e}\right)^{50} \sqrt{100\pi} = 3.03634 \times 10^{64}.$$

The ratio of these two huge numbers is 1.00167, so the error in our approximation is about one-tenth of one percent.

For more on estimating integrals, see DeBruijn (1981).

Exercise 9.1.1. Use calculus to show that for $x > 0$, both the functions $-x^2/2$ and $h(x) = -x + \log(1+x)$ decrease for $x > 0$. Therefore, because both are 0 at $x = 0$, both are negative for $x > 0$. So, for large values of s, both $\exp(-sx^2/2)$ and $\exp(sh(x))$ are close to 0, even though $-x^2/2$ and $h(x)$ are not close to each other.

Exercise 9.1.2. Show that

$$\int_{-\infty}^{\infty} \exp(-y^2)dy = \sqrt{\pi}$$

as follows. Take two copies of the integral multiplied:

$$\int_{-\infty}^{\infty} \exp(-y^2)dy \int_{-\infty}^{\infty} \exp(-x^2)dx = \int_{-\infty}^{\infty}\int_{-\infty}^{\infty} \exp(-x^2 - y^2)\,dxdy.$$

Now, switch to polar coordinates.

Exercise 9.1.3. The beauty of Stirling's formula is that the larger n is the *better* approximation you get, at least in the sense of relative error above. (We first saw an example of this in Exercise 3.2.2.) That is what the symbol $\sim$ means; the ratio of the two sides tends to 1 as n goes to ∞. Use the following factorials and compute the relative error in Stirling's formula for these values of n.

$10! = 3628800$
$20! = 2432902008176640000$
$30! = 265252859812191058636308480000000$
$40! = 815915283247897734345611269596115894272000000000$

For a particular choice of n, one can improve the approximation by using a technique called Euler–Maclaurin summation. The basic idea is to approximate a sum using an integral, just as you did for $\log(n!)$ in Exercise 3.3.1. Euler–Maclaurin summation uses the Bernoulli numbers discussed in Section 6.2 to control the error in this approximation. This was secretly used to approximate the Harmonic numbers, H_n, in Exercises 3.2.3 and 3.2.4.

Exercise 9.1.4. With Euler–Maclaurin summation, one can show that for any m,

$$n! \approx \left(\frac{n}{e}\right)^n \sqrt{2\pi n} \exp\left(\sum_{k=1}^{m} \frac{B_{2k}}{2k(2k-1)n^{2k-1}}\right).$$

The bigger m is, the better the approximation. Use this with $m = 2$ to get a better approximation to $50!$ than the one above. How many of the leading digits of your approximation are correct? In fact, $50!$ divided by this new approximation is 1.000000000002538, so it's a very good approximation indeed. To learn about Euler–Maclaurin summation, see Graham, Knuth, and Patashnik (1994).

9.2. The Transformation Formula

For a real number $t > 0$, we will consider another function, the JACOBI THETA FUNCTION:

$$\Theta(t) = \sum_{k=-\infty}^{\infty} \exp(-\pi k^2 t).$$

The main purpose of this section is to prove a remarkable symmetry, the transformation formula for $\Theta(t)$.

Theorem. *The function* $t^{1/4}\Theta(t)$ *is invariant under the symmetry* $t \to 1/t$. *Another way of saying this is that*

$$\Theta\left(\frac{1}{t}\right) = \sqrt{t}\,\Theta(t). \tag{9.2}$$

The function $\Theta(t)$ has a more general version $\Theta(t, x)$, which is a function of two variables. These functions play an important role not only in number theory, but also in mathematical physics, in the study of the heat equation

$$\frac{\partial u}{\partial t} = \frac{\partial^2 u}{\partial x^2},$$

a partial differential equation (PDE) that describes the diffusion of heat over time t through a metal rod or wire stretched out on the x axis. In fact, Chapter 14 of Bellman (1961) sketches a proof of the transformation formula based on the following physical considerations. One side of the transformation formula arises from a "separation of variables" solution to the PDE, derived from consideration of large values of t. The other side is what physicists call a "similarity solution" to the PDE, which arises from consideration of small values of t. Because a PDE with boundary conditions has a unique solution, the two sides must be equal.

A more standard proof of the transformation formula uses Fourier analysis. It is very beautiful but uses some advanced techniques. We will give yet another proof, attributable to G. Polya (Polya, 1927), that relies only on Stirling's formula (9.1) for $n!$, and a couple of other basic facts. For example,

$$\lim_{n\to\infty} \left(1 + \frac{x}{n}\right)^n = \exp(x),$$

and therefore, because $\exp(x)$ is a continuous function of x,

$$\lim_{n\to\infty} \left(1 + \frac{x_n}{n}\right)^n = \exp(x) \qquad \text{if} \qquad \lim_{n\to\infty} x_n = x. \tag{9.3}$$

We begin the proof by developing some weird-looking identities. First, for any z in $\mathbb{C}$, and for any positive integer m,

$$\left(z^{1/2} + z^{-1/2}\right)^{2m} = \sum_{k=0}^{2m} \binom{2m}{k} z^{k/2} z^{-(2m-k)/2},$$

according to the Binomial Theorem, which becomes

$$= \sum_{j=-m}^{m} \binom{2m}{m+j} z^j$$

after changing the variables $k = m + j$. Now, let l be a positive integer and let $\omega = \exp(2\pi i / l)$. This makes $\omega^l = \exp(2\pi i) = 1$. We take $z = \omega^n$ for some various values of n in the previous formula and add them together, as follows:

$$\sum_{-l/2 \leq n < l/2} \{(\omega^n)^{1/2} + (\omega^n)^{-1/2}\}^{2m} = \sum_{-l/2 \leq n < l/2} \sum_{j=-m}^{m} \binom{2m}{m+j} \omega^{nj}.$$

Now, change the order on the double sum:

$$= \sum_{j=-m}^{m} \binom{2m}{m+j} \sum_{-l/2 \leq n < l/2} \omega^{nj}.$$

Lemma.

$$\sum_{n} \omega^{nj} = \begin{cases} l, & \text{in the case of } j = lk \text{ for some integer } k, \\ 0, & \text{otherwise.} \end{cases}$$

Proof. If $j = lk$, $\omega^{nj} = \omega^{nlk} = (\omega^l)^{nk} = 1$. The sum is just the number of terms, which is l. Otherwise,

$$\omega^j = \exp(2\pi i j / l) = \cos(2\pi j / l) + i \sin(2\pi j / l) \neq 1.$$

Because of the periodicity of cos and sin, the value of ω^{nj} only depends on the value of n modulo l (in the language of Interlude 4); in other words, $\omega^{nj} = \omega^{(n+kl)j}$ for any k. So, we get the same answer if we sum over $0 \leq n < l$ instead. Replacing n with $n + 1$ gives the same terms (modulo l), merely in a different order. So,

$$\sum_{n} \omega^{nj} = \sum_{n} \omega^{(n+1)j} = \omega^j \sum_{n} \omega^{nj}.$$

Because $\omega^j \neq 1$, the sum on n must be 0.   □

So we have

$$\sum_{-l/2 \leq n < l/2} \{(\omega^n)^{1/2} + (\omega^n)^{-1/2}\}^{2m} = l \sum_{k=-[m/l]}^{[m/l]} \binom{2m}{m+lk}. \qquad (9.4)$$

Now, let $t > 0$ be fixed and let $l = [\sqrt{\pi m t}]$, so

$$l^2 \sim \pi m t. \qquad (9.5)$$

Divide both sides of (9.4) by 2^{2m} to get

$$\sum_{-l/2 \le n < l/2} \left\{ \frac{\exp(\pi in/l) + \exp(-\pi in/l)}{2} \right\}^{2m} = \sum_{k=-[m/l]}^{[m/l]} \frac{[\sqrt{\pi mt}]}{2^{2m}} \binom{2m}{m+lk}.$$

(9.6)

So far, this is completely mysterious. We will now take limits as $m \to \infty$ (so $l \to \infty$ also) and show that the two sides of (9.6) give the two sides of the transformation formula (9.2).

Lemma.

$$\lim_{m \to \infty} \sum_{-l/2 \le n < l/2} \left\{ \frac{\exp(\pi in/l) + \exp(-\pi in/l)}{2} \right\}^{2m} = \sum_{n=-\infty}^{\infty} \exp(-\pi n^2/t).$$

Proof. Its clear that as m and l tend to infinity, we are summing *something* from $n = -\infty$ to ∞. We just need to see that each term in the sum tends to $\exp(-\pi n^2/t)$. We are looking at terms such as

$$\cosh(x)^{2m} = \left\{ \frac{\exp(x) + \exp(-x)}{2} \right\}^{2m}, \quad \text{where} \quad x = \pi in/l.$$

But because

$$\cosh(x) = 1 + \frac{x^2}{2} + O(x^4),$$

we see that

$$\cosh(\pi in/l)^{2m} = \left(1 + \frac{(\pi in)^2}{2l^2} + O\left(\frac{1}{l^4}\right) \right)^{2m}$$

$$= \left\{ \left(1 + \frac{-\pi^2 n^2}{2l^2} + O\left(\frac{1}{l^4}\right) \right)^{2l^2} \right\}^{m/l^2}.$$

Now, according to (9.5),

$$\frac{m}{l^2} \sim \frac{1}{\pi t},$$

and according to (9.3),

$$\left(1 + \frac{-\pi^2 n^2}{2l^2} + O\left(\frac{1}{l^4}\right) \right)^{2l^2} \sim \exp(-\pi^2 n^2).$$

So,

$$\cosh(\pi in/l)^{2m} \sim \exp(-\pi n^2/t).$$

$\square$

So we're half done.

Lemma.

$$\lim_{m\to\infty} \sum_{k=-[m/l]}^{[m/l]} \frac{[\sqrt{\pi m t}]}{2^{2m}} \binom{2m}{m+lk} = \sqrt{t} \sum_{k=-\infty}^{\infty} \exp(-\pi k^2 t).$$

Proof. First, note that

$$[\sqrt{\pi m t}] \sim \sqrt{t}\sqrt{\pi m},$$

which gives the $\sqrt{t}$ term we're looking for. According to the definition of binomial coefficients, we get

$$\frac{\sqrt{\pi m}}{2^{2m}} \binom{2m}{m+lk} = \frac{\sqrt{\pi m}}{2^{2m}} \frac{2m!}{m+lk!\, m-lk!}.$$

We apply (9.1) to each of $2m!$, $m+lk!$, and $m-lk!$ to get

$$\sim \frac{\sqrt{\pi m}}{2^{2m}} \left(\frac{2m}{e}\right)^{2m} \sqrt{2\pi 2m} \frac{(e/(m+lk))^{m+lk}}{\sqrt{2\pi(m+lk)}} \frac{(e/(m-lk))^{m-lk}}{\sqrt{2\pi(m-lk)}}.$$

Observe that all the $\sqrt{\pi}$ terms and powers of e and 2 cancel out, giving

$$= \frac{m^{2m}}{(m+lk)^{m+lk}(m-lk)^{m-lk}} \frac{m}{\sqrt{m+lk}\sqrt{m-lk}}.$$

Now,

$$\frac{m}{\sqrt{m+lk}\sqrt{m-lk}} = \frac{m}{\sqrt{m^2-l^2k^2}} = \frac{1}{\sqrt{1-l^2k^2/m^2}} \sim 1,$$

because $l^2 \sim m\pi t$, according to (9.5). So we're left with

$$\frac{m^{2m}}{(m+lk)^{m+lk}(m-lk)^{m-lk}} = \frac{m^{2m}}{(m+lk)^m(m+lk)^{lk}(m-lk)^m(m-lk)^{-lk}}$$

$$= \frac{m^{2m}}{(m^2-l^2k^2)^m} \frac{(m-lk)^{lk}}{(m+lk)^{lk}}.$$

To finish the lemma, we need to show two things: that

$$\frac{m^{2m}}{(m^2-l^2k^2)^m} \sim \exp(\pi k^2 t) \tag{9.7}$$

and that

$$\frac{(m-lk)^{lk}}{(m+lk)^{lk}} \sim \exp(-2\pi k^2 t). \tag{9.8}$$

For (9.7), observe that

$$\frac{m^{2m}}{(m^2 - l^2 k^2)^m} = \frac{1}{(1 - l^2 k^2/m^2)^m} \sim \frac{1}{\exp(-\pi k^2 t)},$$

according to (9.3), because $l^2 \sim \pi m t$. Meanwhile,

$$\frac{(m - lk)^{lk}}{(m + lk)^{lk}} = \left(\frac{1 - lk/m}{1 + lk/m}\right)^{lk}.$$

As $(1 - x)/(1 + x) = 1 - 2x + O(x^2)$, this is

$$(1 - 2lk/m + O(1/l^2))^{lk} \sim \exp(-2\pi t k)^k = \exp(-2\pi k^2 t),$$

according to (9.3), because $l/m \sim \pi t/l$. This proves (9.8). $\qquad\square$

The proof we just finished looks like many mysterious calculations, but it has physical meaning. Here is a quote from Bellman (1961):

Although the foregoing result at first may seem like a tour de force, in actuality it is closely connected with the fact that the continuous diffusion process may be considered to be a limit of a discrete random walk process. Since the random walk is ruled by the binomial distribution, and the diffusion process by the heat equation which gives rise to the Gaussian distribution, we see that it is not at all surprising that a modification of binomial expansions should yield the theta function formula.

9.3. The Functional Equation

In this section, we can finally prove the symmetry relation for $\zeta(s)$, conjectured by Euler and proved by Riemann, that we have been aiming at the whole chapter.

Theorem (Riemann). *Let* $\Lambda(s) = \pi^{-s/2}\Gamma(s/2)\zeta(s)$. *Then* $\Lambda(1 - s) = \Lambda(s)$.

Proof. Our starting point is very similar to that for the partial fractions expansion of $\Gamma(s)\zeta(s)$. First, for an integer n, consider the integral

$$\int_0^\infty \exp(-\pi n^2 t) t^{s/2} \frac{dt}{t} = \frac{1}{n^s}\Gamma(s/2)\pi^{-s/2}.$$

We've written this integral in a slightly different way, by grouping terms $t^{s/2} dt/t$ instead of $t^{s/2-1} dt$. This makes the identity easy to prove when we change the variables by $\pi n^2 t = x$: $dt/t = dx/x$ and $t^{s/2} = \pi^{-s/2} n^{-s} x^{s/2}$.

Suppose now that $\mathrm{Re}(s) > 1$. If we sum both sides over all $n \geq 1$, we get

$$\Lambda(s) = \pi^{-s/2}\Gamma(s/2)\zeta(s) = \sum_{n=1}^{\infty}\int_0^{\infty} \exp(-\pi n^2 t)t^{s/2}\frac{dt}{t}.$$

Now, change the sum and integral to get

$$= \int_0^{\infty} f(t)t^{s/2}\frac{dt}{t},$$

where

$$f(t) = \sum_{n=1}^{\infty} \exp(-\pi n^2 t).$$

Notice that our theta function is just

$$\Theta(t) = 1 + 2f(t)$$
$$= t^{-1/2}(1 + 2f(t^{-1})),$$

according to the transformation formula (9.2) for $\Theta(t)$. So, solving for $f(t)$, we see that

$$f(t) = 1/2(t^{-1/2} - 1) + t^{-1/2}f(t^{-1}).$$

We now split the integral defining $\Lambda(s)$ into two pieces,

$$\Lambda(s) = \int_0^1 f(t)t^{s/2}\frac{dt}{t} + \int_1^{\infty} f(t)t^{s/2}\frac{dt}{t},$$

and use the symmetry for $f(t)$:

$$= \frac{1}{2}\int_0^1 (t^{-1/2} - 1)t^{s/2}\frac{dt}{t} + \int_0^1 t^{-1/2}f(t^{-1})t^{s/2}\frac{dt}{t}$$
$$+ \int_1^{\infty} f(t)t^{s/2}\frac{dt}{t}.$$

The first integral is easy to compute:

$$\frac{1}{2}\int_0^1 t^{s/2-3/2} - t^{s/2-1}dt = \frac{t^{s/2-1/2}}{s-1}\Big|_0^1 - \frac{t^{s/2}}{s}\Big|_0^1 = \frac{1}{s-1} - \frac{1}{s}.$$

In the second integral, change variables by $\tau = 1/t$ therefore $d\tau/\tau = -dt/t$

and $t = 0$ corresponds to $\tau = \infty$. We get

$$\int_0^1 t^{-1/2} f(t^{-1}) t^{s/2} \frac{dt}{t} = -\int_\infty^1 \tau^{1/2} f(\tau) \tau^{-s/2} \frac{d\tau}{\tau}$$

$$= \int_1^\infty \tau^{1/2} f(\tau) \tau^{-s/2} \frac{d\tau}{\tau}$$

$$= \int_1^\infty f(t) t^{(1-s)/2} \frac{dt}{t}$$

when we rename τ as t again. This is legal; it is just a dummy variable anyway. To summarize, we've shown that

$$\Lambda(s) = \frac{1}{s-1} - \frac{1}{s} + \int_1^\infty f(t) t^{(1-s)/2} \frac{dt}{t} + \int_1^\infty f(t) t^{s/2} \frac{dt}{t}$$

$$= \frac{-1}{s(1-s)} + \int_1^\infty f(t)(t^{(1-s)/2} + t^{s/2}) \frac{dt}{t},$$

$$(9.9)$$

where we have combined terms in a way to highlight the symmetry under $s \to 1 - s$.

This is the main idea of the proof. What remains is more technical, and my feelings won't be hurt if you skip it. We still need to check whether it is legal to switch a sum and an integral in saying that

$$\sum_{n=1}^\infty \int_0^\infty \exp(-\pi n^2 t) t^{s/2} \frac{dt}{t} = \int_0^\infty f(t) t^{s/2} \frac{dt}{t}.$$

To justify this, first observe that the series

$$\sum_{n=1}^\infty x^{n^2} = x + O(x^4) \quad \text{for } x \to 0$$

converges absolutely for $|x| < 1$ in comparison to the Geometric series. Set $x = \exp(-\pi t)$ to see that

$$f(t) = \exp(-\pi t) + O(\exp(-4\pi t)), \quad \text{for } t \to \infty,$$
$$\Theta(t) = 1 + O(\exp(-\pi t)).$$

So, $\int_1^\infty |f(t) t^{s/2-1}| dt$ is finite because $\int_1^\infty |\exp(-\pi t) t^{s/2-1}| dt$ is finite. For the rest of the integral, we have to be sneaky. As $t \to 0$, all of the terms in the sum for $f(t)$ tend to 1, so $f(t)$ gets big. How big? As $t \to 0$, $t^{-1} \to \infty$, and so,

$$f(t) = \frac{\Theta(t) - 1}{2} = \frac{t^{-1/2}\Theta(t^{-1}) - 1}{2}$$

$$= \frac{t^{-1/2}(1 + O(\exp(-\pi t^{-1}))) - 1}{2}$$

$$\sim \frac{t^{-1/2}}{2} \quad \text{for } t \to 0.$$

In other words, $f(t)$ goes to ∞ like $t^{-1/2}/2$ as $t \to 0$, because $\exp(-\pi t^{-1})$ goes to 0 and the 1 stays constant. So, $\int_0^1 |f(t)t^{s/2-1}|dt$ is finite in comparison to

$$\frac{1}{2}\int_0^1 |t^{-1/2}t^{s/2-1}|dt = \frac{1}{2}\int_0^1 t^{\mathrm{Re}(s)/2-3/2}dt$$

$$= \frac{t^{(\mathrm{Re}(s)-1)/2}}{s - 1}\Big|_0^1$$

$$= \frac{1}{s - 1} \quad \text{if } \mathrm{Re}(s) > 1.$$

Notice that the improper integral is *not* finite for $\mathrm{Re}(s) < 1$, because $\lim_{t\to 0} t^{(\mathrm{Re}(s)-1)/2}$ is infinite in that case. $\quad\square$

Even though we started by assuming that $\mathrm{Re}(s) > 1$, this new expression for $\Lambda(s)$ is finite for all s except, of course, the poles at $s = 0, 1$. This follows from our preceding analysis of the integral $\int_1^\infty |f(t)t^{s/2-1}|dt$. In other words, this is another proof of the analytic continuation of $\zeta(s)$.

The symmetry of the function $\Theta(t)$ has other uses in number theory. As a first step toward the Riemann Hypothesis, G.H. Hardy proved in 1914 that the function $\zeta(s)$ has infinitely many zeros on the vertical line $\mathrm{Re}(s) = 1/2$. The proof uses the symmetry of $\Theta(t)$. Another application is the proof of the quadratic reciprocity law, mentioned in Section 11.3.

Chapter 10

Explicit Formula

10.1. Hadamard's Product

The theorem in Section 8.4 showed that $\zeta(s)$ could be defined for all complex s except for a simple pole at $s = 1$. So, $(s - 1)\zeta(s)$ has no poles at all. Functions such as this are now called ENTIRE FUNCTIONS. Other simple examples are $\exp(z)$ and $\sin(z)$.

Entire functions are in some ways similar to polynomials. When there are no poles, the Taylor series turns out to have infinite radius of convergence. From the ratio test, one can deduce that the coefficients in the expansion tend to zero very rapidly. In some sense, this is the next best thing to being a polynomial, which has all but finitely many coefficients equal to zero.

In addition to being a sum, any nonconstant polynomial can be factored into a product over its zeros. (The Fundamental Theorem of Algebra says that every polynomial factors reduces to linear factors over the complex numbers.) We have already used the factorization

$$\sin(\pi z) = \pi z \prod_{n=1}^{\infty} \left(1 - \frac{z^2}{n^2} \right),$$

based on the fact that $\sin(\pi z) = 0$ if $z = 0$ or $z = \pm n$. On the other hand, the function $\exp(z)$ is never equal to zero, so it certainly has no factorization. Even when factorizations exist, they may have nonlinear terms. For example, $1/\Gamma(s + 1)$ is zero at $s = -1, -2, -3, \ldots$, and it is a theorem in complex analysis that

$$\frac{1}{\Gamma(s + 1)} = \exp(\gamma s) \prod_{n=1}^{\infty} \left(1 + \frac{s}{n} \right) \exp(-s/n).$$

Each linear factor $1 + s/n$ is paired with a term $\exp(-s/n)$. Incidentally, the γ in the $\exp(\gamma s)$ is still Euler's constant, but for the remainder of the book γ will denote something else: $\gamma = \text{Im}(\rho)$, the imaginary part of a zero of $\zeta(s)$.

It was Riemann's great contribution to the subject to realize that a product formula for $(s - 1)\zeta(s)$ would have great significance for the study of primes. Riemann's proof of the factorization was not complete; this was later fixed by Hadamard. He showed that

Theorem.

$$(s - 1)\zeta(s) =$$

$$\frac{1}{2}\left(\frac{2\pi}{e}\right)^s \prod_{n=1}^{\infty}\left(1 + \frac{s}{2n}\right)\exp(-s/2n)$$

$$\times \prod_{\rho}\left(1 - \frac{s}{\rho}\right)\exp(s/\rho). \tag{10.1}$$

Here, the first infinite product shows the contribution of the trivial zeros at $s = -2, -4, -6, \dots$. We proved in Section 8.4 that these zeros exist. The second infinite product shows the contribution of all remaining zeros ρ. In the graphics of Chapter 8 we saw three examples of zeros ρ: at $1/2 + i14.13473\dots$, $1/2 + i21.02204\dots$, and $1/2 + i25.01086\dots$. We commented earlier that the Euler product over primes cannot ever be zero. So, $\zeta(s)$ is not zero if $\mathrm{Re}(s) > 1$. According to the symmetry under $s \to 1 - s$ of

$$\Lambda(s) = \pi^{-s/2}\Gamma(s/2)\zeta(s),$$

the only zeros for $\mathrm{Re}(s) < 0$ must be cancelled by the poles of $\Gamma(s/2)$. These are the trivial zeros above. In other words, all the nontrivial zeros of $\zeta(s)$ are zeros of $\Lambda(s)$. They satisfy

$$0 \le \mathrm{Re}(\rho) \le 1. \tag{10.2}$$

Furthermore, the symmetry under $s \to 1 - s$ implies that if $\zeta(\rho) = 0$, then $\zeta(1 - \rho) = 0$ also. In other words, the zeros ρ are located symmetrically relative to the vertical line $\mathrm{Re}(s) = 1/2$.

The proof of Hadamard's product is too sophisticated for this book. In fact, Hadamard wrote,

[a]s for the properties for which [Riemann] gave only a formula, it took me almost three decades before I could prove them, all except one.

We are instead interested in the consequences of Hadamard's product. Remember that in Exercise 6.3.1 you computed the logarithmic derivative of the product formula for $\sin(\pi z)$. The logarithmic derivative is useful here.

We see that

$$\frac{\frac{d}{ds}((s-1)\zeta(s))}{(s-1)\zeta(s)} = \frac{1}{s-1} + \frac{\zeta'(s)}{\zeta(s)}.$$

Meanwhile, on the other side of the equation, for every term $f(s)$ in the original product, we get a term $d/ds(\log(f(s))) = f'(s)/f(s)$ in a sum. We see that

$$\frac{\frac{d}{ds}((s-1)\zeta(s))}{(s-1)\zeta(s)}$$
$$= \log(2\pi) - 1 + \sum_{n=1}^{\infty}\left\{\frac{1}{s+2n} - \frac{1}{2n}\right\} + \sum_{\rho}\left\{\frac{1}{s-\rho} + \frac{1}{\rho}\right\}.$$

If we put every term in { } over common denominators and solve for $\zeta'(s)/\zeta(s)$, we get

Theorem.

$$\frac{\zeta'(s)}{\zeta(s)} = \log(2\pi) + \frac{s}{1-s} + \sum_{n=1}^{\infty}\frac{-s}{2n(s+2n)} + \sum_{\rho}\frac{s}{\rho(s-\rho)}.$$
$$(10.3)$$

On the other hand, we also know that for $\mathrm{Re}(s) > 1$,

$$\zeta(s) = \prod_{p\text{ prime}}(1-p^{-s})^{-1}.$$

We compute that

$$\frac{d}{ds}\log((1-p^{-s})^{-1}) = -\frac{\log(p)p^{-s}}{1-p^{-s}} = -\log(p)\sum_{k=1}^{\infty}p^{-ks}$$

by means of the Geometric series. This gives us a second expression.

Theorem. *For* $\mathrm{Re}(s) > 1$,

$$\frac{\zeta'(s)}{\zeta(s)} = -\sum_{p\text{ prime}}\log(p)\sum_{k=1}^{\infty}p^{-ks}.$$
$$(10.4)$$

Another way to write this sum is to invent a new notation, the Von Mangoldt function:

$$\Lambda(n) = \begin{cases}\log(p), & \text{where } n = p^k \text{ is a power of a single prime,} \\ 0, & \text{otherwise.}\end{cases}$$

(Of course, the notation $\Lambda(n)$ is not the same as the function $\Lambda(s)$ discussed in Chapter 8.) So, for example, $\Lambda(6) = 0$, $\Lambda(7) = \log(7)$, $\Lambda(8) = \log(2)$, $\Lambda(9) = \log(3)$, $\Lambda(10) = 0$. With this convention, we can write more conveniently

$$\frac{\zeta'(s)}{\zeta(s)} = -\sum_{n=1}^{\infty} \Lambda(n)n^{-s}.$$

10.2. Von Mangoldt's Formula

Riemann's great idea was that these two expressions encoded arithmetic information about the primes less than x, as x increases. Instead of just counting the prime numbers $p < x$, this first, simpler version gives each prime power p^k the weight $\log(p)$. The other integers are assigned weight 0. In other words, each integer n, prime or not, is counted with weight $\Lambda(n)$. We want to study the growth of the total weight

$$\sum_{n<x} \Lambda(n)$$

as a function of the parameter x.

Exercise 10.2.1. Compute the above sum for $x = 12.99$, and also for $x = 13.01$.

Because this sum, as a function of x, has a jump whenever x is a power of a prime number, we can get a slightly nicer function by averaging. So let

$$\Psi(x) = \frac{1}{2} \left(\sum_{n<x} \Lambda(n) + \sum_{n\leq x} \Lambda(n) \right).$$

Notice that the only difference in the two sums is the $<$ or the $\leq$. $\Psi(x)$ still has jumps, but the value of Ψ at the point is the average of the limits from the right and from the left. If this is confusing, don't worry about it.

Exercise 10.2.2. Compute $\Psi(13)$.

Figure 10.1 shows three plots of $\Psi(x)$, on three different scales. The first thing we notice is that although $\Psi(x)$ is very irregular on the smallest scale, on the largest scale it looks very much like the straight line $y = x$. In some sense, this justifies the rather complicated looking definition of the Von Mangoldt function $\Lambda(n)$. It is just the right weight to attach to each integer so that the

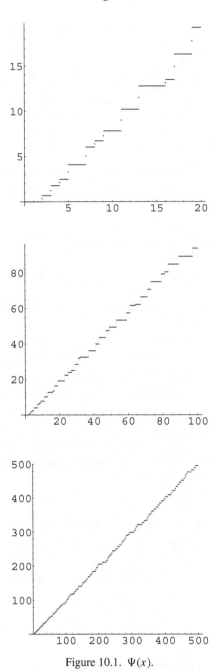

Figure 10.1. $\Psi(x)$.

total $\Psi(x)$ behaves nicely. A. E. Ingham, in his classic book (Ingham, 1990), writes that

[i]t happens that [the function], which arises most naturally from the analytical point of view, is the one most remote from the original problem, namely $\Psi(x)$. For this reason it is usually most convenient to work in the first instance with $\Psi(x)$, and to [then] deduce results about $\pi(x)$. This is a complication which seems inherent in the subject, and the reader should familiarize himself at the outset with the function $\Psi(x)$, which is to be regarded as the fundamental one.

We will follow Ingham's advice and study $\Psi(x)$ further. The miraculous thing is that there is an *exact* formula for this function in terms of the zeros of the zeta function

Theorem (Von Mangoldt's Explicit Formula). *For $x > 1$,*

$$\Psi(x) = x - \sum_{\rho} \frac{x^\rho}{\rho} - \frac{1}{2}\log\left(1 - \frac{1}{x^2}\right) - \log(2\pi), \tag{10.5}$$

where, as usual, ρ are the nontrivial zeros of $\zeta(s)$, $0 < Re(\rho) < 1$.

The idea of the proof is very simple, merely a comparison of two integrals. We have already seen how the Gamma function $\Gamma(s)$, as a function of the variable s, is given by an integral of x^{s-1} times the function $f(x) = \exp(-x)$. We can do this for any function $f(x)$, but because we will soon be interested only in functions that are 0 for $x < 1$, we will make the integral start at 1 instead of 0. Also, it will simplify notation greatly to integrate $f(x)$ against x^{-s-1}, so we will change s in the conventional notation to $-s$. Explicitly, given a function $f(x)$, we define the MELLIN TRANSFORM as

$$\mathcal{M}f(s) = \int_1^\infty f(x)x^{-s-1}dx.$$

Again, it should be emphasized this is a slightly different notation for Mellin transform than is traditional.

Exercise 10.2.3. Show that for $f(x)$, the constant function 1,

$$\mathcal{M}(1)(s) = 1/s.$$

(You may assume that $Re(s) > 0$; so, the improper integral makes sense.)

Exercise 10.2.4. Compute the Mellin transform of x.

The key fact here, which we will not prove, is that the Mellin transform is INJECTIVE as an operator on functions. That is, if $\mathcal{M}f(s) = \mathcal{M}g(s)$ as functions of the s variable, then $f(x) = g(x)$. The Mellin transform, $\mathcal{M}$, is completely analogous to the Laplace transform, $\mathcal{L}$, which you may have seen in differential equations. A standard method is to use the Laplace transform to convert a differential equation for an unknown function $f(x)$ into an algebraic equation for $\mathcal{L}f(s)$ that can be readily solved. The solution to the differential equation is, then, whatever function $f(x)$ has as the Laplace transform, the known function $\mathcal{L}f(s)$.

Proof of Theorem. Observe first that

$$n^{-s} = s \int_n^\infty x^{-s-1} dx,$$

according to calculus. So,

$$\frac{\zeta'(s)}{\zeta(s)} = -\sum_{n=1}^\infty \Lambda(n)n^{-s} = -\sum_{n=1}^\infty \Lambda(n)s \int_n^\infty x^{-s-1} dx.$$

We now want to change the order of the sum and the integral. Because the limits of the integral depend on n, this is tricky, like interchanging a double integral in calculus. Because x is always greater than n in the integral, when we change we get

$$\frac{\zeta'(s)}{\zeta(s)} = -s \int_1^\infty \sum_{n \leq x} \Lambda(n)x^{-s-1} dx.$$

We can replace $\sum_{n \leq x} \Lambda(n)$ with the nicer function $\Psi(x)$; they are equal except at isolated points, which won't change the value of the integral. Again, if this is confusing, don't worry about it. So,

$$\frac{\zeta'(s)}{\zeta(s)} = -s \int_1^\infty \Psi(x)x^{-s-1} dx = -s\mathcal{M}\Psi(s). \tag{10.6}$$

So we know that $-s\mathcal{M}\Psi(s)$ is equal to the right side of (10.4), which is therefore equal to the right side of (10.3). To complete the proof, we need only show that the right side of (10.5) has this same Mellin transform.

Another useful property of the Mellin transform is that it is LINEAR; that is, for two functions $f(x)$ and $g(x)$ and for any constant c,

$$\mathcal{M}(f + g)(s) = \mathcal{M}f(s) + \mathcal{M}g(s) \quad \text{and} \quad \mathcal{M}(cf)(s) = c\mathcal{M}f(s).$$

This is clear because the analogous facts are true for integrals. We can, thus, prove the theorem by computing Mellin transforms term by term in the sum

$$x - \sum_\rho \frac{x^\rho}{\rho} - \frac{1}{2} \log\left(1 - \frac{1}{x^2}\right) - \log(2\pi).$$

You showed in Exercise 10.2.3 that $\mathcal{M}(1)(s) = 1/s$; that is,

$$-s\mathcal{M}(-\log(2\pi))(s) = \log(2\pi),$$

by linearity. This accounts for the first term in (10.3). In Exercise 10.2.4, you computed that $\mathcal{M}(x)(s) = 1/(s-1)$; so,

$$-s\mathcal{M}(x)(s) = \frac{s}{1-s},$$

the second term in (10.3).

Exercise 10.2.5. Show that

$$-s\mathcal{M}\left(-\frac{x^\rho}{\rho}\right)(s) = \frac{s}{\rho(s-\rho)}.$$

Thus, all the terms corresponding to the nontrivial zeros in (10.5) and (10.3) match up correctly. The only term left to compute the Mellin transform of is $-1/2\log(1 - 1/x^2)$, and it must somehow correspond to the infinite sum over all the trivial zeros. If you have read this far, you should know what comes next; stop reading and try to figure it out on your own. Here is a hint if you need one: We know that $1/x^2 < 1$.

The way to compute $\mathcal{M}(-1/2\log(1 - 1/x^2))(s)$ is to first expand the function as a series.

Exercise 10.2.6. Use (I2.8) to show that

$$-\frac{1}{2}\log(1 - \frac{1}{x^2}) = \sum_{n=1}^{\infty} \frac{x^{-2n}}{2n} \qquad \text{for } x > 1.$$

Exercise 10.2.7. Show that

$$-s\mathcal{M}\left(-\frac{1}{2}\log\left(1 - \frac{1}{x^2}\right)\right)(s) = \sum_{n=1}^{\infty} \frac{-s}{2n(s+2n)}.$$

Because the sum of all these terms has the same transform as $\Psi(x)$, they must be equal. This is the main idea the proof. There are some serious issues that we neglected. For example, we interchanged infinite sums and integrals, and this must be justified for a rigorous proof. In fact, the sum over the nontrivial zeros ρ of the zeta function is only conditionally, not absolutely, convergent. The proof that this is valid is beyond the scope of this book. □

What is this theorem saying? What do these function x^ρ/ρ look like? After all, the ρ are complex and the function $\Psi(x)$ is certainly real. If ρ is a zero, then so is $\bar{\rho}$; so, we should group these terms together. (In fact, strictly speaking, we *have* to group these terms together. The sum over the zeros is not absolutely convergent.) We can write each ρ as $\rho = \beta + i\gamma$. Then, $\bar{\rho}$ is $\beta - i\gamma$. We group the terms corresponding to the zeros ρ and $\bar{\rho}$ to get

$$\frac{x^{\beta+i\gamma}}{\beta + i\gamma} + \frac{x^{\beta-i\gamma}}{\beta - i\gamma}.$$

We factor out a term x^β and use

$$x^{\pm i\gamma} = \exp(\pm i\gamma \log(x)) = \cos(\gamma \log(x)) \pm i \sin(\gamma \log(x)).$$

Put everything over a common denominator and do lots of tedious multiplying to get

$$\frac{x^\rho}{\rho} + \frac{x^{\bar{\rho}}}{\bar{\rho}} = \frac{2x^\beta}{\beta^2 + \gamma^2} \left(\beta \cos(\gamma \log(x)) + \gamma \sin(\gamma \log(x))\right).$$

Next, observe that $\beta^2 + \gamma^2$ is $\rho\bar{\rho} = |\rho|^2$. So,

$$\frac{x^\rho}{\rho} + \frac{x^{\bar{\rho}}}{\bar{\rho}} = \frac{2x^\beta}{|\rho|} \left(\frac{\beta}{|\rho|} \cos(\gamma \log(x)) + \frac{\gamma}{|\rho|} \sin(\gamma \log(x))\right).$$

We can put this in what is called "phase-amplitude" form, by finding an angle θ such that

$$\cos(\theta) = \frac{\beta}{|\rho|}, \; -\sin(\theta) = \frac{\gamma}{|\rho|} \quad \Rightarrow \quad \theta = -\arctan(\gamma/\beta).$$

It is not hard to see that θ is just the angle that ρ (thought of as a vector (β, γ) in the plane) makes with the horizontal axis. From the addition formula

$$\cos(\theta) \cos(\gamma \log(x)) - \sin(\theta) \sin(\gamma \log(x)) = \cos(\theta + \gamma \log(x)),$$

we get

$$\frac{x^\rho}{\rho} + \frac{x^{\bar{\rho}}}{\bar{\rho}} = \frac{2x^\beta}{|\rho|} \cos(\gamma \log(x) - \arctan(\gamma/\beta)). \tag{10.7}$$

If we let $t = \log(x)$, we get

$$= \frac{2\exp(\beta t)}{|\rho|} \cos(\gamma t - \arctan(\gamma/\beta)).$$

If we temporarily ignore the term $\exp(\beta t) = x^\beta$, what is left is a purely periodic function. The factor $2/|\rho|$ is the AMPLITUDE. The constant $\theta = -\arctan(\gamma/\beta)$ inside the cosine represents a PHASE SHIFT. It simply shifts the cosine to the right. The PERIOD is $2\pi/\gamma$. Figure 10.2 shows the graph of (10.7), along with its ENVELOPE $\pm 2x^{1/2}/|\rho|$, for the first zero $\rho = 1/2 + $

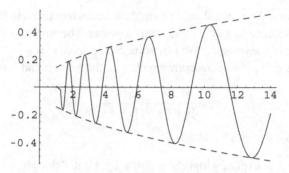

Figure 10.2. Contribution of a single zero to the explicit formula.

$i14.13473\ldots$. The larger $|\rho|$ is the smaller the amplitude and the faster the oscillations.

Exercise 10.2.8. You can use *Mathematica* to help understand Von Mangoldt's Explicit Formula. First, create a function that takes a number

Table 10.1. *Imaginary Part of the First* 100 *Zeros*

14.13473	92.49190	146.00098	193.07973
21.02204	94.65134	147.42277	195.26540
25.01086	95.87063	150.05352	196.87648
30.42488	98.83119	150.92526	198.01531
32.93506	101.31785	153.02469	201.26475
37.58618	103.72554	156.11291	202.49359
40.91872	105.44662	157.59759	204.18967
43.32707	107.16861	158.84999	205.39470
49.77383	111.87466	163.03071	209.57651
52.97032	114.32022	165.53707	211.69086
56.44625	116.22668	167.18444	213.34792
59.34704	118.79078	169.09452	214.54704
60.83178	121.37013	169.91198	216.16954
65.11254	122.94683	173.41154	219.06760
67.07981	124.25682	174.75419	220.71492
69.54640	127.51668	176.44143	221.43071
72.06716	129.57870	178.37741	224.00700
75.70469	131.08769	179.91648	224.98332
77.14484	133.49774	182.20708	227.42144
79.33738	134.75651	184.87447	229.33741
82.91038	138.11604	185.59878	231.25019
84.73549	139.73621	187.22892	231.98724
87.42527	141.12371	189.41616	233.69340
88.80911	143.11185	192.02666	236.52423

x and an integer n as input and returns the main term,

$$x - \frac{1}{2}\log\left(1 - \frac{1}{x^2}\right) - \log(2\pi),$$

plus the contribution of the first n pairs of zeros ρ and $\bar{\rho}$. Use (10.7) to include the contribution of the zeros (observe that they are subtracted from, not added to, (10.5).) Next, create a function that takes an integer n as input and uses *Mathematica*'s Plot to graph the function you created above for $2 \leq x \leq 20$. Finally, you can make a Table of these for various n. The imaginary parts γ of the first 100 zeros are shown in Table 10.1. (The real parts β of the first 100 zeros are all $1/2$.)

10.3. The Logarithmic Integral

In Section 5.1, we introduced the Logarithmic integral function

$$\mathrm{Li}(x) = \int_0^x \frac{dt}{\log(t)};$$

according to the Fundamental Theorem of Calculus, this is just an antiderivative of $1/\log(x)$. To get a better understanding of the primes than Von Mangoldt's formula provides, we need to know more about the Logarithmic integral. One thing that should have been mentioned earlier is that since $\log(1) = 0$, the integral is improper. We have to be a little careful and compute it by the Cauchy principal value. This need not really concern us here, it is enough to say that $\mathrm{Li}(e) = 1.89512\ldots$ is some constant and

$$\mathrm{Li}(x) = \int_0^e \frac{dt}{\log(t)} + \int_e^x \frac{dt}{\log(t)} = \mathrm{Li}(e) + \int_e^x \frac{dt}{\log(t)}.$$

To better understand this function in terms of more elementary functions, we can integrate by parts with $u = 1/\log(t)$, $dv = dt$, $du = -dt/(\log(t)^2 t)$, and $v = t$ to get

$$\int \frac{dt}{\log(t)} = \frac{t}{\log(t)} + \int \frac{dt}{\log(t)^2}.$$

So,

$$\mathrm{Li}(x) = \int_e^x \frac{dt}{\log(t)} + \mathrm{Li}(e) = \frac{t}{\log(t)}\Big|_e^x + \int_e^x \frac{dt}{\log(t)^2} + \mathrm{Li}(e)$$

$$= \frac{x}{\log(x)} + \int_e^x \frac{dt}{\log(t)^2} + C_1,$$

where $C_1 = \mathrm{Li}(e) - e$

Exercise 10.3.1. Integrate by parts again to show that

$$\text{Li}(x) = \frac{x}{\log(x)} + \frac{x}{\log(x)^2} + 2\int_e^x \frac{dt}{\log(t)^3} + C_2,$$

where C_2 is some other constant. By repeating this process, you can show that for some constant C_k,

$$\text{Li}(x) = \frac{x}{\log(x)} + \frac{x}{\log(x)^2} + 2\frac{x}{\log(x)^3} + \cdots$$

$$+ (k-1)!\frac{x}{\log(x)^k} + k!\int_e^x \frac{dt}{\log(t)^{k+1}} + C_k.$$

At some point, we get sick of this and want to know how big the integral is that is left over, that is, the "remainder."

Lemma.

$$k!\int_e^x \frac{dt}{\log(t)^{k+1}} + C_k = O\left(\frac{x}{\log(x)^k}\right).$$

This means that for fixed k, as x gets big, the left side is bounded by some constant (depending on k) times $x/\log(x)^k$. By making the constant a little bigger, we can absorb the $k!$ and neglect the C_k, because $x/\log(x)^k$ tends to infinity; it is certainly bigger than C_k.

Proof. First, break the integral into two pieces:

$$\int_e^x \frac{dt}{\log(t)^{k+1}} = \int_e^{\sqrt{x}} \frac{dt}{\log(t)^{k+1}} + \int_{\sqrt{x}}^x \frac{dt}{\log(t)^{k+1}}.$$

Because $e \leq t$, $1 < \log(t)^{k+1}$; so, $1/\log(t)^{k+1} < 1$ and

$$\int_e^{\sqrt{x}} \frac{dt}{\log(t)^{k+1}} < \int_e^{\sqrt{x}} 1\, dt = \sqrt{x} - e < \sqrt{x}.$$

We want to show that

$$\sqrt{x} = O\left(\frac{x}{\log(x)^k}\right).$$

To see this, write $y = \log(x)$. We know that $y^{2k}/2k! \leq \exp(y)$, by comparison to the series for exp(y). So,

$$y^{2k} \leq 2k!\exp(y) \Rightarrow \log(x)^{2k} \leq 2k!x$$
$$\Rightarrow \log(x)^k \leq \sqrt{2k!}\sqrt{x}$$
$$\Rightarrow \log(x)^k\sqrt{x} \leq \sqrt{2k!}\,x$$
$$\Rightarrow \sqrt{x} \leq \sqrt{2k!}\frac{x}{\log(x)^k}.$$

Meanwhile, in the second integral, $t < x$, so $1 < x/t$ and

$$\int_{\sqrt{x}}^{x} \frac{dt}{\log(t)^{k+1}} < x\int_{\sqrt{x}}^{x} \frac{dt}{t\,\log(t)^{k+1}}.$$

Now, integrate by substitution with $u = \log(t)$, $du = dt/t$ to get

$$= x\left.\left(\frac{\log(t)^{-k}}{-k}\right)\right|_{\sqrt{x}}^{x}$$
$$= x\left(\frac{1}{k\log(\sqrt{x})^k} - \frac{1}{k\log(x)^k}\right)$$
$$< x\frac{1}{k\log(\sqrt{x})^k} = \frac{2^k x}{k\log(x)^k} = O\left(\frac{x}{\log(x)^k}\right).$$

$\square$

What this means is that for any k,

$$\mathrm{Li}(x) = \frac{x}{\log(x)} + \frac{x}{\log(x)^2} + 2\frac{x}{\log(x)^3} + \cdots$$
$$+ (k-2)!\frac{x}{\log(x)^{k-1}} + O\left(\frac{x}{\log(x)^k}\right); \qquad (10.8)$$

that is, the error made in this approximation is bounded by a constant times the first term omitted. In particular, this says that because

$$\mathrm{Li}(x) = \frac{x}{\log(x)} + O\left(\frac{x}{\log(x)^2}\right),$$

we can divide both sides by $x/\log(x)$ to get

$$\frac{\mathrm{Li}(x)}{x/\log(x)} = 1 + O\left(\frac{1}{\log(x)}\right).$$

And because $1/\log(x) \to 0$ as $x \to \infty$,

$$\mathrm{Li}(x) \sim \frac{x}{\log(x)}.$$

Formula (10.8) is an example of an ASYMPTOTIC EXPANSION. This is similar
to what we did in Section 9.1 when we approximated the Euler integral for
$\Gamma(s + 1)$ to get Stirling's formula. But this idea is more subtle than it at first
appears. In Exercise I2.6.9, you showed that the series $\sum_{k=0}^{\infty} k! y^k$ diverges
for every y except $y = 0$. If we substitute in $y = 1/\log(x)$, we see that the
series

$$\frac{x}{\log(x)} \sum_{k=0}^{\infty} \frac{k!}{\log(x)^k}$$

diverges for every $x < \infty$, so it is certainly not equal to Li(x). Nonetheless,
divergent series are still useful. The point here is in some sense the opposite
of that in Interlude 2. There, we took a fixed x and got better approximations
as the order k increased. Here, we take a fixed order approximation; it gets
better as x approaches some limiting value.

Riemann's version of the explicit formula will also involve the functions
Li(x^ρ), where the ρ are complex zeros of $\zeta(s)$. This function's precise definition will involve integration of $1/\log(t)$ along paths in the complex plane
ending at the point x^ρ and, thus, is beyond the scope of this book. But, again,
according to the Fundamental Theorem of Calculus and the chain rule,

$$\frac{d}{dx}\text{Li}(x^\rho) = \frac{1}{\log(x^\rho)} \cdot \frac{d}{dx}(x^\rho) = \frac{\rho x^{\rho-1}}{\rho \log(x)} = \frac{x^{\rho-1}}{\log(x)}. \qquad (10.9)$$

And it has the same asymptotic expansion, in particular

$$\text{Li}(x^\rho) \sim \frac{x^\rho}{\rho \log(x)}.$$

10.4. Riemann's Formula

Von Mangoldt's formula is very nice and, in some sense, very natural. But if
we want to count primes and not just prime powers weighted by $\log(p)$, we
must work a little harder. Riemann's formula gets rid of the $\log(p)$ weight
factor. As before, $\pi(x)$ is the function that counts the number of primes less
than x, but averaged (like $\Psi(x)$ was.) When x happens to be prime, the value
at the jumps is the average of the limits from the right and left. The formula is

$$\pi(x) = \frac{1}{2}\left(\sum_{p \text{ prime } < x} 1 + \sum_{p \text{ prime } \leq x} 1 \right).$$

And we define

$$\Pi(x) = \pi(x) + \frac{1}{2}\pi(x^{1/2}) + \frac{1}{3}\pi(x^{1/3}) + \frac{1}{4}\pi(x^{1/4}) \cdots = \sum_{k=1}^{\infty} \frac{1}{k}\pi(x^{1/k}).$$

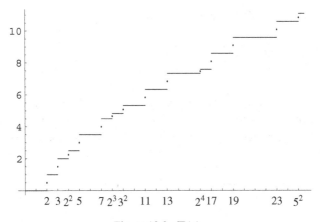

Figure 10.3. $\Pi(x)$.

What does this function count? The first term in the sum just counts primes below x. A prime p is less than $x^{1/2}$ when p^2 is less than x. So, the second term counts squares of primes below x, with weight $1/2$. Similarly, the third term counts cubes of primes below x with weight $1/3$, and so forth. Figure 10.3 shows a plot of $\Pi(x)$. Observe that there is a jump by 1 at each prime, and that there are smaller jumps at larger powers of primes.

Surprisingly, $\Pi(x)$ and $\pi(x)$ are about the same size. In fact,

Lemma.

$$\Pi(x) = \pi(x) + O(x^{1/2}\log(x)). \tag{10.10}$$

Because Chebyshev's estimate $x/\log(x) \ll \pi(x)$ from Chapter 5 shows that $\pi(x)$ (and thus $\Pi(x)$) is much bigger than the error $x^{1/2}\log(x)$, (10.10) shows that

$$\Pi(x) \sim \pi(x).$$

Proof. First, notice that the sum defining $\Pi(x)$ is actually finite, because $\pi(x^{1/k}) = 0$ when $x^{1/k} < 2$. This happens when $\log(x)/k < \log(2)$, or $\log(x)/\log(2) < k$. Let $n = [\log(x)/\log(2)]$ be the index of the last nonzero term. Then,

$$\Pi(x) - \pi(x) = \frac{\pi(x^{1/2})}{2} + \cdots + \frac{\pi(x^{1/n})}{n}.$$

Because $\pi(t) < t$ for any t, we get the estimate

$$\Pi(x) - \pi(x) < \frac{x^{1/2}}{2} + \cdots + \frac{x^{1/n}}{n} < (n-1) \cdot \frac{x^{1/2}}{2},$$

which proves (10.10). $\qquad\qquad\square$

Theorem (Riemann's Explicit Formula). *For $x > 1$,*

$$\Pi(x) = Li(x) - \sum_\rho Li(x^\rho) - \log(2) + \int_x^\infty \frac{dt}{t(t^2 - 1)\log(t)}. \qquad (10.11)$$

Proof of First Main Idea. We begin by comparing $\log(\zeta(s))$ to the Mellin transform of $\Pi(x)$. This is very similar to the proof of the Von Mangoldt formula, but the review is worthwhile. From the Euler product over primes, we have

$$\log(\zeta(s)) = -\sum_{p \text{ prime}} \log(1 - p^{-s}) = \sum_{p \text{ prime}} \sum_{k=1}^\infty \frac{1}{k} p^{-ks},$$

according to the series expansion (I2.8) for $-\log(1 - x)$. Next, we get

$$= \sum_{p \text{ prime}} \sum_{k=1}^\infty \frac{s}{k} \int_{p^k}^\infty x^{-s-1} dx,$$

according to calculus. Now, we interchange the integral and the sum as before to get

$$= s \int_1^\infty \sum_{\substack{p,k \\ \text{with } p^k < x}} \frac{1}{k} x^{-s-1} dx,$$

We can replace the sum $\sum_{p^k < x} 1/k$ with $\Pi(x)$, as they are equal except at isolated points, which does not change the value of the integral. We see that

$$\log(\zeta(s)) = s \mathcal{M}\Pi(s). \qquad (10.12)$$

$\square$

We also know that $\log(\zeta(s))$ can be written as a sum over the zeros of $\zeta(s)$, coming from taking the logarithm of the Hadamard product. We could further imitate the proof of (10.5) by comparing integrals again. This is more or less what Riemann himself did. But most of the integrals in this approach are too hard to compute without advanced knowledge of complex analysis. For example, one can find in Gradshteyn and Ryzhik (1979) that

$$\mathcal{M}(Li)(s) = \int_1^\infty Li(x)\, x^{-s-1} dx = -\frac{\log(s - 1)}{s}.$$

This means that the main term, $Li(x)$ in Riemann's formula, corresponds to the $(s - 1)$ term in the Hadamard product coming from the pole at $s = 1$.

Instead, we will give a simpler proof, which has a weaker conclusion: The two sides of the equation are equal up to a constant. We need two more facts about the Mellin transform.

Lemma. *Let $\mathcal{D}$ denote d/ds, a derivative with respect to the s variable. Then,*

$$\mathcal{D}\mathcal{M}f(s) = \mathcal{M}(-f \cdot \log)(s). \tag{10.13}$$

Proof. This is easy. Because $x^{-s} = \exp(-s\log(x))$, we get

$$\mathcal{D}\mathcal{M}f(s) = \frac{d}{ds}\int_1^\infty f(x)x^{-s}\frac{dx}{x} = \int_1^\infty f(x)\frac{d}{ds}x^{-s}\frac{dx}{x}$$

$$= \int_1^\infty -f(x)\log(x)x^{-s}\frac{dx}{x} = \mathcal{M}(-f \cdot \log)(s).$$

$\square$

Lemma. *Let $\mathcal{E}$ denote the Euler operator on functions, given by $\mathcal{E}f(x) = xf'(x)$. Suppose that $f(1) = 0$. Then,*

$$\mathcal{M}\mathcal{E}f(s) = s\mathcal{M}f(s). \tag{10.14}$$

Proof. Again, this is not so hard. We have

$$\mathcal{M}\mathcal{E}f(s) = \int_1^\infty xf'(x)x^{-s}\frac{dx}{x} = \int_1^\infty f'(x)x^{-s}dx.$$

Now, integrate by parts with $u = x^{-s}$, $du = -sx^{-s-1}$, $dv = f'(x)dx$. So, with $v = f(x)$, we get

$$= x^{-s}f(x)\big|_1^\infty + s\int_1^\infty f(x)x^{-s-1}dx.$$

But $f(1) = 0$. And it is implicit that $x^{-s}f(x) \to 0$ as $x \to \infty$ or the integral will not make sense. So, this is just

$$= s\int_1^\infty f(x)x^{-s-1}dx = s\mathcal{M}(f)(s).$$

$\square$

We saw the Euler operator $\mathcal{E}$ in Section 8.2. Notice that the operators $\mathcal{D}$ and $\mathcal{E}$ live on different sides of the Mellin transform. $\mathcal{D}$ only makes sense after the transform; $\mathcal{E}$ only makes sense before the transform.

Proof of Second Main Idea. We will give a proof that is "formal" in the sense that we ignore all analytical difficulties. (In this sense of the word, formal actually means *not* rigorous.) Because $\Psi(1) = 0$, we can say that

$$\mathcal{M}\mathcal{E}\Psi(s) = s\mathcal{M}\Psi(s), \qquad\qquad\qquad\qquad \text{by (10.14)},$$

$$= -\frac{\zeta'(s)}{\zeta(s)}, \qquad\qquad\qquad\qquad \text{by (10.6)},$$

$$= -\mathcal{D}\log(\zeta(s)), \qquad\qquad\qquad \text{chain rule, by (10.12)},$$

$$= -\mathcal{D}(s\mathcal{M}\Pi(s)),$$

$$= -\mathcal{M}\Pi(s) - s\mathcal{D}\mathcal{M}\Pi(s), \qquad\qquad \text{product rule},$$

$$= -\mathcal{M}\Pi(s) + s\mathcal{M}(\Pi \cdot \log)(s), \quad \text{by (10.13)},$$

$$= -\mathcal{M}\Pi(s) + \mathcal{M}\mathcal{E}(\Pi \cdot \log)(s),$$

according to (10.14) again, because $\Pi(1) \cdot \log(1) = 0$. We've shown that

$$\mathcal{M}\mathcal{E}\Psi(s) = -\mathcal{M}\Pi(s) + \mathcal{M}\mathcal{E}(\Pi \cdot \log)(s).$$

According to the injectivity of the Mellin transform, $\mathcal{M}$, we see that

$$\mathcal{E}\Psi(x) = -\Pi(x) + \mathcal{E}(\Pi(x) \cdot \log(x)).$$

Both $\mathcal{D}$ and $\mathcal{E}$ are *linear* operators, as $\mathcal{M}$ is. Now, we need only show that for each term $f(x)$ on the right side of Riemann's formula, $-f(x) + \mathcal{E}(f(x) \cdot \log(x))$ gives the the Euler operator $\mathcal{E}f(x)$ of the corresponding term in Von Mangoldt's formula.

Exercise 10.4.1. For $\alpha = 1, \rho$, or $-2n$, show that

$$x\frac{d}{dx}\left(\frac{x^\alpha}{\alpha}\right) = -\mathrm{Li}(x^\alpha) + x\frac{d}{dx}\left(\mathrm{Li}(x^\alpha) \cdot \log(x)\right).$$

Use the formula (10.9) for the derivative of $\mathrm{Li}(x^\alpha)$. Don't forget the product rule.

This means that the main term, $\mathrm{Li}(x)$, corresponds to the main term, x, in the Von Mangoldt formula. And the terms $\mathrm{Li}(x^\rho)$ correspond to the terms x^ρ/ρ from before. What about the contribution of the trivial zeros? We saw before that

$$-\frac{1}{2}\log\left(1 - \frac{1}{x^2}\right) = -\sum_{n=1}^{\infty}\frac{x^{-2n}}{-2n} \qquad \text{for } x > 1.$$

We need to check that

$$\int_x^\infty \frac{dt}{t(t^2-1)\log(t)} = -\sum_{n=1}^\infty \mathrm{Li}(x^{-2n}) \quad \text{for } x > 1.$$

In the integral, change variables by $u = \log(t)$, $\exp(u) = t$, $du = dt/t$ to see that it becomes

$$\int_{\log(x)}^\infty \frac{1}{\exp(2u)-1} \frac{du}{u} = \int_{\log(x)}^\infty \frac{\exp(-2u)}{1-\exp(-2u)} \frac{du}{u}$$

after multiplying the numerator and denominator by $\exp(-2u)$. Now, expand the integrand as a Geometric series to get

$$\int_{\log(x)}^\infty \sum_{n=1}^\infty \exp(-2nu) \frac{du}{u} = \sum_{n=1}^\infty \int_{\log(x)}^\infty \exp(-2nu) \frac{du}{u}.$$

In each term of the series, change the variables again, with $t = \exp(-2nu)$, $dt = -2n\exp(-2nu)\,du$. So, $\log(t) = -2nu$ and $dt/\log(t)$ is the same as $\exp(-2nu)/u\,du$. Note that as $u \to \infty$, $t \to 0$, so we get

$$\sum_{n=1}^\infty \int_{x^{-2n}}^0 \frac{dt}{\log(t)} = -\sum_{n=1}^\infty \int_0^{x^{-2n}} \frac{dt}{\log(t)} = -\sum_{n=1}^\infty \mathrm{Li}(x^{-2n}).$$

Using this method of proof, we cannot keep track of the constants, because the Euler operator $\mathcal{E}$ is not quite injective. If $\mathcal{E}f(x) = \mathcal{E}g(x)$, then $xf'(x) = xg'(x)$. This means that $f'(x) = g'(x)$, and so $f(x)$ and $g(x)$ can differ by a constant.

What was not rigorous about this? Well, our functions $\Psi(x)$ and $\Pi(x)$ are not even continuous at the jumps $x = p^k$, so they are certainly not differentiable there. The sum over the zeros ρ in Riemann's formula is only conditionally convergent, just as in Von Mangoldt's formula. We are claiming that it is legal to interchange the operations of infinite sum $\sum_\rho$ and differential operator $\mathcal{E} = xd/dx$, but this needs to be justified for a fully rigorous proof. $\qquad\qquad\Box$

What about an exact formula for the actual count of primes $\pi(x)$? This follows from Riemann's explicit formula. Because we defined

$$\Pi(x) = \pi(x) + \frac{1}{2}\pi(x^{1/2}) + \frac{1}{3}\pi(x^{1/3}) + \frac{1}{4}\pi(x^{1/4})\cdots = \sum_{n=1}^\infty \frac{1}{n}\pi(x^{1/n}),$$

by Möbius inversion (a generalization of Section 2.3), we get that

$$\pi(x) = \Pi(x) - \frac{1}{2}\Pi(x^{1/2}) - \frac{1}{3}\Pi(x^{1/3}) - \frac{1}{5}\Pi(x^{1/5}) + \frac{1}{6}\Pi(x^{1/6})\ldots$$

$$= \sum_{n=1}^{\infty} \frac{\mu(n)}{n}\Pi(x^{1/n}).$$

Remember that for a fixed value of x, each sum is actually a finite sum; both $\pi(x^{1/n})$ and $\Pi(x^{1/n})$ are 0 when $x^{1/n} < 2$. Because the Möbius inversion formula is linear, we can do this with each term in the explicit formula for $\Pi(x)$. Let

$$R(x) = \sum_{n=1}^{\infty} \frac{\mu(n)}{n}\mathrm{Li}(x^{1/n}), \qquad R(x^\rho) = \sum_{n=1}^{\infty} \frac{\mu(n)}{n}\mathrm{Li}(x^{\rho/n}).$$

So, from Riemann's explicit formula for $\Pi(x)$, we deduce that

Theorem.

$$\pi(x) = R(x) + \sum_{\rho} R(x^\rho) + \sum_{n=1}^{\infty} \frac{\mu(n)}{n} \int_{x^{1/n}}^{\infty} \frac{dt}{t(t^2 - 1)\log(t)}.$$

The $-\log(2)$ disappears from this theorem, because a consequence of the Prime Number Theorem says that $\sum_n \mu(n)/n = 0$. The logarithmic integral has a series expansion:

$$\mathrm{Li}(x) = \gamma + \log(\log(x)) + \sum_{k=1}^{\infty} \frac{\log(x)^k}{k!k} \qquad \text{for } x > 1.$$

Using this and another deep result,

$$\sum_{n=1}^{\infty} \frac{\mu(n)\log(n)}{n} = -1,$$

one can show that

$$R(x) = 1 + \sum_{k=1}^{\infty} \frac{\log(x)^k}{k!k\zeta(k+1)}.$$

This is called Gram's series.

   What are some of the consequences of Riemann's formula? It can be proved that

Theorem. *The Prime Number Theorem is true:*

$$\pi(x) \sim Li(x) \sim \frac{x}{\log(x)}.$$

We won't give the full proof here; you can find it in any other book on analytic number theory. But here are some of the ideas. We saw in the previous section that $Li(x) \sim x / \log(x)$. And (10.10) shows that that $\pi(x) \sim \Pi(x)$.

The main idea of the theorem is that the first term written in Riemann's explicit formula for $\Pi(x)$, the $Li(x)$ term, is actually the most significant. It dominates all the others. We already mentioned the fact that this term comes from the pole of $\zeta(s)$ at $s = 1$; that is, the Harmonic series diverges. In fact, it is known that the Prime Number Theorem is equivalent to the following theorem, which is an improvement on (10.2).

Theorem. *The nontrivial zeros ρ of $\zeta(s)$ satisfy*

$$0 < Re(\rho) < 1. \tag{10.15}$$

Proof. This deep result begins with the trivial trig identity that for any angle θ,

$$3 + 4\cos(\theta) + \cos(2\theta) = 2(1 + \cos(\theta))^2 \geq 0.$$

We will make use of this in the series expansion for $Re(\log(\zeta(s)))$:

$$Re\left(\log\left(\prod_p (1 - p^{-s})^{-1}\right)\right) = Re\left(\sum_p \sum_{k=1}^{\infty} \frac{1}{k} p^{-ks}\right)$$

$$= \sum_p \sum_{k=1}^{\infty} \frac{1}{k} p^{-k\sigma} \cos(kt \log(p)).$$

We apply this to $\zeta(1 + \epsilon)$, to $\zeta(1 + \epsilon + it)$, and to $\zeta(1 + \epsilon + 2it)$ and take the following clever linear combination:

$$3Re(\log(\zeta(1 + \epsilon))) + 4Re(\log(\zeta(1 + \epsilon + it))) + Re(\log(\zeta(1 + \epsilon + 2it)))$$

$$= \sum_p \sum_{k=1}^{\infty} \frac{1}{k} p^{-k(1+\epsilon)} \left(3 + 4\cos(kt \log(p)) + \cos(k2t \log(p))\right).$$

According to the trig identity, this mess is ≥ 0. But the facts about logarithms in Interlude 3 imply that the left side is actually

$$3\log(|\zeta(1 + \epsilon)|) + 4\log(|\zeta(1 + \epsilon + it)|) + \log(|\zeta(1 + \epsilon + 2it)|).$$

So, when we exponentiate,

$$|\zeta(1+\epsilon)|^3|\zeta(1+\epsilon+it)|^4|\zeta(1+\epsilon+2it)| \geq 1.$$

So far, we have not assumed anything about t. But now suppose that t is the imaginary part of a zero of the form $1 + i\gamma$. Then, a Taylor series expansion at $1 + i\gamma$ has no constant term:

$$\zeta(1+\epsilon+i\gamma) = O(\epsilon) \quad \text{as } \epsilon \to 0, \quad \text{so}$$
$$\zeta(1+\epsilon+i\gamma)^4 = O(\epsilon^4).$$

Meanwhile, from (7.1),

$$\zeta(1+\epsilon) = \frac{1}{\epsilon} + O(1), \quad \text{so}$$
$$\zeta(1+\epsilon)^3 = \frac{1}{\epsilon^3} + O(\epsilon^{-2}).$$

And because $\zeta(s)$ has no poles except at $s = 1$,

$$\zeta(1+\epsilon+2it) = O(1).$$

Combining these, we see that

$$\zeta(1+\epsilon)^3\zeta(1+\epsilon+it)^4\zeta(1+\epsilon+2it) = O(\epsilon)$$

goes to zero as $\epsilon \to 0$, which contradicts the fact that the absolute value must be ≥ 1. $\qquad\square$

From this theorem and from (10.16), it is clear that each individual term $\text{Li}(x^\rho)$ in Riemann's explicit formula (10.11) is smaller than the first term $\text{Li}(x)$. But because there are infinitely many such terms, one still needs to make sure that the contribution of all of them together is not larger than $\text{Li}(x)$. Once this is done, one has $\Pi(x) \sim \text{Li}(x)$.

Figure 10.4 shows a graph of $\Pi(x)$ together with $\text{Li}(x) - \log(2)$, the terms in the explicit formula *except* those coming from the zeros. (Actually, it is clear that the contribution of the trivial zeros, that is,

$$\int_x^\infty \frac{dt}{t(t^2 - 1)\log(t)},$$

is practically negligible. For $x = 2$, it is 0.14001, and it is a decreasing function of x because as x increases, the integral is over a smaller region.) Figure 10.4 indicates how the main term captures the size of the function $\Pi(x)$. This is

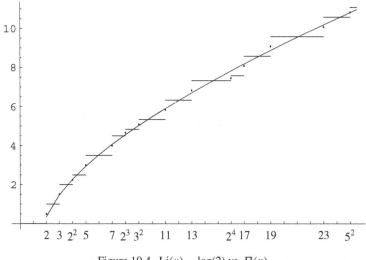

Figure 10.4. $\text{Li}(x) - \log(2)$ vs. $\Pi(x)$.

the analog, for Riemann's formula, of what you did in Exercise 10.2.8 for Von Mangoldt's formula.

Riemann does not get the credit for proving the Prime Number Theorem. For one thing, a rigorous proof of the factorization formula (10.1) did not come until after his death. It is interesting that Riemann's paper never mentions the Prime Number Theorem as a goal. He was interested in deeper questions about the distribution of primes. The miracle is that we can see not just the size of $\Pi(x)$, but the actual *value* of the function, by including the contribution of the nontrivial zeros ρ. Remember that in (10.7) we showed that

$$\frac{x^{\rho}}{\rho} + \frac{x^{\bar\rho}}{\bar\rho} = \frac{2x^{\beta}}{|\rho|} \cos(\gamma \log(x) - \arctan(\gamma/\beta)).$$

Because $\text{Li}(x^{\rho}) \sim x^{\rho}/\rho \log(x)$, we get an analogous formula for the contribution of the pair of zeros ρ, $\bar\rho$:

$$\text{Li}(x^{\rho}) + \text{Li}(x^{\bar\rho}) \sim \frac{2x^{\beta}}{|\rho| \log(x)} \cos(\gamma \log(x) - \arctan(\gamma/\beta)). \qquad (10.16)$$

The contributions to the explicit formula from some of the zeros are shown in the eight graphs of Figure 10.5. The more zeros we include, the more fine detail of the function $\Pi(x)$ we see. This is analogous to what you did in Exercise 10.2.8.

At the beginning of Chapter 9, I made a big deal about how important symmetry is. We had to work very hard to prove the transformation law for $\Theta(t)$. And we had to work just as hard to prove the functional equation for

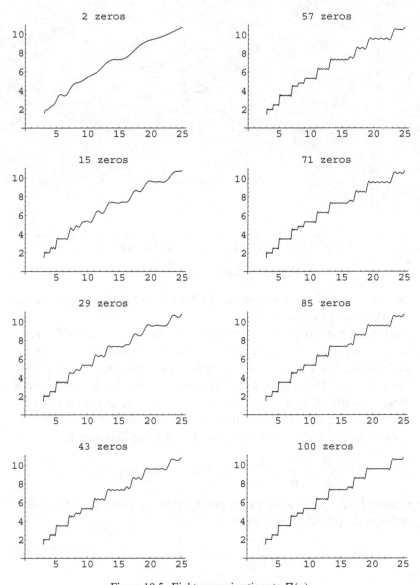

Figure 10.5. Eight approximations to $\Pi(x)$.

$\Lambda(s)$. It is not hard to see that the zeros of $\Lambda(s)$ are precisely the nontrivial zeros ρ of $\zeta(s)$. (The trivial zeros of $\zeta(s)$ are cancelled out by the poles of $\Gamma(s/2)$.) Because of the symmetry, Riemann was led to conjecture the Riemann Hypothesis, the conjecture that all the zeros ρ are actually on the vertical line fixed under the transformation $s \to 1 - s$, that is, on the line

Re(s) = 1/2. This is only a conjectural symmetry, but if it is true, does it propagate further? What does the Riemann Hypothesis say about the primes?

In the explicit formula, the contribution of the nontrivial zeros is more or less equal to

$$\Pi(x) - \text{Li}(x);$$

you should think of this as variation of $\Pi(x)$ around the average $\text{Li}(x)$. From Eq. (10.16), we can deduce that if *all* the nontrivial zeros have $\beta = \text{Re}(\rho) = 1/2$, then $\bar{\rho} = 1 - \rho$ and

$$\text{Li}(x^\rho) + \text{Li}(x^{1-\rho}) \sim \frac{2x^{1/2}}{|\rho| \log(x)} \cos(\gamma \log(x) - \arctan(2\gamma)).$$

The point is that we can factor out the term $x^{1/2} / \log(x)$ from the contribution of every zero ρ, as a sort of "universal amplitude." What is left over is a sum of purely *periodic* functions. The Riemann Hypothesis would mean that the distribution of the primes is amazingly regular. A physicist will think of a sum of periodic functions as a superposition of waves, a vibration or sound. This is what the physicist Sir Michael Berry means in (Berry, Keating, 1991) by the following:

[W]e can give a one-line nontechnical statement of the Riemann hypothesis: The primes have music in them.

Exercise 10.4.2. *Mathematica* has a package called Audio. It includes a function ListWaveform, which takes a table of amplitudes and frequencies as input and produces a sound with those characteristics. Read about it in the Help, and then make the sound that corresponds to the contribution of the first few zeros to the explicit formula.

Interlude 4

Modular Arithmetic

Traditional books on number theory that emphasize the algebraic aspects begin with the subject of modular arithmetic. This is a very powerful technique, based on the work of Fermat, Euler, and Gauss. The idea is to fix an integer, n, and group all the rest of the integers into one of n classes, depending on what remainder you get when you divide by n. A more elegant way of saying this is that two integers, a and b, are in the same class if n divides $b - a$ or, in other words, $b - a$ is an integer multiple of n. In this case, we write $a \equiv b \bmod n$. (This is pronounced "a is congruent to b modulo n.") For example, with $n = 6$, $5 \equiv 17 \bmod 6$, $-2 \equiv 4 \bmod 6$, and even $6 \equiv 0 \bmod 6$. Because there are seven days in a week, the fact that $3 \equiv 10 \bmod 7$ means that the 3rd of the month and the 10th of the month fall on the same weekday.

Exercise I4.0.1. The following steps show that $\equiv$ is an equivalence relation.

1. Show that $\equiv$ is reflexive, that is, that $a \equiv a \bmod n$.
2. Show that $\equiv$ is symmetric; that is, if $a \equiv b \bmod n$, then $b \equiv a \bmod n$.
3. Show that $\equiv$ is transitive; that is, if $a \equiv b \bmod n$ and $b \equiv c \bmod n$, then $a \equiv c \bmod n$.

Using long division, we can write any integer a as $a = k \cdot n + r$ for some integer k, and remainder r with $0 \leq r \leq n - 1$. (The remainder r, if it exists, must be less than n; otherwise, we would subtract off another n and increase k by 1.) So, n divides $a - r = k \cdot n$ and, thus, $a \equiv r \bmod n$. This shows that there are n equivalence classes, and that they have representatives 0, 1, ..., $n - 1$. We can define arithmetic modulo n by doing the operations of addition and multiplication on equivalence classes instead of on individual integers. What this means in practice is that we take representatives of the classes and add them (or multiply), and the answer is the equivalence class of whatever integer we get. An example will clarify. To multiply 4 times 5

254

modulo 6, we use that $4 \cdot 5 = 20 = 3 \cdot 6 + 2$; so, we write $4 \cdot 5 \equiv 2 \bmod 6$. And $4 + 5 = 9 = 6 + 3$, so $4 + 5 \equiv 3 \bmod 6$.

Exercise I4.0.2. Write out the complete multiplication table modulo 6. Do the same for addition.

For any b, $-b \equiv (n-1)b \bmod n$, and this makes subtraction easy: $a - b$ is just $a + (-b)$. What about division? We expect that a/b is just $a \cdot (1/b)$. What is $1/b$? It is just whatever class, when multiplied by b, gives 1. Because $5 \cdot 5 \equiv 1 \bmod 6$, then $5 \equiv 1/5 \bmod 6$! But from your multiplication table, you see that 2, 3, and 4 have no inverse. What is worse, $2 \cdot 3 \equiv 0 \bmod 6$, and $3 \cdot 4 \equiv 0 \bmod 6$. (Actually, this is not "worse"; it is the same fact stated in a different way.) The problem is that 2 and 6 have a common factor, as do 3 and 6 and 4 and 6.

Here is a fact based on the Euclidean algorithm that we will find very useful. If the greatest common divisor of a and n is d, then there are integers b and c such that

$$b \cdot a + c \cdot n = d.$$

(You can find the Euclidean algorithm in most books on number theory.) If a and n are relatively prime, that is, the greatest common divisor is 1, then there are b and c such that

$$b \cdot a + c \cdot n = 1.$$

And this says that $b \cdot a \equiv 1 \bmod n$, so $b \equiv 1/a \bmod n$. The classes that are relatively prime to the modulus n form a GROUP. The product of any two such class is another class prime to the modulus. The argument above says that every class a prime to the modulus has an inverse class b, so $ab \equiv 1 \bmod n$.

Most of our troubles go away if our modulus is a prime. So, we will assume that and write q instead of n now. If the greatest common divisor d of a and q is not 1, it must be q because it divides q and q is prime. Because it also divides a, this says that $a \equiv 0 \bmod q$. So, for a prime modulus, every class except the 0 class is invertible.

Exercise I4.0.3. Write out the addition and multiplication tables modulo 7. For each nonzero class a, identify $1/a$.

Exercise I4.0.4. Using your multiplication table, compute all the powers $3, 3^2, 3^3, \ldots$ modulo 7. Do the same for powers of 2.

The point of modular arithmetic is that it is a convenient notation for stating many results in number theory.

Exercise I4.0.5. Assume that x and y are integers. Show (by considering all the cases) that $x^2 + y^2 \equiv 0, 1$, or $2 \bmod 4$. This means that no prime number $p \equiv 3 \bmod 4$ can be the sum of two squares. The converse is true but a little harder to prove: Every prime $p \equiv 1 \bmod 4$ is the sum of two squares. For example, $5 = 1^2 + 2^2$, $13 = 2^2 + 3^2$, and $17 = 1^2 + 4^2$.

Exercise I4.0.6. Which primes p do you think can be written as $x^2 + 3y^2$, where x and y are integers?

There is a lot more that can be done with congruences; we will just prove a couple of theorems indicating why Fermat, Euler, and Gauss were interested in congruences.

Theorem (Fermat's Little Theorem). *If q is a prime and a is some integer not divisible by q, then*

$$a^{q-1} \equiv 1 \bmod q.$$

Of course, if q does divide a, then $a \equiv 0 \bmod q$, and so is a^{q-1}.

Proof. The nonzero residue classes modulo p are represented by $1, 2, \ldots, q - 1$. If we multiply each of these by a, we get the residue classes $a, 2a, \ldots, (q - 1)a$. These are all distinct, because if $ca \equiv da \bmod q$, we would be able to multiply by the class of a^{-1} to get $c \equiv d \bmod q$. (This is the only place we use the fact that a is not $\equiv 0 \bmod q$.) So, we have just written down the original $q - 1$ classes in a different order. (In your multiplication table for Exercise I4.0.3, each row except the one for 0 contains all the classes modulo 7, in a different order.) Thus, if we multiply all the classes together, we get that

$$1 \cdot 2 \cdot \ldots \cdot (q - 1) \equiv a \cdot 2a \cdot \ldots \cdot (q - 1)a \bmod q,$$

because they are the same numbers, just in a different order. If we group the terms, we get that

$$(q - 1)! \equiv a^{q-1} \cdot (q - 1)! \bmod q.$$

And because $(q - 1)!$ is not $\equiv 0 \bmod q$, we can cancel it out from both sides, thereby proving the theorem. □

Lemma. *Suppose that q is a prime and that a is some integer not divisible by q. Let e denote the least positive integer such that $a^e \equiv 1$ mod q. For any integer k, if $a^k \equiv 1$ mod q, then e divides k.*

Proof. Suppose that $a^k \equiv 1$ mod q. If we divide e into k, it will go some number of times, say d, with some remainder, r. Then,

$$k = e \cdot d + r, \qquad \text{with } 0 \le r < e.$$

(Again, the remainder, r, if it exists, must be less than e, otherwise we would subtract off another e and increase d by 1.) We are done if we can show that $r = 0$, which means e divides k. But if $r > 0$, we get a contradiction from

$$1 \equiv a^k \equiv a^{ed+r} \equiv (a^e)^d \cdot a^r \equiv 1^d \cdot a^r \equiv a^r \text{ mod } q,$$

because we assumed that e was the least positive integer with this property, and $r < e$. So, necessarily, $r = 0$.   □

Exercise I4.0.7. The converse is true as well. Try to prove it.

Theorem (Fermat). *If a prime number q divides a Mersenne number $M_p = 2^p - 1$, then p divides $q - 1$. So, $q = kp + 1$ for some k.*

Proof. If q divides $2^p - 1$, then $2^p \equiv 1$ mod q. Let e be the least positive integer such that $2^e \equiv 1$ mod q. Then, according to the lemma, e divides p. Because p is a prime and $e \ne 1$, $e = p$. Now, q is an odd prime because it divides M_p, which is an odd number. So, q does not divide 2. According to Fermat's Little Theorem, we also know that $2^{q-1} \equiv 1$ mod q. Again, according to the lemma, we know that $p = e$ divides $q - 1$.   □

In fact, we can say a little more. The integer k has to be even, because k and p are both odd, thus $kp + 1$ is even and, thus, not prime.

Exercise I4.0.8. Is $M_{29} = 536870911$ a prime? According to Fermat's theorem, the only possible prime divisors are of the form $58n + 1$. Try this with $n = 1, 2, 3, 4, \ldots$. Of course, when $58n + 1$ is composite, you don't even have to do the trial division, as the smallest divisor is certainly a prime.

A couple more theorems about congruences will be useful later on, when we want to relate congruences to different moduli. Unlike the preceding paragraphs, we will no longer restrict ourself to prime modulus.

Theorem (Hensel's Lemma). *Suppose that p is an odd prime, a is not 0 modulo p, $j \geq 1$, and we have a solution, x_j, to the equation*

$$x_j^2 \equiv a \bmod p^j.$$

Then, there is a unique solution, x_{j+1}, to the equation

$$x_{j+1}^2 \equiv a \bmod p^{j+1},$$

which satisfies $x_{j+1} \equiv x_j \bmod p^j$.

A few examples will give an idea of what Hensel's Lemma says. First, take $p = 5$, $a = -1$, $j = 1$. It so happens that $x_1 = 2$ satisfies the equation $2^2 \equiv -1 \bmod 5$. Now, Hensel says that there is exactly one class x_2 modulo 25 with $x_2 \equiv 2 \bmod 5$ and $x_2^2 \equiv -1 \bmod 25$. In fact, $x_2 = 7$ works, and if you want, you may check that there is no other solution by doing a tedious search. We can repeat the process with $j = 2$; $x_3 = 57$ satisfies $57 \equiv 7 \bmod 25$ and $57^2 \equiv -1 \bmod 125$, and so on. With $j = 3$, $x_4 = 182 \equiv 57 \bmod 125$ and $182^2 \equiv -1 \bmod 625$. How am I getting the solutions $x_2, x_3, x_4, \ldots$? The proof of Hensel's Lemma will indicate the algorithm.

Proof. According to hypothesis, we have $x_j^2 = a + cp^j$ for some integer c. If a solution x_{j+1} exists, it must be of the form $x_{j+1} = x_j + kp^j$ for some k, and the solution is unique if k is. Computing modulo p^{j+1}, we need

$$
\begin{aligned}
a \equiv x_{j+1}^2 &= (x_j + kp^j)^2 \\
&\equiv x_j^2 + 2kx_j p^j + p^{2j} \\
&\equiv a + cp^j + 2kx_j p^j + p^{2j} \bmod p^{j+1}.
\end{aligned}
$$

The a cancels out on both sides, and $j \geq 1$ means that $2j \geq j + 1$, so $p^{2j} \equiv 0 \bmod p^{j+1}$. So, all we need to be true is that

$$
\begin{aligned}
0 &\equiv cp^j + 2kx_j p^j \\
&\equiv (c + 2kx_j)p^j \bmod p^{j+1},
\end{aligned}
$$

or $0 \equiv c + 2kx_j \bmod p$. This equation has a unique solution in the unknown k; namely, $k \equiv -c/(2x_j) \bmod p$. Here, we are using the fact that p is odd, so 2 is invertible, and $x_j \equiv a$ is different from 0 modulo p^j and, thus, is also modulo p. Thus, x_j is also invertible modulo p. $\square$

Exercise 14.0.9. $x_1 = 3$ also satisfies $x_1^2 \equiv -1 \bmod 5$. Find an $x_2 \equiv x_1 \bmod 5$ such that $x_2^2 \equiv -1 \bmod 25$. Find an $x_3 \equiv x_2 \bmod 25$ such that $x_3^2 \equiv -1 \bmod 125$.

The next theorem is complementary to Hensel's Lemma, in that it relates solutions to equations when the moduli are relatively prime.

Theorem (Chinese Remainder Theorem). *If the moduli m and n are relatively prime, and if a and b are given, then there is a unique x modulo mn such that $x \equiv a$ mod m and $x \equiv b$ mod n.*

For example, with $m = 3$ and $n = 4$, the theorem claims that there is an x that is unique modulo 12 with $x \equiv 2$ mod 3 and $x \equiv 3$ mod 4. The solution is that $x \equiv 11$ mod 12. Again, the proof of the theorem will show how I did this.

Proof. First, observe that we have a map π between classes modulo mn and pairs of classes modulo m and n, respectively:

$$\pi : y \text{ mod } mn \to (y \text{ mod } m, y \text{ mod } n).$$

What the theorem is saying is that the map π is a bijection. All we have to do is write down the inverse map, π^{-1} going the other direction. The key fact is that because m and n are relatively prime, there are integers c and d such that $cm + dn = 1$, according to the Euclidean Algorithm. Reducing this equation first modulo m and then modulo n, we deduce that

$$d \equiv n^{-1} \text{ mod } m \qquad \text{and} \qquad c \equiv m^{-1} \text{ mod } n.$$

The map π^{-1} works as follows. Given a mod m and b mod n, we let $x = adn + bcm$ mod mn define $\pi^{-1}(a, b)$. This is a bijection because

$$\begin{aligned}
\pi(x) &= (x \text{ mod } m, x \text{ mod } n) \\
&= (adn + bcm \text{ mod } m, adn + bcm \text{ mod } n) \\
&= (adn + 0 \text{ mod } m, 0 + bcm \text{ mod } n) \\
&= (a \cdot 1 \text{ mod } m, b \cdot 1 \text{ mod } n).
\end{aligned}$$

$\square$

In the preceding example, we had $c = -1$ and $d = 1$ because $-1 \cdot 3 + 1 \cdot 4 = 1$. With $a = 2$ and $b = 3$, we take $x = 2 \cdot 1 \cdot 4 + 3 \cdot (-1) \cdot 3 = -1 \equiv 11$ mod 12.

Exercise I4.0.10. Find which x mod 15 satisfies $x \equiv 2$ mod 3 and $x \equiv 4$ mod 5.

Chapter 11
Pell's Equation

In Chapters 6 and 8, we saw how Euler computed values of the functions $\zeta(s)$ for various integer values of s. The comparison between them led him to conjecture the symmetry under $s \to 1 - s$ that was proved in Chapter 9. But it is natural to ask what is the *arithmetic* significance of an identity such as $\sum_n 1/n^2 = \pi^2/6$? This is hard to explain precisely because the Riemann zeta function is so fundamental in the study of the integers. It turns out there are other examples that are easier to understand.

Number theorists now define an Euler product like that of $\zeta(s)$ for almost every kind of arithmetic object they can think of. These new Euler products are called *L*-functions. Understanding the connection between the values these functions take on at particular values of the variable s, and the underlying arithmetic, is one of the basic goals of number theory. In the last three chapters, we will look at a couple of specific examples of *L*-functions attached to Diophantine equations.

11.1. The Cattle Problem of Archimedes

One fact that everyone insists on telling you about Pell's equation is that Pell had nothing to do with it. Euler apparently made the mistake of attributing Brouckner's work to Pell. But the name still applies to the study of integer solutions (x, y) to the equation

$$x^2 - Ny^2 = 1$$

for various integers N. If N is a square, or if $N < 0$, it is not hard to see that there are only the trivial solutions $(\pm 1, 0)$. Pell's equation with $N = 8$ appeared in Exercise 1.1.3, in connection with the question of which triangular numbers are also squares.

Fermat was the first mathematician to conjecture that for $N > 0$, and not a square, there are infinitely many solutions. He posed the special cases of $N = 61$ and 109 to his colleagues, saying that he had chosen quite small numbers *pour ne vous donner pas trop de peine* ("so as not to give you too much trouble"). But, really, he was showing off; the smallest solutions are

$$1766319049^2 - 61 \cdot 226153980^2 = 1 \quad \text{and}$$
$$158070671986249^2 - 109 \cdot 15140424455100^2 = 1.$$

In fact, the study of Pell's equations is much older than this. Archimedes, in *The Measurement of the Circle*, used the fact that $1351/780$ was a very good approximation to $\sqrt{3}$. The reason is that

$$1351^2 - 3 \cdot 780^2 = 1, \quad \text{so} \quad \frac{1351^2}{780^2} - 3 = \frac{1}{780^2}.$$

How much did Archimedes know about Pell's equation? In 1773, a manuscript was discovered in the Wolfenbüttel library. It described a problem posed by Archimedes to colleagues Eratosthenes and Apollonius, written as a poem in 22 couplets:

If thou are diligent and wise, O stranger, compute the number of cattle of the Sun, who once upon a time grazed on the fields of Thrinacia, divided into four herds of different colors, one milk white, another glossy black. . . .

Thrinacia is Sicily, the "three-sided" island. The problem goes on to describe the relations between the sizes of the herds of different colors. In all, there are seven linear equations and eight unknowns: the number of cows and bulls of each color. So far, this is a straightforward linear algebra problem, although the numbers involved are large. The smallest solution is a herd of 50389082 cattle. Archimedes says that if you can solve this much, that

. . . thou wouldst not be called unskilled or ignorant of numbers, but not yet shalt thou be numbered among the wise.

He goes on to add two more conditions: that the number of white bulls and black bulls be a square number and the number of yellow bulls and spotted bulls be a triangular number.

If thou art able, O stranger, to find out all these things . . . thou shalt depart crowned in glory and knowing that thou hast been adjudged perfect in this species of wisdom.

After some algebraic manipulation, the extra conditions require solving the Pell equation

$$x^2 - 4729494y^2 = 1,$$

with the side condition that y be divisible by 9314. A representation of the solution, described below, was found by Amthor in 1880. The size of the herd is about $7.76 \cdot 10^{206544}$ cattle. In 1895, the Hillsboro, Illinois, Mathematical Club got an answer correct to 32 decimal places. They spent four years on the computation and announced proudly (Bell, 1895) that the final answer was one-half mile long. The exact solution was obtained by Williams, German and Zarnke in 1965 (Williams, German, Zarnke, 1965), on an IBM 7040 with just 32K memory. For a good discussion of the problem, see Vardi (1998). Weil's book (Weil, 1983) has more on the history of Pell's equation.

11.2. Group Structure

Solutions to Pell's equation can be viewed geometrically as lattice points on a hyperbola. But an algebraic interpretation is also very useful. We can factor Pell's equation as

$$(x + \sqrt{N}y)(x - \sqrt{N}y) = 1.$$

This suggests looking at numbers of the form $x + \sqrt{N}y$ where x and y are integers. In some ways, these are like complex numbers, with $\sqrt{N}$ playing the role of i. We can multiply such numbers with the obvious rule:

$$(x + \sqrt{N}y) \cdot (a + \sqrt{N}b) = ax + \sqrt{N}ay + \sqrt{N}bx + Nby$$
$$= (ax + Nby) + \sqrt{N}(bx + ay).$$

Exercise 11.2.1. Show that if (x, y) is one solution to Pell's equation, and if (a, b) is another, then defining $(a, b) \cdot (x, y)$ as equal to $(ax + Nby, bx + ay)$ is a new solution. Observe that the operation $\cdot$ is a *new* way of combining pairs of points.

With this observation, we can say that the solutions to Pell's equation form a group. The multiplication rule for combining two solutions is the one given by $\cdot$ above. The identity element of the group is the trivial solution $(1, 0)$, corresponding to the number $1 = 1 + \sqrt{N}0$, and the inverse of the solution (x, y) is the solution $(x, -y)$.

Exercise 11.2.2. Show that if (x, y) is a solution to Pell's equation, then $(1, 0) \cdot (x, y) = (x, y)$ and $(x, y) \cdot (x, -y) = (1, 0)$.

We've seen different examples of groups before. In Section 2.3, the group operation there was convolution $*$; the identity element of the group was the function $e(n)$ (see Exercise 2.3.2.) In Interlude 4, the classes modulo n that are relatively prime to n form a group under multiplication.

From now on, we can just identify a solution (x, y) to Pell's equation with the number $x + \sqrt{N}y$. This lets us generate new solutions from old ones and write them in a compact format.

Exercise 11.2.3. For example, $2 + \sqrt{3}$ solves Pell's equation for $N = 3$. Use the Binomial Theorem to write $(2 + \sqrt{3})^6$ in the form $a + b\sqrt{3}$ to get Archimedes approximation to $\sqrt{3}$, mentioned earlier.

Amthor's solution of the Pell equation that arises from Archimedes' cattle problem can be expressed as

$$(109931986732829734979866232821433543901088049 +$$
$$50549485234315033074477819735540408986340\sqrt{4729494})^{2329}.$$

There is quite a lot I haven't said about the algebra of Pell's equation; it is beyond the scope of this book. The point here is that it is a hard problem with a long history.

11.3. Solutions Modulo a Prime

Because it is hard to find solutions with integers to Pell's equation, we might change the problem to a different, related one. If we fix a prime number p, are there solutions modulo p? In fact, there must be because Lagrange proved that there are solutions to the original problem. These integers reduced modulo p give a solution modulo p. More interesting is the question of how many solutions there are modulo p.

In fact, in this section, we will prove that

Theorem. *If p is an odd prime that does not divide N, then the number of solutions is*

$$\#\{(x, y)|x^2 - Ny^2 \equiv 1\} = \begin{cases} p - 1, & \text{if some } a^2 \equiv N \text{ mod } p, \\ p + 1, & \text{otherwise.} \end{cases} \quad (11.1)$$

We need a lemma.

Lemma. *Whether or not N is a square modulo p, there are $p - 1$ solutions to the equation $z^2 - w^2 \equiv N$ mod p.*

Proof. We can introduce new variables, u and v, related to z and w by the equations

$$u = z - w, \quad v = z + w \quad \Leftrightarrow \quad z = \frac{u + v}{2}, \quad w = \frac{u - v}{2}.$$

(We can divide by 2 because p is odd.) Then,

$$u \cdot v \equiv N \bmod p \quad \Leftrightarrow \quad z^2 - w^2 \equiv N \bmod p.$$

But the equation in u and v has exactly $p - 1$ solutions, because for each residue class $u = 1, 2, \ldots, p - 1$, we get a solution by taking $v = N/u$. $\quad \square$

Proof of Theorem. Suppose first that $N \equiv a^2 \bmod p$. From the lemma, we know that the equation $x^2 - w^2 \equiv 1 \bmod p$ has exactly $p - 1$ solutions (x, w). (Here, the number 1 plays the role of N in the lemma, but so what?) Letting $y = w/a$, we get $p - 1$ solutions (x, y) to the equation $x^2 - Ny^2 \equiv 1 \bmod p$. (We know that a is not $\equiv 0 \bmod p$ because p does not divide N either.) This proves the first case.

Next, suppose that N is not congruent to any square modulo p. We know from the lemma that there are $p - 1$ solutions (z, w) to the equation $z^2 - w^2 \equiv N \bmod p$. Furthermore, none of the w is $\equiv 0 \bmod p$, otherwise $N \equiv z^2$. Write the equation as $z^2/w^2 - N/w^2 \equiv 1 \bmod p$. With the change of variables $x = z/w$, $y = 1/w$, we get $p - 1$ solutions to

$$x^2 - Ny^2 \equiv \frac{z^2}{w^2} - \frac{N}{w^2} \equiv 1 \bmod p.$$

Furthermore, none of the y is 0, as they are of the form $1/w$. There are precisely two more solutions that do have $y \equiv 0 \bmod p$, namely $(\pm 1, 0)$. This makes $p + 1$ solutions in total. $\quad \square$

This is nice so far, but it requires solving a new congruence for each prime p, to find out whether N is congruent to a square modulo p. Luckily, there is a theorem called Quadratic Reciprocity, which is spectacularly useful in number theory. We won't prove it here; it can be found in most books. The reciprocity law relates the question of whether N is a square modulo p to the question of whether p is a square modulo N (hence "reciprocity"). There are various ways to state the theorem; this version will be sufficient for us.

Theorem (Quadratic Reciprocity). *Suppose that N and p are both odd primes. If either is congruent to 1 modulo 4, then N and p have the same "parity"; that is, either both are congruent to squares or neither is. If both N*

and p are congruent to 3 modulo 4, then they have opposite parity; exactly one of the two is congruent to a square.

Meanwhile, 2 is a square modulo an odd prime p exactly when p is congruent to ±1 modulo 8. If p is congruent to 1 modulo 4, then −1 is a square modulo p, but it is not if p is congruent to 3 modulo 4.

Gauss proved the Quadratic Reciprocity Theorem at age 17, after unsuccessful attempts by Euler and Legendre. He called it his *theorema fundamentale* and went on to prove it eight different ways. Remarkably, this symmetry between primes p and N can also be deduced from the symmetry property of the theta function $\Theta(t)$ of Section 9.2. This seems to have first been noticed by Polya (Polya, 1927). You can find an exposition in McKean and Moll (1997).

By using Quadratic Reciprocity, we can convert the infinitely many congruences $N \equiv a^2 \bmod p$ as p varies into the question of which of finitely many residue classes modulo N does p fall into. Some examples will help clarify.

Example ($N = 2$). We define a function $\chi_8(n)$ by

$$\chi_8(n) = \begin{cases} +1, & \text{if } n \equiv 1 \text{ or } 7 \bmod 8, \\ -1, & \text{if } n \equiv 3 \text{ or } 5 \bmod 8, \\ 0, & \text{otherwise.} \end{cases} \tag{11.2}$$

Then, for all odd primes p, the Pell equation $x^2 - 2y^2 \equiv 1$ has $p - \chi_8(p)$ solutions modulo p, according to (11.1) and Quadratic Reciprocity.

Example ($N = 3$). We define a function $\chi_{12}(n)$ by

$$\chi_{12}(n) = \begin{cases} +1, & \text{if } n \equiv 1 \text{ or } 11 \bmod 12, \\ -1, & \text{if } n \equiv 5 \text{ or } 7 \bmod 12, \\ 0, & \text{otherwise.} \end{cases} \tag{11.3}$$

Then, for all primes $p \neq 3$, the Pell equation $x^2 - 3y^2 \equiv 1$ has $p - \chi_{12}(p)$ solutions modulo p.

Proof. For $p = 2$, we've defined $\chi_{12}(2) = 0$. And

$$x^2 - 3y^2 \equiv x^2 + y^2 \bmod 2.$$

There are two solutions to $x^2 + y^2 \equiv 1 \bmod 2$, namely, $(1, 0)$ and $(0, 1)$. So, $p = 2$ checks out. For odd primes $p \neq 3$, we will need to use Quadratic Reciprocity, and so, we will need to know the squares modulo 3. Because

$2^2 \equiv 1 \equiv 1^2 \bmod 3$, 1 is a square and 2 is not. According to (11.1), there are $p - 1$ solutions to the Pell equation exactly when 3 is a square modulo p. This happens in two cases only. One is when $p \equiv 1 \bmod 3$ and $p \equiv 1 \bmod 4$; according to the Chinese Remainder Theorem, this means that $p \equiv 1 \bmod 12$. The other case is when $p \equiv 2 \bmod 3$ and $p \equiv 3 \bmod 4$; this happens when $p \equiv 11 \bmod 12$. In the other two cases, 3 is not a square. Specifically, this happens if $p \equiv 2 \bmod 3$ and $p \equiv 1 \bmod 4$, because the primes that are 5 mod 12 satisfy this congruence, or if $p \equiv 1 \bmod 3$ and $p \equiv 3 \bmod 4$, which happens for primes 7 mod 12. $\square$

11.4. Dirichlet *L*-Functions

Now that we've done all this, we can get back to analysis. This section will continue to look at the special cases of $N = 3$ and 2, but much of this is true more generally. We can take the function χ_{12}, defined earlier, and use it to define an Euler product by

$$L(s, \chi_{12}) = \prod_{p \text{ prime}} (1 - \chi_{12}(p)p^{-s})^{-1}.$$

For $\mathrm{Re}(s) > 1$ and any prime p, we know that $|\chi_{12}(p)p^{-s}| < 1$. So, we can use the Geometric series to write

$$L(s, \chi_{12}) = \prod_{p \text{ prime}} \sum_{k=0}^{\infty} \frac{\chi_{12}(p)^k}{p^{ks}}.$$

By considering the cases in (11.3), one can show that

$$\chi_{12}(n)\chi_{12}(m) = \chi_{12}(nm).$$

So, we can multiply the terms corresponding to different primes in the Euler product to get

$$L(s, \chi_{12}) = \sum_{n=1}^{\infty} \frac{\chi_{12}(n)}{n^s}$$

$$= 1 - \frac{1}{5^s} - \frac{1}{7^s} + \frac{1}{11^s} + \frac{1}{13^s} - \frac{1}{17^s} - \frac{1}{19^s} + \frac{1}{23^s} \cdots.$$

Using the comparison test with $\zeta(s)$, we can show that $L(s, \chi_{12})$ is absolutely convergent for $\mathrm{Re}(s) > 1$. And, similarly to the function $\phi(s)$ in Section 8.1, $L(s, \chi_{12})$ converges conditionally, but not absolutely, for $0 < \mathrm{Re}(s) \leq 1$. This suggests that something interesting is happening at $s = 1$. What does the Euler product look like at $s = 1$? Formally, if we plug in $s = 1$, we get

$$L(1, \chi_{12}) = \prod_{p \text{ prime}} \frac{1}{1 - \chi_{12}(p)/p} = \prod_{p \text{ prime}} \frac{p}{p - \chi_{12}(p)}.$$

The terms in the denominator are exactly the number of solutions to Pell's equation modulo p. This is not too miraculous; we more or less defined $L(s, \chi_{12})$ so that this would happen. What is surprising is that this number gives a solution to the original, integer Pell equation.

Theorem.

$$L(1, \chi_{12}) = \frac{\log(7 + 4\sqrt{3})}{\sqrt{12}} = \frac{\log((2 + \sqrt{3})^2)}{\sqrt{12}}.$$

Observe that $1 = 7^2 - 3 \cdot 4^2$. You should be amazed; this is a miracle, with a capital M.

Proof. We want to compute

$$1 - \frac{1}{5} - \frac{1}{7} + \frac{1}{11} + \frac{1}{13} - \frac{1}{17} - \frac{1}{19} + \frac{1}{23} \cdots,$$

using Abel's theorems from Section I2.7. The partial sums of the series $\sum_n \chi_{12}(n)$ are all either 1, 0, or -1 because χ_{12} is periodic. So Abel's Theorem II (I2.20) applies with $\sigma = 1$; the series $L(1, \chi_{12})$ converges. Now we use Abel's Theorem I (I2.19). That is, we define a Taylor series for $|x| < 1$ by

$$f_{12}(x) = \sum_{n=1}^{\infty} \frac{\chi_{12}(n)}{n} x^n$$

$$= x - \frac{x^5}{5} - \frac{x^7}{7} + \frac{x^{11}}{11} + \frac{x^{13}}{13} - \frac{x^{17}}{17} - \frac{x^{19}}{19} + \frac{x^{23}}{23} \cdots.$$

If we can find a closed-form expression for $f_{12}(x)$ that is continuous at $x = 1$, then the series we are interested in is just $f_{12}(1)$. Because $\chi_{12}(n)$ has period 12, we can write $f_{12}(x)$ as

$$\sum_{k=0}^{\infty} \left\{ \frac{x^{12k+1}}{12k + 1} - \frac{x^{12k+5}}{12k + 5} - \frac{x^{12k+7}}{12k + 7} + \frac{x^{12k+11}}{12k + 11} \right\}.$$

The function $f_{12}(x)$ is complicated. So, we take a derivative to get that

$$f_{12}'(x) = \sum_{k=0}^{\infty} \{x^{12k} - x^{12k+4} - x^{12k+6} + x^{12k+10}\}$$

$$= \{1 - x^4 - x^6 + x^{10}\} \sum_{k=0}^{\infty} x^{12k}$$

$$= \frac{1 - x^4 - x^6 + x^{10}}{1 - x^{12}}$$

$$= \frac{(1-x)(1+x)}{1 - x^2 + x^4}.$$

Here, in the last step, we used that

$$1 - x^4 - x^6 + x^{10} = (1-x)(1+x)(1 - x^2 + x^6 + x^8)$$

and

$$1 - x^{12} = (1 - x^2 + x^4)(1 - x^2 + x^6 + x^8).$$

The function $f_{12}(x)$ we are looking for is the antiderivative

$$\int \frac{(1-x)(1+x)}{1 - x^2 + x^4} dx,$$

which is 0 at $x = 0$ (because the constant term of the Taylor series $f_{12}(x)$ is 0). The denominator factors as

$$1 - x^2 + x^4 = (1 + \sqrt{3}x + x^2)(1 - \sqrt{3}x + x^2).$$

So, the antiderivative can be done by partial fractions. After some tedious algebra, we find that

$$\frac{(1-x)(1+x)}{1 - x^2 + x^4} = \frac{1}{\sqrt{12}} \left\{ \frac{2x + \sqrt{3}}{1 + \sqrt{3}x + x^2} - \frac{2x - \sqrt{3}}{1 - \sqrt{3}x + x^2} \right\}.$$

So,

$$f_{12}(x) = \frac{1}{\sqrt{12}} \left\{ \int \frac{2x + \sqrt{3}}{1 + \sqrt{3}x + x^2} dx - \int \frac{2x - \sqrt{3}}{1 - \sqrt{3}x + x^2} dx \right\}.$$

This is now an easy substitution integral. We find that

$$f_{12}(x) = \frac{1}{\sqrt{12}}\{\log(1 + \sqrt{3}x + x^2) - \log(1 - \sqrt{3}x + x^2)\}$$

$$= \frac{1}{\sqrt{12}}\log\left(\frac{1 + \sqrt{3}x + x^2}{1 - \sqrt{3}x + x^2}\right).$$

All this, so far, is legal for $|x| < 1$. But the right side above is defined and continuous for $x = 1$. So, the sum of the series is

$$f_{12}(1) = \frac{1}{\sqrt{12}}\log\left(\frac{2 + \sqrt{3}}{2 - \sqrt{3}}\right)$$

$$= \frac{1}{\sqrt{12}}\log((2 + \sqrt{3})^2).$$

$\square$

We can define a similar L-function for $N = 2$:

$$L(s, \chi_8) = \prod_{p \text{ prime}} (1 - \chi_8(p)p^{-s})^{-1}$$

$$= \sum_{n=1}^{\infty} \frac{\chi_8(n)}{n^s}$$

$$= 1 - \frac{1}{3^s} - \frac{1}{5^s} + \frac{1}{7^s} + \frac{1}{9^s} - \frac{1}{11^s} - \frac{1}{13^s} + \frac{1}{15^s} \cdots.$$

Again,

$$L(1, \chi_8) = \prod_{p \text{ prime}} \frac{p}{p - \chi_8(p)}$$

is an infinite product measuring the number of solutions to

$$x^2 - 2y^2 \equiv 1 \bmod p.$$

Theorem.

$$L(1, \chi_8) = \frac{\log(3 + 2\sqrt{2})}{\sqrt{8}}.$$

And $1 = (3 - 2\sqrt{2})(3 + 2\sqrt{2}) = 3^2 - 2 \cdot 2^2$. So, the L-function value encodes a solution to Pell's equation.

Proof in Exercises. We want to sum the series

$$1 - \frac{1}{3} - \frac{1}{5} + \frac{1}{7} + \frac{1}{9} - \frac{1}{11} - \frac{1}{13} + \frac{1}{15} \cdots .$$

Abel's theorems apply just as before. Let

$$f_8(x) = \sum_{n=1}^{\infty} \frac{\chi_8(n)}{n} x^n.$$

Exercise 11.4.1. Show that for $|x| < 1$,

$$f_8'(x) = \frac{(1-x)(1+x)}{1+x^4}.$$

Exercise 11.4.2. Verify that

$$f_8'(x) = \frac{1}{2\sqrt{2}} \left\{ \frac{2x + \sqrt{2}}{x^2 + \sqrt{2}x + 1} - \frac{2x - \sqrt{2}}{x^2 - \sqrt{2}x + 1} \right\}.$$

Exercise 11.4.3. Do the integration to find $f_8(x)$, and plug in $x = 1$ to show that

$$L(1, \chi_8) = f_8(1) = \frac{1}{\sqrt{8}} \log(3 + 2\sqrt{2}).$$

□

What we have been looking at so far are examples of binary quadratic forms. The general format is a function of the shape

$$F(x, y) = ax^2 + bxy + cy^2$$

for relatively prime integers a, b, c. The discriminant $d = b^2 - 4ac$, just as in the quadratic formula. Notice that $x^2 - 2y^2$ has discriminant 8, and $x^2 - 3y^2$ has discriminant 12. We can get variations of the Pell equation by looking at other binary quadratic forms. For example, if $d \equiv 1 \bmod 4$, then $(d-1)/4$ is an integer and $x^2 + xy - \frac{d-1}{4}y^2$ has discriminant d. We can factor

$$x^2 + xy - \frac{d-1}{4}y^2 = \left(x + \frac{1 + \sqrt{d}}{2}y \right) \left(x + \frac{1 - \sqrt{d}}{2}y \right).$$

Integer solutions (x, y) correspond, as before, to numbers of the form $x + \frac{1+\sqrt{d}}{2}y$, and we still have a group structure, as before:

Exercise 11.4.4. Verify that

$$\left(x + \frac{1+\sqrt{d}}{2}y\right)\left(a + \frac{1+\sqrt{d}}{2}b\right) =$$

$$\left(ax + \frac{d-1}{2}by\right) + \frac{1+\sqrt{d}}{2}(ay + 2by + bx).$$

So, the multiplication law on pairs is

$$(x, y) \cdot (a, b) = \left(ax + \frac{d-1}{2}by, \, ay + 2by + bx\right).$$

Verify that if (x, y) and (a, b) are solutions to the Pell equation, so is $(x, y) \cdot (a, b)$.

Theorem. *If p is an odd prime that does not divide d, then, as before, the number of solutions modulo p is*

$$\#\left\{(x, y) \mid x^2 + xy - \frac{d-1}{4}y^2 \equiv 1\right\} =$$

$$\begin{cases} p - 1, & \text{if some } a^2 \equiv d \bmod p, \\ p + 1, & \text{otherwise.} \end{cases} \qquad (11.4)$$

Proof. We can complete the square in the equation modulo p to write it as

$$\left(x + \frac{y}{2}\right)^2 - d\left(\frac{y}{2}\right)^2 \equiv 1 \bmod p.$$

We can divide by 2 because p is odd. Now, change the variables with $z = x + y/2$ and $w = y/2$ to write this as

$$z^2 - dw^2 \equiv 1 \bmod p.$$

From (11.1), we know that there are $p - 1$ solutions (z, w) (respectively $p + 1$ solutions) if d is a square modulo p (resp. not a square). Changing the variables back by $y = 2w$, $x = z - w$ gives the correct number of solutions (x, y) to the original equation. $\qquad\square$

Example ($d = 5$). We define a function $\chi_5(n)$ as

$$\chi_5(n) = \begin{cases} +1, & \text{if } n \equiv 1 \text{ or } 4 \bmod 5, \\ -1, & \text{if } n \equiv 2 \text{ or } 3 \bmod 5, \\ 0, & \text{if } n \equiv 0 \bmod 5. \end{cases} \qquad (11.5)$$

Then, for all primes $p \neq 5$, the Pell equation $x^2 + xy - y^2 \equiv 1$ has $p - \chi_5(p)$ solutions modulo p.

Exercise 11.4.5. Prove this statement. For odd primes $p \neq 5$, you need to use Quadratic Reciprocity. Which classes modulo 5 are squares? For $p = 2$, you need to explicitly find three solutions.

The Dirichlet L-function for $d = 5$ is, as you might expect,

$$L(s, \chi_5) = \prod_{p \text{ prime}} (1 - \chi_5(p)p^{-s})^{-1}$$

$$= \sum_{n=1}^{\infty} \frac{\chi_5(n)}{n^s}$$

$$= 1 - \frac{1}{2^s} - \frac{1}{3^s} + \frac{1}{4^s} + \frac{1}{6^s} - \frac{1}{7^s} - \frac{1}{8^s} + \frac{1}{9^s} \dots.$$

As before,

$$L(1, \chi_5) = \prod_{p \text{ prime}} \frac{p}{p - \chi_5(p)}$$

is an infinite product measuring the number of solutions to

$$x^2 + xy - y^2 \equiv 1 \bmod p.$$

Theorem. *The value of the L-function is*

$$L(1, \chi_5) = \frac{1}{\sqrt{5}} \log \left(\frac{3 + \sqrt{5}}{2} \right) = \frac{1}{\sqrt{5}} \log \left(1 + 1 \cdot \frac{1 + \sqrt{5}}{2} \right),$$

encoding the solution $(1, 1)$ *to Pell's equation.*

Proof in Exercises. We want to sum the series

$$1 - \frac{1}{2} - \frac{1}{3} + \frac{1}{4} + \frac{1}{6} - \frac{1}{7} - \frac{1}{8} + \frac{1}{9} \dots.$$

Abel's theorems apply just as before. Let

$$f_5(x) = \sum_{n=1}^{\infty} \frac{\chi_5(n)}{n} x^n.$$

Exercise 11.4.6. Show that for $|x| < 1$,

$$f_5'(x) = \frac{(1 - x)(1 + x)}{1 + x + x^2 + x^3 + x^4}.$$

Exercise 11.4.7. For convenience, let

$$\epsilon_+ = (1 + \sqrt{5})/2, \quad \text{and} \quad \epsilon_- = (1 - \sqrt{5})/2.$$

Verify that

$$f_5'(x) = \frac{1}{\sqrt{5}} \left\{ \frac{2x + \epsilon_+}{x^2 + \epsilon_+ x + 1} - \frac{2x + \epsilon_-}{x^2 + \epsilon_- x + 1} \right\}.$$

You will need the identities $\epsilon_+ + \epsilon_- = 1$, $\epsilon_+\epsilon_- = -1$.

Exercise 11.4.8. Do the integration to find $f_5(x)$, and plug in $x = 1$ to show that

$$L(1, \chi_5) = f_5(1) = \frac{1}{\sqrt{5}} \log((3 + \sqrt{5})/2).$$

$\square$

Without more background in abstract algebra, the techniques available to us are not powerful enough to do other examples with positive discriminant d. But the phenomenon we have observed occurs generally. The L-function value at $s = 1$ always encodes a solution to Pell's equation, but which solution? Here is a rough answer. Lagrange proved that for each discriminant d, there is a fundamental solution ϵ to Pell's equation, one that generates all the others by taking powers of ϵ. Gauss, in his work on binary quadratic forms, defined an invariant h, called the CLASS NUMBER, depending on the discriminant d. Dirichlet proved that

Theorem (Analytic Class Number Formula). *For d, a positive discriminant*

$$L(1, \chi_d) = \frac{1}{\sqrt{d}} \log(\epsilon^h) = \frac{h}{\sqrt{d}} \log(\epsilon). \tag{11.6}$$

There is an analogous formula for negative discriminants but without the logarithms, because the corresponding Pell equation has only trivial solutions. In fact, Gregory's series for $\pi/4$, Exercise I2.6.4, is the case of $d = -4$. The Analytic Class Number Formula in the case of negative discriminants is proved in Chapter 13.

Chapter 12

Elliptic Curves

In the previous chapter, we looked at a quadratic equation, Pell's equation. Now we will look at something more complicated: cubic equations in x and y. By some changes of variable, we can reduce the degree of one of the variables, and simplify things. So, we will look at equations of the form

$$y^2 = x^3 + Ax + B.$$

The numbers A and B are parameters we can play with, like the d in Pell's equation. These equations are called ELLIPTIC CURVES. This is not the same thing as ellipses; the name comes from connection to certain integrals, elliptic integrals, which determine arc length on an ellipse.

12.1. Group Structure

The points on an elliptic curve, like the solutions to Pell's equation in Chapter 11, have an algebraic structure; they form a group. Again, this lets us get new solutions from old ones. The rule for combining points or solutions is geometric and very elegant. Given two points, P and Q, on E, we draw the line through them. This line meets the curve at a third point, R. Figure 12.1 shows the points $P = (-2, 1)$ and $Q = (0, -1)$ on the curve $E : y^2 = x^3 - 4x + 1$. This gives a new point on the curve, in this example $R = (3, -4)$. This is not quite the group law, however; it is not complicated enough. Because if we try to combine R and P, the third point on the line between them is just Q, a point we already know about. To liven things up, after finding the third point, R, on the line PQ, we intersect the vertical line through R with the curve E. The curve is symmetric around the x-axis, this amounts to changing the sign of the y-coordinate of R. This gives another point on the curve, which we define as $P + Q$. In the example of Figure 12.1, $P + Q = (3, 4)$.

The points P and Q are also points in the plane; they can be added together with the usual vector addition. But vector addition has no inherent connection to the curve; the vector sum of P and Q will probably not even be a point on

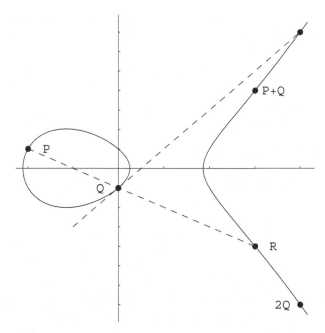

Figure 12.1. $E : y^2 = x^3 - 4x + 1$.

E. When we write $P + Q$, we always mean the addition rule defined above.
It is probably worth mentioning that this group law is written with an additive
notation, $+$, whereas the group operation $\cdot$ on the solutions to Pell's equation
is written multiplicatively.

A group needs an identity element, something that does not change any
point it is added to. For E, the identity element is a point, O, thought of as
being "at infinity," infinitely far up on the vertical axis. The line through O
and Q is, thus, the vertical line through Q. The third point on this line is by
definition $-Q$. The vertical line through $-Q$ meets the curve at Q again.
So, $Q + O = Q$. The point R in Figure 12.1 is actually $-(P + Q)$. A very
elegant way of describing the group law on E is that three points, P, Q, and
R, add up to the identity O exactly when they are colinear.

A special case of the group law describes how to add a point Q to itself,
that is, how to compute $2Q = Q + Q$. We must find the tangent line to E at
Q; this line has a "double" intersection with E at Q and hits E at exactly one
other point, which is $-2Q$. The point $2Q$ is the other point on the vertical line
through $-2Q$. So, in the example $Q = (0, -1)$, it turns out that $2Q = (4, -7)$.

This description of the group law is sometimes called the method of secants
and tangents. A line through two distinct points P and Q is called a secant line

in geometry. The geometric view described here is apparently attributable to Isaac Newton, although he did not have the abstract concept of group to work with.

So far, all you have is my word that these various lines hit E where I said. We should do some examples, with the equations written out explicitly. Let us add $P = (-2, 1)$ and $Q = (0, -1)$. See that the line between them has slope $(-1 - 1)/(0 + 2) = -1$. The line through $(0, -1)$ with slope -1 is

$$y + 1 = -(x - 0) \quad \text{or} \quad y = -x - 1.$$

Plug this into the equation for E to get the cubic equation

$$(-x - 1)^2 = x^3 - 4x + 1,$$

which simplifies to

$$x^3 - x^2 - 6x = 0.$$

Factoring a cubic is hard, but we know two roots already: the x-coordinates of the two points $(-2, 1)$ and $(0, -1)$ that we started with. So, $(x + 2)(x - 0)$ divides the cubic, and tedious synthetic division (or a computer algebra package) shows that the quotient is $x - 3$. So, the x-coordinate is 3, and we get the y-coordinate, -4, by plugging $x = 3$ into the equation of the line $y = -x - 1$. This says that $R = (3, -4)$ and $P + Q = (3, 4)$.

Here's another explicit example: doubling a point. From the equation $y^2 = x^3 - 4x + 1$ for E, we can use implicit differentiation to say that

$$2y \, dy = (3x^2 - 4)dx \quad \text{or} \quad \frac{dy}{dx} = \frac{3x^2 - 4}{2y}.$$

At $P = (-2, 1)$, the slope of the tangent line is thus 4, and the equation of the tangent line at P is $y - 1 = 4(x + 2)$, or $y = 4x + 9$. Plugging into the equation for E, we see that the line meets E where

$$(4x + 9)^2 = x^3 - 4x + 1 \quad \text{or} \quad x^3 - 16x^2 - 76x - 80 = 0.$$

Again, cubics are hard to factor, but we know that $(x + 2)^2$ divides it, because $x = -2$ is a double root from the tangent line at P. The quotient is $x - 20$. So, the x-coordinate is 20, and the y-coordinate of the point $-2P$ comes from the equation of the line: $y = 4 \cdot 20 + 9 = 89$. So, $2P = (20, -89)$.

Exercise 12.1.1. Show that $P - Q$, that is, that $P + (-Q)$ is $(2, -1)$. Show that $2Q = (4, -7)$. Compute $-P + 2Q$, $2P - Q$, and $3Q = Q + 2Q$.

For a positive integer n, we use the notation

$$nP = \underbrace{P + P + \cdots + P}_{n \text{ copies of } P}$$

for repeated addition. And $(-n)P$ is just $-(nP)$. For this particular curve, every point on E with rational number coordinates is of the form $nP + mQ$ for some integers n and m. Thus, the algebraic structure of E is like that of two copies of the integers. This setup is somewhat similar to a vector space, where P and Q play the role of a basis for the vector space and the expression $nP + mQ$ is like a linear combination of basis vectors with coefficients n and m. The big difference is that we allow only integer coefficients, not real numbers.

In fact, a very beautiful theorem by Mordell says that this is always the case: For any elliptic curve E, there is a finite list of points, $P_1, P_2, \ldots, P_r$, such that every rational point on E can be written uniquely as

$$n_1 P_1 + n_2 P_2 + \cdots + n_r P_r + \text{ a torsion point}$$

for integers $n_1, n_2, \ldots, n_r$. The number of generators r is called the RANK of E; it is the analog of the dimension of a vector space. A point P is a torsion point if $nP = O$ for some n; there are only finitely many possibilities. So, the elliptic curve $E : y^2 = x^3 - 4x + 1$ we've been studying has rank 2. Determining the rank of a given elliptic curve, and finding the generators, is a hard problem. While we are talking about rank, it is worth recalling that the solutions of Pell's equation in the previous chapter formed a group with rank 1. Every solution is a power of a single generator, up to an annoying factor of ± 1. But in Pell's equation, the notation is multiplicative, not additive.

12.2. Diophantine Equations

Roughly speaking, Diophantine equations are polynomial equations for which we seek integer, or sometimes rational, solutions. Pell's equation in the previous chapter is one example; an elliptic curve is another. The Diophantine tradition of number theory is distinct from that of the Pythagoreans: The latter studies the inherent properties of numbers (e.g., prime or perfect), whereas the former studies solutions of equations. It is surprising how many Diophantine problems are connected to elliptic curves. This section will sketch some of these connections, and the history of how Diophantus came to be lost and then found again.

Diophantus of Alexandria (circa 250 A.D.). Virtually nothing of his life is known. A reference to Diophantus in an eleventh century manuscript by Michael Psellus led historians to believe that Diophantus lived in the middle of the third century. Of his works, only four are known, even by title.

1. *Moriastica*. This is mentioned only once, in a scholium to Iamblichus' commentary on Nicomachus' *Arithmetica*. It seems to have treated computations with fractions.
2. *Porismata*. This may have been a collection of lemmas used in *Arithmetica*. It is referred to in that work.
3. *On Polygonal Numbers*. This work survives only in an incomplete form and contains little that is new. It does contain the result of Exercise 1.2.4 for the nth a polygonal number (a vertices). The fragment ends in the middle of an investigation of the number of ways an integer can be a polygonal number.
4. *Arithmetica*. Originally in thirteen books, only the first six survive. The *Arithmetica* was Diophantus' major work, but it was lost to scholars for a thousand years.

The Pythagorean tradition of number theory survived because of Boethius's translation of Nicomachus. But Boethius did not know of the work of Diophantus. We would not have any record of the *Arithmetica* but for Hypatia.

Hypatia (370–415 A.D.). Hypatia was the daughter and pupil of Theon of Alexandria. She was the first woman to lecture and write on mathematics. She wrote commentaries on Ptolemy's *Almagest* and on the *Conics* of Apollonius. She also wrote a commentary on the first six books of Diophantus, and according to Heath (1981), that is the reason these six were not lost. All subsequent manuscripts seem to be derived from her commentary. Hypatia became the head of the Neo-Platonic School in Alexandria. Because she was a pagan, her visibility and influence threatened the Christians. She was murdered, probably at the instigation of St. Cyril. The *Chronicle* of John, Bishop of Nikiû (Charles, 1916), says the following:

And in those days there appeared in Alexandria a female philosopher, a pagan named Hypatia, and she was devoted at all times to magic, astrolabes and instruments of music, and she beguiled many people through Satanic wiles. And the governor of the city honored her exceedingly; for she had beguiled him through her magic. And he ceased attending church as had been his custom.... And the Christians mustered all together and went and marched in wrath to the synagogues of the Jews and took possession of them, and purified them and converted them into churches... And as for the Jewish assassins they expelled them from the city, and pillaged all their possessions and drove them forth wholly despoiled.... And thereafter a multitude of believers in God arose under the guidance of Peter the magistrate—now this Peter was a perfect believer in all respects in Jesus Christ—and they proceeded to seek for the pagan woman who had beguiled the people of the city and the prefect through her enchantments. And when they learnt the place where she was, they proceeded to her and found her seated on a chair; and having made her descend they dragged her along till they brought her to the great church, named Caesarion. Now this was in the days of the fast. And they tore off her clothing and dragged her through the streets of the city till she died. And they carried her to a place named Cinaron, and they burned her body with fire. And all the people surrounded the patriarch Cyril and named him 'the new Theophilus'; for he had destroyed the last remains of idolatry in the city.

Many consider Hypatia to be the last mathematician of note in the ancient world. In 529 A.D., Emperor Justinian closed the Platonic School in Athens in an effort to eradicate paganism. In 641 A.D., the Caliph Omar ordered what remained of the library at Alexandria to be destroyed. The scrolls were burned in the public baths of the city. It took six months to destroy them all.

According to Tannery, as cited in Heath (1964), all existing Greek manuscripts by Diophantus are derived from a single copy of Hypatia's commentary. This original was destroyed when the Crusaders sacked Constantinople in 1204, but it was copied by Michael Psellus.

Michael Psellus (1017–1078). Psellus was head of the imperial secretariat in Constantinople, and later head of the philosophical faculty of the University. He wrote on diverse topics and helped to preserve ancient knowledge. What we know of Iamblichus comes through him. He wrote on demonology, marvels, alchemy, and astrology, too. Unfortunately, he also arranged for his close friend and former student to become Emperor Constantine X. Norwich writes (Norwich, 1997) about Psellus, "His burden of guilt must be heavy indeed; for there is no Emperor... whose accession had more disastrous consequences... [and] rendered inevitable the first of the two great catastrophes that were ultimately to bring about the downfall of Byzantium."

12.2.1. Congruent Number Problem

Which positive integers n are the area of a right triangle with rational sides? We will say that n is a CONGRUENT NUMBER if there are rational numbers (a, b, c) such that

$$n = \frac{ab}{2} \quad \text{and} \quad a^2 + b^2 = c^2.$$

This use of the word congruent is totally different from that in Interlude 4. It is just a historical accident that the same word is used. The Pythagorean triple $3^2 + 4^2 = 5^2$ tells us that $6 = 3 \cdot 4/2$ is a congruent number. This problem appears in an anonymous Arab manuscript written before 972, according to Dickson (1999). Fermat showed that $n = 1, 2$, and 3 are not congruent numbers; no solution exists. On the other hand, 5 is a congruent number because

$$\left(\frac{3}{2}\right)^2 + \left(\frac{20}{3}\right)^2 = \left(\frac{41}{6}\right)^2 \qquad 5 = \frac{3}{2} \cdot \frac{20}{3} \cdot \frac{1}{2}.$$

As you might guess from this last example, it is not easy to tell whether or not an integer n is a congruent number. But we may as well assume that n is square free because n is congruent if and only if nd^2 is.

In the tenth century, Muhammad ibn Alhocain realized that n is a congruent number exactly when there is a rational number t such that $t - n, t$, and $t + n$ are all squares of rational numbers. Like many theorems, this is easier to prove than it was to discover in the first place. If $a^2 + b^2 = c^2$ and $ab/2 = n$, then $t = c^2/4$, $t - n = (a - b)^2/4$, and $t + n = (a + b)^2/4$ are all squares. On the other hand, if $t - n, t$, and $t + n$ are all squares, then

$$a = \sqrt{t+n} + \sqrt{t-n}, \; b = \sqrt{t+n} - \sqrt{t-n}, \; c = 2\sqrt{t}$$

are all rational numbers, $a^2 + b^2 = c^2$, and $ab/2 = n$.

Leonardo of Pisa, also known as Fibonacci, discussed congruent numbers in his book *Liber Quadratorum* of 1225.

Leonardo of Pisa (1170–1240). He used the name "Bigollo" for himself, from the Tuscan slang *bighellone*, variously translated as loafer or absentminded. Perhaps this is a joke about his abstract mathematical work. As a young man, he traveled around the Mediterranean with his father, the Secretary of Pisa, where he learned mathematics from Arabic sources far in advance of what was known in the West. Problems in the *Liber Quadratorum* are very similar to those discussed by Diophantus. They may have come from Arabic writings from the tenth century based on work by Diophantus, manuscripts which have not survived. Another of Leonardo's works, *Liber Abbaci*, contains the following puzzle. Seven old women are traveling to Rome, and each has seven mules. On each mule there are seven sacks, in each sack there are seven loaves of bread, in each loaf there are seven knives, and each knife has seven sheaths. Find the total of all of them. You can compare this to a problem that is 3,000 years older, Exercise 1.2.11.

He held a disputation on congruent numbers at the court of Holy Roman Emperor Frederick II. Frederick spoke six languages and was knowledgeable in mathematics, science, medicine, and philosophy. He was known as *Stupor Mundi*, the wonder of the world. Frederick II had an eclectic mixture of Islamic, Jewish, Byzantine Greek, and Norman French scholars at his court in Sicily. One was the court astrologer Michael Scot, who had studied the quadrivium at the University of Paris, where he was known (Brown, 1897) as "Michael the Mathematician." Leonardo's *Liber Abbaci* is even dedicated to Scot, but he is remembered as a magician in Boccaccio's *Decameron*, and he appears in the eighth circle of Hell in Dante's *Inferno*:

The next is Michael Scot... who could by magic artistry Against the demons' subtlest wiles prevail.

What is the connection between congruent numbers and elliptic curves? If n is a congruent number, then with t as above, $(t - n)t(t + n) = t^3 - n^2t$ is the product of three squares. So, it is the square of a rational number, and there is a rational point $(t, \sqrt{t^3 - n^2t})$ on the elliptic curve

$$y^2 = x^3 - n^2x.$$

Furthermore it cannot be any of the trivial (obvious) ones, namely $(0, 0)$ or $(\pm n, 0)$. Because t cannot equal $\pm n$ because t is a square and n is square free. And t cannot be 0 because $t + n$ is also a square.

In the example $n = 6$ with $(a, b, c) = (3, 4, 5)$, we see that $t = 25/4$, $t - 6 = 1/4$, and $t + 6 = 49/4$ are all squares. And $(t - 6)t(t + 6) = 1225/64 = (35/8)^2$, so $(25/4, 35/8)$ is a point on the curve $y^2 = x^3 - 36x$.

Exercise 12.2.1. Find three rational squares that differ by 5, by starting with the Pythagorean triple $(3/2, 20/3, 41/6)$ with area 5. Now, use this to find a nontrivial point on the curve $y^2 = x^3 - 25x$.

How does this help solve the congruent number problem? We know that if the curve $y^2 = x^3 - n^2x$ has only trivial solutions, the n is not a congruent number. And although we will not prove it, the converse is true: If $y^2 = x^3 - n^2x$ has nontrivial solutions, then n is a congruent number. In fact, the group law we defined in the previous section shows that the trivial points $P = (-n, 0)$, $Q = (0, 0)$ and $R = (n, 0)$ are all of order 2; that is, $2P = 2Q = 2R = O$. One can show by reducing the curve modulo primes p (see the next section) that these are the only points of finite order. So, our nontrivial point, if it exists, has infinite order. An integer n is a congruent number if and only if the rank of $y^2 = x^3 - n^2x$ is greater than or equal to 1.

12.2.2. The Rediscovery of Diophantus in the Renaissance

In Rome, the Renaissance began with the election of Pope Nicholas V in 1447. He used papal resources to put together a spectacular library, intended to be the greatest library since that of Alexandria. The fall of Constantinople to the Ottoman Turks in 1453 led to a great influx of refugees, and they brought with them ancient Greek manuscripts. The inventory of Bibliotheca Vaticana done after Nicholas' death in 1455 includes the first reference to a manuscript by Diophantus, as well as those by Euclid and Ptolemy. The 1484 inventory by Pope Sixtus IV included two more Diophantus manuscripts. But few mathematicians consulted them at the time; they were not allowed out on loan.

Cardano claimed that Leonardo of Pisa's *Liber Quadratorum* was also rediscovered at this time, after centuries of neglect, when a manuscript was found in the library of San Antonio di Castello.

Regiomontanus (1436–1476). Regiomontanus was the pseudonym of Johannes Müller. He was an astronomer at the University of Vienna. The papal legate to the Holy Roman Empire in Vienna, Cardinal Bessarion, was very interested in arranging the translation of ancient Greek authors. Regiomontanus traveled to Rome with Bessarion in 1461. He later reported in a letter that he had discovered in Venice a manuscript of the first six books of Diophantus' *Arithmetica* and he asked for help in locating the other seven books. He said that if he could locate a complete version, he would translate it into Latin.

No one has yet translated from the Greek into Latin the thirteen books of Diophantus, in which the very flower of arithmetic lies hidden. (Rose, 1975)

He never found the rest of the manuscript and never wrote a translation, but he was the first modern mathematician to read and lecture on Diophantus.

Rafael Bombelli (1526–1572). Bombelli was what would today be called a hydraulic engineer. During a break in the draining of the Val di Chiana marshes, Bombelli wrote a treatise on algebra at his patron's villa in Rome. He had the chance to see a manuscript of Diophantus' *Arithmetica* in the Vatican library and, together with Antonio Pazzi, set out to translate the manuscript.

[A] Greek manuscript in this science was found in the Vatican Library, composed by Diophantus. . . . This was shown to me by Antonio Maria Pazzi, public professor of mathematics in Rome. . . . We set to translating it and have already done five of the seven extant books. The rest we have not been able to finish because of our other commitments. (Rose, 1975)

The project was never finished, and the Bombelli–Pazzi translation is now lost. But Bombelli revised his own *Algebra*, including many abstract problems from Diophantus. This was a significant change from the style of "practical arithmetics" then current. Bombelli helped to raise the status of abstract algebra, and to raise awareness of the work of Diophantus.

Wilhelm Holzmann (1532–1576). Holzmann, who wrote under the pseudonym Xylander, was more of a scholar of the ancient world than a mathematician, but he made the first translation of Diophantus from Greek into Latin in 1575.

Françoise Viète (1540–1603). Viète gave up a career as a lawyer to be come tutor to the princess Catherine de Parthenay, supervising her education and remaining her friend and adviser for the rest of his life. His connections to the royal court led him into and out of favor under the various kings at the end of the Valois dynasty. He was in favor under Henry IV, serving as a cryptographer against the Spanish. He was so successful that the king of Spain complained to the pope that the French were using sorcery against him, contrary to good Christian morals. (The pope was not impressed; his own cryptographers had also broken the Spanish code.)

The most important of Viète's works is *Introduction to the Analytic Art* in 1591, quoted in the Preface on page x. This was the first work on symbolic algebra. Bombelli's *Algebra* and Holzmann's translation of Diophantus both influenced Viète. Ironically, the efforts to restore the lost analytical arts of the ancient world led to the development of a completely new form of abstract mathematics, algebra. Viète's *Zeteticorum Libri Quinque* ("Five Books of Research") explicitly parallels the work of Diophantus, treating the same problems as Diophantus did but by modern methods.

12.2.3. Sums of Two Cubes

It was Bachet who made the first accurate mathematical translation of Diophantus' *Arithmetica*.

Claude-Gaspar Bachet de Méziriac (1581–1638). Bachet was from a noble family. He wrote and translated poetry and was a member of the Académie Française. He is remembered most significantly for his translation of Diophantus' *Arithmetica*, based on the work of Bombelli and Holzmann. Bachet printed the Greek text along with a Latin translation based on Holzmann's. But he was able to correct mistakes and fill in details better than his predecessors.

His exposition was much clearer. Bachet's Diophantus, printed in 1621, became the standard edition and marks the beginning of number theory in the modern world.

In *Arithmetica*, one finds the simple problem of finding a square and a cube that differ by 2, for example $5^2 + 2 = 3^3$. This says that $P = (3, 5)$ is a point on the elliptic curve $E : y^2 = x^3 - 2$, and Bachet gave another solution to the problem, which amounts to the duplication formula for $2P$ coming from the tangent line to E at P discussed in the previous section. The elliptic curve

$$y^2 = x^3 + B$$

with the parameter $A = 0$ is sometimes called the Bachet equation because of this. It is also known as Mordell's equation. A change of variables,

$$X = \frac{36D + y}{6x}, \quad Y = \frac{36D - y}{6x}, \tag{12.1}$$

converts the Bachet equation $y^2 = x^3 - 432D^2$ with $B = -432D^2$ to

$$X^3 + Y^3 = D.$$

The inverse map going back is

$$x = \frac{12D}{X + Y}, \quad y = 36D\frac{X - Y}{X + Y}. \tag{12.2}$$

So, we say that these two curves are BIRATIONAL. The existence of rational points on this curve answers the very old question of whether an integer D can be written as a sum of two rational cubes. G. H. Hardy tells the following story (Hardy, 1978):

I remember going to see [Ramanujan] once when he was lying ill in Putney. I had ridden in taxi-cab No. 1729, and remarked that the number seemed to me a rather dull one, and that I hoped that it was not an unfavorable omen. 'No,' he replied, 'it is a very interesting number; it is the smallest number expressible as a sum of two cubes in two different ways'.

In the language of elliptic curves, Ramanujan was observing that $D = 1729$ was the smallest number such that the curve $X^3 + Y^3 = D$ had two integer (as opposed to merely rational) points, namely (1, 12) and (9, 10).

12.2.4. Fermat's Last Theorem

In the Introduction, I mentioned Andrew Wiles' proof of Fermat's Last Theorem: For an integer $n > 2$, there are no integer solutions to

$$a^n + b^n = c^n$$

except the obvious ones, where one of a, b, or c is 0. Fermat was reading his copy of Bachet's Diophantus when he conjectured this and made his famous note that the margin was too small for his proof.

Pierre de Fermat (1601–1665). Fermat came from a middle class family; he was a lawyer and civil servant. Perhaps his passion for mathematics detracted from his professional work. According to *Dictionary of Scientific Biography* (1970–1980), "... a confidential report by the *intendant* of Languedoc to Colbert in 1664 refers to Fermat in quite deprecatory terms." Fermat was strongly influenced by the development of symbolic algebra in Viète's work. Fermat saw his work as a continuation of Viète's and, like him, tried to use the modern methods to restore lost texts of the ancient Greeks, including Archimedes and others, as well as Diophantus.

Fermat refused to publish any of his discoveries; instead, he wrote letters to colleagues. His *Observations on Diophantus* was only published after his death by his son, as part of a second edition of Bachet's Diophantus. Because of this, he was somewhat isolated later in life from the current trends in mathematics, and his reputation declined until a revival of interest occurred in the nineteenth century.

The "Fermat curves" are not themselves elliptic curves except in a few special cases. For $n = 3$, an integer solution (a, b, c) to $a^3 + b^3 = c^3$ is the same as a rational number point $(x = a/c, y = b/c)$ on the elliptic curve $x^3 + y^3 = 1$. It was Euler who proved that there are no integer solutions besides the trivial ones $(1, 0)$ and $(0, 1)$.

For a general exponent, the situation is more complicated. In Section 7.3, we mentioned that it is enough to prove Fermat's Last Theorem for an odd prime exponent q. It turns out that a solution $a^q + b^q = c^q$ in integers would mean that the elliptic curve

$$y^2 = x(x - a^q)(x + b^q)$$

would have some very exotic properties; in fact, Wiles showed that such curves can't exist!

12.2.5. Euler's Conjecture

One of the problems in Diophantus' *Arithmetica*, restated in modern notation, is to find a solution to the equation

$$x^4 + y^4 + z^4 = t^2.$$

Fermat noted the following in his margin:

Why does Diophantus not ask for a sum of *two* biquadrates [i.e., fourth powers] to be a square? This is, indeed, impossible. ...

Fermat went on to show more, that there is no rational number solution to

$$x^4 + y^4 = t^4,$$

and because any fourth power would automatically be a square ($t^4 = (t^2)^2$), this proves his above claim. (This is, of course, the $n = 4$ case of Fermat's Last Theorem. The fact that Fermat wrote up the special case of $n = 4$ shows that he eventually realized he did not have a valid proof of the general theorem.)

In 1769, Euler made a conjecture that was a generalization of this. He claimed that

it is certain that it is impossible to exhibit three biquadrates [i.e., fourth powers] whose sum is a biquadrate. ... In the same manner it would seem to be impossible to exhibit four fifth powers whose sum is a fifth power, and similarly for higher powers.

In fact, Euler was wrong this time. In 1966, Lander and Parkin gave a counterexample for the case of fifth powers, found by computer search:

$$27^5 + 84^5 + 110^5 + 133^5 = 144^5.$$

Surprisingly, no counterexample was found for the case of fourth powers. But in 1988, Elkies was able to produce infinitely many counterexamples. An integer solution to

$$A^4 + B^4 + C^4 = D^4$$

is the same as a rational point ($x = A/D$, $y = B/D$, $z = C/D$) on the surface

$$x^4 + y^4 + z^4 = 1.$$

Elkies realized that this surface was a family of elliptic curves with a varying parameter. He found the simplest curve in the family that could possibly have a rational point, and a search on that curve was successful. The simplest counterexample is

$$95800^4 + +217519^4 + 414560^4 = 422481^4.$$

12.3. Solutions Modulo a Prime

As with the Pell equation, we can get more information by looking at the equation of an elliptic curve modulo a prime number p. For convenience, we will write the equation of the curve as

$$E : y^2 = x^3 + Ax + B = f(x).$$

There is a trivial upper bound on the number of points on the curve modulo p. Namely, for each of the p possible choices of x-coordinate, $f(x)$ has, at most, two square roots. So there are, at most, two choices of y-coordinate that satisfy $y^2 = f(x)$. This gives, at most, $2p$ points, or $2p + 1$ when we include the point at infinity, which we always do. We will define

$$N_p = \#\left\{(x, p) \bmod p \,|\, y^2 \equiv f(x) \bmod p\right\} + 1$$

as the number of points on the curve reduced modulo p, including the point at infinity (the $+1$).

If fact, one can do better than the trivial bound. Hasse proved that

Theorem. *For a prime number p, define $a(p)$ as $p + 1 - N_p$. Then, $|a(p)| < 2\sqrt{p}$. In other words, the number of points is bounded between $p - 2\sqrt{p} + 1$ and $p + 2\sqrt{p} + 1$.*

This is the first example of a general phenomenon for curves and is, in fact, a version of the Riemann Hypothesis.

The preceding trivial bound leads to a very simple approach to actually computing the number of points. We simply list all possible x-coordinates, compute $f(x)$, and see whether or not it is a square. This last step can be done by looking in a table of the squares of all the classes modulo p, or by using Quadratic Reciprocity. We get one, two, or zero points according to whether or not $f(x) \equiv$ zero, a nonzero square, or a nonzero nonsquare, respectively. At the end, we have to remember the point at infinity.

For example, suppose that we want to count the number of points $y^2 \equiv x^3 + 5 \bmod 7$. From Table 12.1, we see that the only nonzero squares modulo 7 are $1 \equiv 1^2 \equiv 6^2$, $4 \equiv 2^2 \equiv 5^2$, and $2 \equiv 3^2 \equiv 4^2$. We see that $x^3 + 5$ is never zero, and that it is a nonzero square if $x \equiv 3$, 5, or 6. So, there are seven points modulo 7, including the one at infinity. Because $8 - \sqrt{7} < 7 < 8 + \sqrt{7}$, this answer satisfies Hasse's bound, a good way to check that the arithmetic is correct.

Table 12.1. *Finding Solutions to* $y^2 \equiv x^3 + 5 \bmod 7$

x	x^2	x^3	$x^3 + 5$	Solutions
0	0	0	5	None
1	1	1	6	None
2	4	1	6	None
3	2	6	4	$(3, \pm 2)$
$4 \equiv -3$	2	1	6	None
$5 \equiv -2$	4	6	4	$(5, \pm 2)$
$6 \equiv -1$	1	6	4	$(6, \pm 2)$

Exercise 12.3.1. Count the points on the curve $y^2 \equiv x^3 + 5 \bmod p$ for $p = 11, 13,$ and 17. In doing the arithmetic, it helps to remember that $p - 1 \equiv -1$, $p - 2 \equiv -2$, and so forth. Make sure your answers satisfy Hasse's bound.

This also gives us some clue as to why there should be about $p + 1$ points on the curve. We need the notation of the LEGENDRE SYMBOL for an odd prime p:

$$\left(\frac{a}{p}\right) = \begin{cases} +1, & \text{if } a \equiv \text{square} \bmod p, \\ -1, & \text{if } a \equiv \text{nonsquare} \bmod p, \\ 0, & \text{if } a \equiv 0 \bmod p. \end{cases} \tag{12.3}$$

We saw an example of the Legendre symbol in Chapter 11, Section 11.4; there, $\chi_5(n)$ is actually $\left(\frac{n}{5}\right)$. For each choice of x-coordinate modulo p, there are $1 + \left(\frac{f(x)}{p}\right)$ choices of y-coordinate to give a point (x, y) on the curve. So,

$$N_p = 1 + \sum_{x \bmod p} \left\{ 1 + \left(\frac{f(x)}{p}\right) \right\}$$

$$= 1 + p + \sum_{x \bmod p} \left(\frac{f(x)}{p}\right).$$

Earlier, we considered the example of the curve $y^2 \equiv x^3 + 5$ modulo 7. Corresponding to x-values $\{0, 1, 2, 3, 4, 5, 6\}$, the values of $x^3 + 5$ were $\{5, 6, 6, 4, 6, 4, 4\}$. The nonzero square modulo 7 are 1, 2 and 4; in particular, 5 and 6 are not congruent to squares. So, in this case,

$$N_7 = 1 + 7 + (-1 - 1 - 1 + 1 - 1 + 1 + 1) = 7.$$

We expect that the mapping $x \to f(x)$ behaves more or less randomly, so we should get a $+1$ about as often as a -1. So, the contribution of the sum should be close to 0.

In fact, Bachet's curve $y^2 = x^3 + B$ is so very nice, that we can do better.

Theorem. *If $p \equiv 2$ mod 3, then $a(p) = 0$; that is,*

$$N_p = \#\{(x, y) \bmod p | y^2 \equiv x^3 + B \bmod p\} + 1 = p + 1. \qquad (12.4)$$

Notice that the answer does not depend on what B is.

Proof. Because $p \equiv 2$ mod 3, we know that 3 divides $p - 2$ and, thus, cannot divide $p - 1$. (The next multiple of 3 is $p + 1$.) Because 3 is prime, this means that the greatest common divisor of 3 and $p - 1$ is just 1. So, there are integers a and b such that $3a + b(p - 1) = 1$.

This implies that there is no element of order 3 modulo p. For if $x^3 \equiv 1$ mod p, then because $x^{p-1} \equiv 1$ mod p (see Interlude 4), we get that

$$x \equiv x^1 \equiv x^{3a+b(p-1)} \equiv (x^3)^a \cdot (x^{p-1})^b \equiv 1 \bmod p.$$

From this, we deduce that the mapping $x \to x^3$ is injective (one to one). That is, if $x_1^3 \equiv x_2^3$ mod p, then $(x_1/x_2)^3 \equiv 1$ mod p. So, $x_1/x_2 \equiv 1$ mod p, and thus, $x_1 \equiv x_2$ mod p. Because the set of equivalence class modulo p is finite, the mapping $x \to x^3$ must be surjective (onto) as well. This is sometimes called the pigeonhole principle: If you have p pigeons and p pigeonholes and no pigeonhole has more than 1 pigeon in it, then every pigeonhole has at least one pigeon in it. So, the mapping $x \to x^3$ is a bijection. The mapping $z \to z + B$ is also a bijection. The inverse map is easy to write: $z \to z - B$. So, the composition of mappings $x \to f(x) = x^3 + B$ is also a bijection. Because the numbers x mod p consist of 0, $(p - 1)/2$ squares, and $(p - 1)/2$ nonsquares, the same is true for the numbers $f(x)$ mod p: $f(x) \equiv 0$ mod p once; $f(x) \equiv$ a square $(p - 1)/2$ times; and $f(x) \equiv$ a nonsquare $(p - 1)/2$ times. Including the point at infinity, we get

$$1 + 2 \cdot \frac{p-1}{2} + 0 \cdot \frac{p-1}{2} + 1 = p + 1$$

points. □

Except for $p = 3$, all other primes are 1 mod 3. In this case, there is also a formula, surprising and elegant. The proof is beyond the scope of this book; you should look at Ireland and Rosen (1990). In fact, to even state the formula in simple language, we will make the further assumption that $B = -432D^2$ for some D. According to (12.1) and (12.2), this curve E is birationally equivalent to $X^3 + Y^3 = D$.

For primes $p \equiv 1 \mod 3$, it turns out that there are always integers L and M such that

$$4p = L^2 + 27M^2.$$

This is a beautiful theorem in its own right. The integer L can't be divisible by 3; otherwise, p is divisible by 9. So, L and $-L$ are $\equiv \pm 1 \mod 3$, and L is uniquely determined if we choose the sign ± 1 such that $L \equiv 1 \mod 3$.

Theorem. *For $p \equiv 1 \mod 3$ and $p > 13$, choose a between $-p/2$ and $p/2$ such that*

$$D^{(p-1)/3} L \equiv a \mod p.$$

Then,

$$\#\{(X, Y) \mod p \,|\, X^3 + Y^3 \equiv D \mod p\} + 1 = p + a - 1.$$

For example, take $p = 19$ and $D = 41$, then $4 \cdot 19 = 7^2 + 27 \cdot 1^2$, and $7 \equiv 1 \mod 3$. Furthermore, $41^6 \cdot 7 \equiv 11 \mod 19$. But $11 > 19/2$, so we instead choose the representative $a = -8 \equiv 11 \mod 19$. The theorem says that there are $19 - 8 - 1 = 10$ solutions to $X^3 + Y^3 \equiv 41 \mod 19$, including the point at infinity. And this is the right answer. The solutions are $(5, 5), (16, 5), (17, 5)$, $(5, 16), (16, 16), (17, 16), (5, 17), (16, 17), (17, 17)$, and the point at infinity.

In the special case where D is congruent to a cube modulo p, $D \equiv C^3 \mod p$, then $D^{(p-1)/3} \equiv C^{p-1} \equiv 1 \mod p$. Then, we have $L \equiv a \mod p$, but the equation $4p = L^2 + 27M^2$ forces L to be between $-2\sqrt{p}$ and $2\sqrt{p}$. So, $L = a$. The fact that, in this case, the number of solutions is just $p + L - 1$ was known to Gauss.

Exercise 12.3.2. Use the theorem to find the number of points on $X^3 + Y^3 = 5$ modulo 31. To write $4 \cdot 31 = 124$ as $L^2 + 27M^2$, compute $124 - 27 \cdot 1^2$, $124 - 27 \cdot 2^2$, $124 - 27 \cdot 3^2$ until you find a square. To compute 5^{10} modulo 31, write 10 as a sum of powers of 2: $10 = 8 + 2$. Thus, you need to compute $5^2, 5^4 = (5^2)^2, 5^8$, and $5^8 \cdot 5^2$ modulo 31.

The elliptic curve $E : y^2 = x^3 - 432 \cdot D^2$ does not have the same number of points, because the map (12.1) is not defined when $x = 0$. But the discrepancy is not too bad, because for $p \equiv 1 \mod 3$, p is a square modulo 3 because $1^2 \equiv 2^2 \equiv 1 \mod 3$. According to Quadratic Reciprocity, -3 is always a square modulo p, so $-432 \cdot D^2 = -3 \cdot 12^2 \cdot D^2$ is always a square modulo

p. Thus, with two choices of the square root, $x = 0$ always contributes two points to the curve. So, E has two extra points modulo p.

Theorem. *Suppose that* $p \equiv 1 \bmod 3$ *and that* $p > 13$. *Choose* L *and the representative* $a \equiv D^{(p-1)/3} L \bmod p$ *as above. Then,* $a(p) = -a$. *That is,*

$$\#\{(x, y) \bmod p | y^2 \equiv x^3 - 432 \cdot D^2 \bmod p\} + 1 =$$
$$N_p = p + a + 1. \quad (12.5)$$

12.4. Birch and Swinnerton-Dyer

Because this is a book about analytic number theory, all the discussion of elliptic curves so far has been leading up to this section. We will construct a new function, an Euler product over all the primes p. The Euler factor at each prime should have something to do with the number of points on the curve reduced modulo p, just as it did in the case of Pell's equation. So, we define

$$L(s, E) = \prod_p (1 - a(p)p^{-s} + p^{1-2s})^{-1},$$

where, as before

$$a(p) = p + 1 - N_p$$
$$= p - \#\{(x, p) \bmod p | y^2 \equiv f(x) \bmod p\}.$$

This looks very complicated compared to the Riemann zeta function or Dirichlet L-function, but it does what we want. Because at $s = 1$,

$$(1 - a(p)p^{-1} + p^{-1})^{-1} = \frac{p}{p + 1 - a(p)} = \frac{p}{N_p}.$$

Just as in the Pell equation examples in the previous chapter, the Euler factor at p, when evaluated at $s = 1$, encodes the number of solutions modulo p.

I have misled you slightly. The discriminant of the cubic polynomial $x^3 + Ax + B$ is the number $\Delta = -16(4A^3 + 27B^2)$; for the finitely many primes p dividing Δ, the Euler factor at p has a different definition. The infinite product means that there is also a corresponding infinite series

$$L(s, E) = \sum_{n=1}^{\infty} \frac{a(n)}{n^s},$$

and from Hasse's theorem in the previous section, one can show that the Euler product, and the series, converge for $\mathrm{Re}(s) > 3/2$. According to the work of Andrew Wiles, we know that $L(s, E)$ has an analytic continuation, like that of $\zeta(s)$ proved in Section 8.4. And there is a functional equation, like the one for $\zeta(s)$ in Chapter 9, but this time the symmetry is $s \to 2 - s$. The zeros of this function are conjectured to lie on the centerline $\mathrm{Re}(s) = 1$.

In the 1960s, Birch and Swinnerton-Dyer began a series of numerical experiments on elliptic curves. They were looking for phenomena similar to what we saw for Pell's equation in the previous chapter. Roughly speaking, if the curve E has "lots" of rational points, then the same should be true of the curve reduced modulo p for each p. More precisely, the larger the rank r of the group of rational points is, the larger the ratios N_p/p should be. They found, in fact, that for large values of x,

$$\prod_{p<x} \frac{N_p}{p} \approx \log(x)^r.$$

But there is much oscillation as well, as you can see in Figure 12.2. The three graphs in the left column represent data for the congruent number curves $y^2 =$

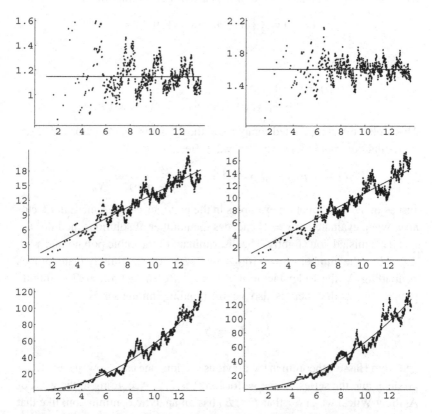

Figure 12.2. Numerical evidence for the Birch Swinnerton-Dyer conjecture.

$x^3 - n^2 x$, with $n = 1, 3$, and 17 and rank $r = 0, 1$, and 2, respectively. The three graphs in the right column represent the data from the curves $X^3 + Y^3 = D$, with $D = 1, 6$ and 19. These curves have rank 0, 1, and 2, respectively. You can see in the first row that $\prod_{p<x} N_p / p$ seems to be bounded, whereas in the second row it grows linearly in $\log(x)$; there is quadratic growth in the third row. (Compare the scales on the vertical axis.)

These examples may remind you of Mertens' Formula (7.10); they should. The proof of (7.10) relied on (7.1), which says that $\zeta(s)$ has a simple pole at $s = 1$. Birch and Swinnerton-Dyer made a similar connection between their experimental data and the function $L(s, E)$. As with Mertens' Formula, we introduce an ϵ such that $x = \exp(1/\epsilon)$ and $s = 1 + \epsilon$. So, $\log(x) = 1/(s - 1)$. In these variables, Birch and Swinnerton-Dyer's numerical evidence is that

$$\prod_{p<x} \frac{p}{N_p} \approx \log(x)^{-r} = (s - 1)^r,$$

where we have inverted because p/N_p is the value of the Euler factor for p at $s = 1$. This lead them to make the

Conjecture (Birch Swinnerton-Dyer). *If the elliptic curve E has rank r, then there is a zero at $s = 1$ of order exactly r; that is,*

$$L(s, E) = c(s - 1)^r + O(s - 1)^{r+1},$$

for some nonzero c.

There are more-precise, complicated versions that describe the constant c, extending the analogy with the Analytic Class Number Formula (11.6). The Birch Swinnerton-Dyer conjecture is now known to be true in the cases when $L(1, E) \neq 0$, or $L(1, E) = 0$ but $L'(1, E) \neq 0$. In these cases, E has rank 0 or 1, respectively, as predicted. This follows from the work of a quite a few mathematicians, including that of Wiles and others on Fermat's Last Theorem, and that of Gross and Zagier, discussed at the end of Chapter 13. The general case is still an open problem.

The Birch Swinnerton-Dyer conjecture has implications for the various Diophantine equations considered in Section 12.2. The L-function has a symmetry under $s \to 2 - s$ as either an even or an odd function, depending on the parameters A and B in a way too complicated to describe here. But if it is an odd function, it must be 0 at the center of symmetry $s = 1$. Birch Swinnerton-Dyer tells us that the curve must, therefore, have rank ≥ 1; there are points of infinite order. From this, one can show that any $n \equiv 5, 6$, or $7 \bmod 8$ is a congruent number, as discussed in Section 12.2.1. And any prime $D \equiv 4, 7$, or 8 mod 9 must be the sum of two cubes of rational numbers, as discussed in Section 12.2.3.

Exercise 12.4.1. Use *Mathematica* to reproduce the numerical experiments of Birch and Swinnerton-Dyer for the curves $E_5 : y^2 = x^3 - 432 \cdot 5^2$ and $E_7 : y^2 - 432 \cdot 7^2$. Use (12.4) and (12.5) and compute $\prod_{p<x} N_p/p$ for various values of x. Does the product grow like $\log(x)^0$, like $\log(x)^1$, or like some higher power of $\log(x)$?

Chapter 13

Analytic Theory of Algebraic Numbers

The Pell equation has some natural generalizations. We can replace the equation $x^2 - Ny^2 = 1$ with

$$x^2 - Ny^2 = \text{a prime}.$$

This is an interesting problem even for N negative, where the original Pell equation has only trivial solutions. Fermat was the first to consider these problems, showing that for an odd prime p,

$$x^2 + y^2 = p \quad \text{has a solution} \Leftrightarrow p \equiv 1 \bmod 4.$$

This is a beautiful theorem in the Pythagorean tradition, for it says that for primes $p \equiv 1 \bmod 4$, there are integers x and y such that

$$(x^2 - y^2)^2 + (2xy)^2 = (x^2 + y^2)^2 = p^2.$$

In other words, p is the hypotenuse of a right triangle with sides $x^2 - y^2$, $2xy$, and p. Examples are 3, 4, 5; and 5, 12, 13; and 8, 15, 17; and 20, 21, 29; and so on. Fermat called this "the fundamental theorem on right triangles" in a letter to Frenicle in 1641. Fermat also showed that

$$x^2 + 2y^2 = p \quad \text{has a solution} \Leftrightarrow p \equiv 1 \text{ or } 3 \bmod 8.$$

For $p \neq 3$, Fermat showed that

$$x^2 + 3y^2 = p \quad \text{has a solution} \Leftrightarrow p \equiv 1 \bmod 3.$$

Similarly, the statement from Section 12.3, that there exist integers L and M such that

$$L^2 + 27M^2 = 4p$$

if and only if $p \equiv 1 \bmod 3$, is part of this theory.

The expressions on the left side above are more examples of binary quadratic forms mentioned in Section 11.4. In general, a binary quadratic form is a function

$$F(x, y) = ax^2 + bxy + cy^2$$

with integers a, b, c that are relatively prime. The discriminant d is defined as $b^2 - 4ac$. Sometimes it is easier to omit the variables and simply write

$$F = \{a, b, c\}.$$

In Chapter 11, we considered some examples with $d > 0$ in connection with Pell's equation. In this chapter, we will give a brief sketch of the theory for $d < 0$. The important algebraic invariant is the class number, introduced by Gauss. With the restriction $d < 0$, we will be able to prove that the values of Dirichlet L-functions at $s = 1$ are related to the class number h. Finally, we will see how the location of the zeros of the Dirichlet L-function is related to the size of the class number.

The title for this final chapter is stolen from a talk by Harold Stark (Stark, 1975). The point is to emphasize the interplay between the algebraic and analytic sides of number theory.

13.1. Binary Quadratic Forms

The question of whether a prime number p can be written, for example, as $2x^2 + 3y^2$ is, obviously, the same question as whether it can be written as $3x^2 + 2y^2$. All we have done is switch the roles of x and y. Somewhat less obviously, it is the same as considering $2x^2 + 4xy + 5y^2$. The reason is that this is just $2(x + y)^2 + 3y^2$. We have just changed the variables (x, y) to $(x + y, y)$, and we can just as easily change them back: $2(x - y)^2 + 4(x - y)y + 5y^2 = 2x^2 + 3y^2$. A solution (x, y) to $2x^2 + 3y^2 = p$ is equivalent to a solution $(x' = x - y, y' = y)$ to $2x'^2 + 4x'y' + 5y'^2 = p$. To avoid this kind of redundancy, Gauss invented an equivalence relation for forms. The forms

$$F = \{a, b, c\}, \qquad F' = \{a', b', c'\}$$

are equivalent, $F \sim F'$, if there is a change of variables that converts F to F'. Specifically, Gauss defined F as equivalent to F' if there is a 2 by 2 integer matrix M with integer inverse such that

$$F'(x, y) = F((x, y)M),$$

where $(x, y)M$ denotes matrix multiplication:

$$(x, y) \begin{bmatrix} r & s \\ t & u \end{bmatrix} \quad \text{is defined as} \quad (rx + ty, sx + uy).$$

You may remember from linear algebra that the inverse of

$$M = \begin{bmatrix} r & s \\ t & u \end{bmatrix} \quad \text{is} \quad \frac{1}{ru - st} \begin{bmatrix} u & -s \\ -t & r \end{bmatrix}.$$

The inverse is an integer matrix when $1/(ru - st)$ is an integer; so, the determinant $ru - st$ must be ± 1. It will simplify things if, in our equivalence relation $\sim$, we allow only matrices M with determinant $+1$. The set of all integer matrices with determinant 1 forms another example of a group. In this example, the group operation, matrix multiplication, is not commutative in general.

The change of variables that makes $2x^2 + 3y^2 \sim 3x^2 + 2y^2$ corresponds to

$$M = \begin{bmatrix} 0 & 1 \\ -1 & 0 \end{bmatrix}, \quad \text{which changes} \quad (x, y) \to (-y, x).$$

The change of variables that makes $2x^2 + 3y^2 \sim 2x^2 + 4xy + 5y^2$ corresponds to

$$M = \begin{bmatrix} 1 & 0 \\ 1 & 1 \end{bmatrix}, \quad \text{which changes} \quad (x, y) \to (x + y, y).$$

Exercise 13.1.1. Check that $\sim$ really is an equivalence relation. What matrix makes $F \sim F$? If M makes $F \sim G$, what matrix makes $G \sim F$? How does the transitive property work? By the way, don't be confused that we have recycled the symbol $\sim$ to mean equivalent instead of asymptotic. We won't need to refer to asymptotics for a while.

The whole point of the equivalence relation is that equivalent forms take on the same values; they have the same range as functions. We will say that F REPRESENTS an integer n if n is in the range of F, that is, if there exists integers x and y such that $n = F(x, y)$. Multiplication shows that

$$4a(ax^2 + bxy + cy^2) = (2ax + by)^2 + (4ac - b^2)y^2.$$

If $d = b^2 - 4ac < 0$, the expression on the right is always positive. This means that $ax^2 + bxy + cy^2$ always has the same sign as the constant a. In other words, a form with a negative discriminant represents only positive numbers or only negative numbers. There is, obviously, a close connection between the cases; the range of values of $\{a, b, c\}$ is the negative of the range of values of $\{-a, -b, -c\}$. For this reason, we will now consider only forms that represent positive integers. Such forms are called POSITIVE DEFINITE; they have $a > 0$. Because $\{a, b, c\}$ is equivalent to $\{c, -b, a\}$ according to $(x, y) \rightarrow (-y, x)$, this latter form also represents only positive integers, and so its first coefficient c is also positive.

In matrix language,

$$F(x, y) = \begin{bmatrix} x & y \end{bmatrix} \begin{bmatrix} a & b/2 \\ b/2 & c \end{bmatrix} \begin{bmatrix} x \\ y \end{bmatrix},$$

and if $F' \sim F$ via a matrix M, then

$$F'(x, y) = \begin{bmatrix} x & y \end{bmatrix} M \begin{bmatrix} a & b/2 \\ b/2 & c \end{bmatrix} {}^{tr}M \begin{bmatrix} x \\ y \end{bmatrix}$$

$$= \begin{bmatrix} x & y \end{bmatrix} M \begin{bmatrix} a & b/2 \\ b/2 & c \end{bmatrix} {}^{tr}(\begin{bmatrix} x & y \end{bmatrix} M),$$

where ${}^{tr}M$ denotes the transposed matrix.

Exercise 13.1.2. Show that equivalent forms have the same discriminant d. (Hint: How does the discriminant of F relate to the determinant of the matrix above?)

As an aside, we remark that Gauss also proved that the equivalence classes of binary quadratic forms for a fixed discriminant also form a group. The composition law is a little too complicated to explain here (see Cox, 1989).

Exercise 13.1.3. Show that the forms

$$F(x, y) = 2x^2 + 3y^2 \quad \text{and} \quad G(x, y) = x^2 + 6y^2$$

both have discriminant -24. Show that $G(x, y)$ represents 1, that $F(x, y)$ does not, and therefore F cannot be equivalent to G.

Which integers d can occur as the discriminant of a binary quadratic form? By reducing $d = b^2 - 4ac$ modulo 4, we see that d must be 0 or 1 mod 4. This necessary condition is also sufficient. If $d \equiv 0 \bmod 4$, then $x^2 - d/4y^2$ has discriminant d. On the other hand, if $d \equiv 1 \bmod 4$, then $x^2 + xy + (1 - d)/4y^2$ has discriminant d. We saw these two forms in Chapter 11; they are called the PRINCIPAL FORM in each case.

How many binary quadratic forms can have discriminant d? Because the change of variables does not change the discriminant, and because there are infinitely many matrices to choose from, the answer is infinite. But this was the wrong question, because we are really only interested in equivalence classes of forms. We should ask how many equivalence classes there are with discriminant d. In fact, there are only finitely many, and we define h, the CLASS NUMBER, as the number of equivalence classes.

Theorem. *For each $d < 0$, the class number h is finite. In fact, every form is equivalent to a form $\{a, b, c\}$ with*

$$|b| \leq a \leq c.$$

Proof. The proof of the theorem consists of showing that if the inequality fails to hold, we can find an equivalent form that reduces the sum of the first and last coefficients. This process can be repeated only a finite number of times because there are only finitely many positive integers less than $a + c$.

So, let $\text{sgn}(b) = \pm 1$ be the sign of b; then, $\text{sgn}(b)b = |b|$. If $a < |b|$, the matrix

$$\begin{bmatrix} 1 & 0 \\ -\text{sgn}(b) & 1 \end{bmatrix} \quad \text{changes } (x, y) \text{ to } (x - \text{sgn}(b)y, y).$$

The corresponding form is

$$a(x - \text{sgn}(b)y)^2 + b(x - \text{sgn}(b)y)y + cy^2 =$$
$$ax^2 + (b - 2\text{sgn}(b)a)xy + (a + c - |b|)y^2.$$

We have that $a + (a + c - |b|) < a + c$, because $a < |b|$.

Exercise 13.1.4. Show that in the case of $c < |b|$, the matrix

$$\begin{bmatrix} 1 & -\text{sgn}(b) \\ 0 & 1 \end{bmatrix}$$

similarly reduces the sum of the first and last coefficients.

Eventually, both a and c are $\geq |b|$, and the matrix $\begin{bmatrix} 0 & 1 \\ -1 & 0 \end{bmatrix}$ interchanges a and c if necessary. This proves the inequality claimed above.

Next, we show that there are only finitely many such triples with discriminant d. The inequalities for a, $|b|$, and c imply that

$$3a^2 = 4a^2 - a^2 \leq 4ac - b^2 = -d = |d|.$$

This means that $a \leq \sqrt{|d|/3}$ and $|b| \leq a \leq \sqrt{|d|/3}$. Also, as observed earlier, $b^2 - 4ac = d$ implies that $b^2 \equiv d \bmod 4$ and, thus, $b \equiv d \bmod 2$. In other words, b is odd if and only if d is. There are only finitely many choices for

a and b, and c is then completely determined by the discriminant equation: $c = (b^2 - d)/(4a)$. $\qquad\qquad\qquad\qquad\qquad\qquad\qquad\qquad\qquad\qquad$ □

The theorem not only proves that the class number is finite, it also gives an upper bound. Here is an example with $d = -35$. We have $\sqrt{|d|/3} = 3.41565\ldots$, so $|b| \le 3$ and $1 \le a \le 3$. Also, b must be odd, as d is; so, b is restricted to $-3, -1, 1,$ or 3. With $b = \pm 1$, $b^2 - d$ is 36. We only get a form when $c = (b^2 - d)/(4a)$ is an integer. The choice $a = 1$ gives rise to the forms $\{1, \pm 1, 9\}$. The choice $a = 2$ gives $c = 36/8$, which is not an integer. The choice $a = 3$ gives rise to the forms $\{3, \pm 1, 3\}$. Meanwhile, if $b = \pm 3$, then $a \ge |b|$ must be 3, and $c = 44/12$ is not an integer. The class number is less than or equal to 4.

Exercise 13.1.5. Carry out this same analysis with discriminant -23 to get a bound on the class number.

In fact, the proof above gives even more. It actually gives an algorithm for finding a representative of a class that satisfies the inequalities. For example, the form $\{33, -47, 17\}$ has discriminant -35. But $47 > 33$, so the theorem says to replace (x, y) with $(x + y, y)$, which gives $\{33, 19, 3\}$. We chose the sign "+" because b was negative. Now, $19 > 3$, so the theorem says to change (x, y) to $(x, y - x)$, which gives $\{17, 13, 3\}$. Again, $13 > 3$, so the same variable change produces $\{7, 7, 3\}$ and then $\{3, 1, 3\}$, which can be reduced no further because the inequality $1 \le 3 \le 3$ is satisfied. Notice that the sum of the first and last entry decreases at each step:

$$33 + 17 > 33 + 3 > 17 + 3 > 7 + 3 > 3 + 3.$$

Exercise 13.1.6. Carry out this algorithm with the form

$$F(x, y) = 12x^2 + 11xy + 3y^2,$$

which has discriminant -23. Also, at each step of the reduction, compute the sum of the first and last coefficients to see that it really does decrease at each step. Do the same for

$$G(x, y) = 39x^2 + 43xy + 12y^2 \quad \text{and}$$
$$H(x, y) = 93x^2 + 109xy + 32y^2,$$

which also both have discriminant -23.

Exercise 13.1.7. Write a *Mathematica* function to carry out this reduction. You can do a single stage or the complete reduction.

We now have a bound on the class number, but we want to know it exactly. The question is, can two forms with the same discriminant satisfying the inequality of the theorem be equivalent to each other? To answer this, we will need the following lemma.

Lemma (Fundamental Inequalities). *Suppose that the quadratic form* $F(x, y) = ax^2 + bxy + cy^2$ *satisfies* $|b| \leq a \leq c$. *Then, a is the minimum of F; that is, for all* $(x, y) \neq (0, 0)$,

$$F(x, y) \geq a.$$

Furthermore, ac is the minimum of products of values of F. In other words, for all pairs of lattice points (x, y) and (u, v) with (x, y) and (u, v) not colinear,

$$F(x, y)F(u, v) \geq ac.$$

Proof. It is easy to see that F actually does represent a, because $a = F(1, 0)$. Similarly, $F(1, 0)F(0, 1) = ac$. If $x \neq 0$, then

$$F(x, 0) = ax^2 \geq a. \tag{13.1}$$

Similarly, if $y \neq 0$, then

$$F(0, y) = cy^2 \geq c,$$

and $c \geq a$. In the general case, neither x nor y is 0. We see that

$$
\begin{aligned}
F(x, y) &= ax^2 + bxy + cy^2 \\
&\geq ax^2 - |b||x||y| + cy^2 \\
&\geq ax^2 - |b||x||y| + cy^2 - a(x - y)^2 \\
&= (2a - |b|)|x||y| + (c - a)y^2 \\
&\geq (2a - |b|) + (c - a) = a + c - |b| \geq c,
\end{aligned}
\tag{13.2}
$$
$$\tag{13.3}$$

because $a \geq |b|$. This proves that a is the minimum value. Suppose that we have two lattice points that are not colinear. They cannot both lie on the horizontal axis. Thus, one of the inequalities (13.2) or (13.3) must hold, and the product $F(x, y)F(u, v) \geq ac$. □

Now, we are ready to determine whether forms satisfying the usual inequalities are equivalent. Almost always they are distinct; the only ambiguity is that $\{a, b, c\} \sim \{a, -b, c\}$ if $b = 0$ (which is trivial), if $|b| = a$ or if $a = c$. We will state this more precisely.

Theorem. *Every form with discriminant d is equivalent to* exactly one *form satisfying the inequalities*

$$|b| \leq a \leq c \quad and \quad b \geq 0 \text{ if either } |b| = a \quad or \quad a = c. \tag{13.4}$$

Proof. First, observe that if $b = 0$, then $\{a, b, c\} \sim \{a, -b, c\}$ is trivial. If $a = c$, then changing (x, y) to $(-y, x)$ shows that $\{a, b, c\} \sim \{c, -b, a\} = \{a, -b, c\}$. Finally, if $|b| = a$, then we saw in the proof of the last theorem that changing (x, y) to $(x - \text{sgn}(b)y, y)$ makes

$$\{a, b, c\} \sim \{a, b - 2\text{sgn}(b)a, a + c - |b|\} = \{a, -b, c\}.$$

Together with the previous theorem this shows that every form is equivalent to at least one form satisfying the inequalities.

We need to show "at most one," that is, that the forms satisfying the inequalities are in distinct classes. Suppose now that $F = \{a, b, c\}$ and $F' = \{a', b', c'\}$ are equivalent, and that both satisfy the inequalities (13.4). Because they are equivalent, they represent the same integers and, thus, have the same minimum. The lemma says that the minimum of F is a and the minimum of F' is a', so $a' = a$. Similarly, the minimum for products of pairs are equal, and so $a'c' = ac$ and thus $c' = c$. According to the fact that $b^2 - 4ac = b'^2 - 4a'c'$, we get that $b' = \pm b$. If $b' = b$, we are done. (Why?) If $b' = -b$, one of the two is negative. Without loss of generality, it is b. Then, the inequalities imply that

$$0 < |b| < a < c.$$

In this case, the proof of (13.3) shows that we have strict inequalities, that is,

$$F(x, y) > c > a \quad \text{if neither } x \text{ nor } y \text{ is } 0. \tag{13.5}$$

Because F and F' are equivalent, there is a change of variables,

$$F'(x, y) = F(rx + ty, sx + uy).$$

Then, $a = F'(1, 0) = F(r, s)$. From (13.5), we deduce that $s = 0$ and $r = \pm 1$. Similarly, $c = F'(0, 1) = F(t, u)$ gives $t = 0$ and $u = \pm 1$. To get determinant 1, the matrix must be either $\begin{bmatrix} 1 & 0 \\ 0 & 1 \end{bmatrix}$ or $\begin{bmatrix} -1 & 0 \\ 0 & -1 \end{bmatrix}$, which means that F' actually was equal to F all along, not just equivalent. □

A form that satisfies the inequalities (13.4) is called REDUCED. The theorem says that every equivalence class contains exactly one reduced form. In our preceding example with discriminant -35, we see that the reduced forms are precisely $\{1, 1, 9\}$ and $\{3, 1, 3\}$, and thus, the class number is 2.

The theorem actually leads to an algorithm to enumerate all the reduced forms and, thus, to compute the class number. We observed above that for a reduced form $\{a, b, c\}$ of discriminant d, we have $|b| \leq \sqrt{|d|/3}$ and $b \equiv d$ mod 2. The first step of the algorithm is to list all possible b values. Next, we notice that $4ac = b^2 - d$ implies that the only possible a choices are divisors of $(b^2 - d)/4$. Furthermore, we need only consider divisors a with $|b| \le a$. Because the c value will be $(b^2 - d)/(4a)$, we will have $a \le c$ exactly when $a \le \sqrt{(b^2 + d)/4} \le c$. The second step is to list all possible a and c values for each b. For each triple a, b, c we count one or two forms according to whether or not $\{a, b, c\} \sim \{a, -b, c\}$, as determined using the second part of (13.4).

Exercise 13.1.8. Compute the class number for the discriminants -19, -20, and -23. Do the same for -424, -427, and -431.

Exercise 13.1.9. Write a *Mathematica* program to compute class numbers, based on the preceding algorithm.

Now that we can compute exactly the number of classes of forms, we return to the question of which integers are represented. A reduction of the problem will be useful. If a form F represents an integer n, then there are integers r and s such that $F(r, s) = n$. If g is an integer that divides both r and s, then we see that g^2 divides n, and $F(r/g, s/g) = n/g^2$, so F represents n/g^2. It will be easier to take out this common factor; then we say that F PROPERLY REPRESENTS an integer n if there are relatively prime integers r and s such that $F(r, s) = n$.

We know that equivalent forms represent the same integers, that is, the point of equivalence. The point of the preceding definition is that we get the following very nice converse result.

Theorem. *If a form $F(x, y)$ properly represents an integer n, then F is equivalent to a form $\{n, m, l\}$ for some m and l, that is, a form with first coefficient n.*

One might ask further if two forms that properly represent the same integer must be equivalent. This is not true. For example, any time the inequalities (13.4) allow inequivalent forms $\{a, b, c\}$ and $\{a, -b, c\}$, where $0 < |b| < a < c$, then the two inequivalent forms both properly represent a.

Proof. Suppose that r and s are relatively prime and that $F(r, s) = n$. We can find integers $-t$ and u such that $ru - ts = 1$; this says that the matrix

$$M = \begin{bmatrix} r & s \\ t & u \end{bmatrix}$$

has determinant 1. The form $G(x, y) = F((x, y)M)$ is equivalent to $F(x, y)$ by definition. But

$$G(1, 0) = F((1, 0)M) = F(r, s) = n.$$

The first coefficient of the form G is $G(1, 0)$, so $G = \{n, m, l\}$ for some m and l.   □

Because the equivalent form G of the theorem has the same discriminant d as F, we see that $m^2 - 4nl = d$, or $m^2 - d = 4nl$. This says that $m^2 \equiv d$ mod $4n$. This immediately gives a powerful necessary condition to test whether *any* form of discriminant d can properly represent an integer n; we must have d congruent to a square modulo $4n$. For example, $n = 2$ is not represented by any form of discriminant -35, because $-35 \equiv 5$ mod 8, and 5 is not a square modulo 8; $1^2 \equiv 3^2 \equiv 5^2 \equiv 7^2 \equiv 1$ mod 8. On the other hand, $n = 5$ is represented because $-35 \equiv 5 \equiv 5^2$ mod 20. There are two m that work for $n = 9$ because $-35 \equiv 1$ mod 36 and $1^2 \equiv 1 \equiv 17^2$ mod 36. So, 9 is represented at least twice.

If we have a particular form F in hand, then given any solution m to the congruence $m^2 \equiv d$ mod $4n$, we find the corresponding $l = (m^2 - d)/(4n)$. If the reduced class corresponding to $\{n, m, l\}$ is the same as the reduced class of F, then F properly represents n. But the congruence class of any one solution m modulo $4n$ contains infinitely many integers $k \equiv m$. We don't want to compute the reduced class for them all. It turns out that we don't have to; it suffices to check the integers m satisfying $0 \le m < 2n$. Here's why. We know that if F properly represents n, it is equivalent to some form

$$F \sim G = \{n, m, l\}.$$

For any integer t, we can change the variables with the matrix

$$T = \begin{bmatrix} 1 & 0 \\ t & 1 \end{bmatrix}.$$

So,

$$
\begin{aligned}
G((x, y)T) &= G(x + ty, y) \\
&= n(x + ty)^2 + m(x + ty)y + ly^2 \\
&= nx^2 + (m + 2nt)xy + (nt^2 + mt + l)y^2.
\end{aligned}
$$

Now, $k = m + 2nt \equiv m$ mod $2n$. Running this argument in reverse, we see that any form $\{n, k, *\}$ of discriminant d with $k \equiv m$ mod $2n$ is in the same class as G. We will summarize this as a

Theorem. *A form F of discriminant d properly represents an integer n if and only if F is equivalent to a form {n, m, l} where*

$$m^2 \equiv d \bmod 4n \quad and \quad 0 \le m < 2n.$$

Exercise 13.1.10. Show that 5 is not represented by any form of discriminant -23 but that 3 is. Show that 6 is represented more than one way.

The next, slightly more sophisticated question one may ask is how many different ways can a form F properly represent an integer n? We suppose, first, that two different pairs, (r, s) and (r', s'), define the same form $\{n, m, l\}$ with $0 \le m < 2n$ through the process described above. So, we have two matrices,

$$M = \begin{bmatrix} r & s \\ t & u \end{bmatrix} \quad \text{and} \quad M' = \begin{bmatrix} r' & s' \\ t' & u' \end{bmatrix},$$

such that

$$F((x, y)M) = nx^2 + mxy + ly^2 = F((x, y)M').$$

If we apply the change of variables M^{-1} to both sides above, this says that

$$F(x, y) = F((x, y)M'M^{-1});$$

the change of variables defined by the matrix $N = M'M^{-1}$ leaves the form F fixed. Such a matrix is called an AUTOMORPHISM of F. The automorphisms of F form a group, a subgroup of the group of all integer determinant 1 matrices, and one can prove the following theorem.

Theorem. *For a binary quadratic form $F = \{a, b, c\}$ of discriminant d, the automorphisms N of F are in one-to-one correspondence with solutions (t, u) of the Pell equation $t^2 - du^2 = 4$ via*

$$(t, u) \leftrightarrow N = \begin{bmatrix} \frac{t-bu}{2} & -cu \\ au & \frac{t+bu}{2} \end{bmatrix}.$$

When $d > 0$, there are infinitely many solutions. For $d = -3$ there are six solutions, for $d = -4$ there are four solutions, and for $d < -4$ there are only the two trivial solutions $(t = \pm 2, u = 0)$.

The proof is largely tedious algebraic calculations, so it is omitted. For $d < -4$, the only automorphisms of F are then $\begin{bmatrix} \pm 1 & 0 \\ 0 & \pm 1 \end{bmatrix}$. This means that if we have a proper representation $F(r, s) = n$ leading to a form $\{n, m, l\}$, then the only other possibility leading to the same $m \bmod 2n$ is the obvious one,

$F(-r, -s) = n$. Of course, there might be other representations corresponding to other solutions m' of the congruence $m'^2 \equiv d \bmod 4$.

With this and the previous theorem, we see that the number of ways n can be properly represented by a form F of discriminant $d < 0$ is finite, and we define the REPRESENTATION NUMBER of n by F as

$$r_F(n) = \frac{1}{w} \#\{(x, y) \text{ relatively prime with } F(x, y) = n\},$$

where $w = 2, 4$, or 6. Of course, $r_F(n)$ only depends on the class of F. The individual $r_F(n)$ are still somewhat mysterious, but if we define

$$r_d(n) = \sum_{\text{classes } F} r_F(n),$$

then the previous two theorems say that

Theorem. *For $d < 0$, $r_d(n)$ is the number of m that satisfy*

$$0 \le m < 2n \quad and \quad m^2 \equiv d \bmod 4n. \tag{13.6}$$

Exercise 13.1.11. Compute the representation numbers $r_d(p)$ for the primes $p = 2, 3, 5, \ldots, 37$ and for discriminants $d = -159, -163$, and -164. Do the same for discriminants $-424, -427$, and -431.

Exercise 13.1.12. Write a *Mathematica* program to compute $r_d(p)$, based on (13.6).

We still aim to connect this with L-functions and analysis later on, so a multiplicative version of this will be helpful. To give it the simplest possible form, we will assume that d is what is called a FUNDAMENTAL DISCRIMINANT. This means that if $d \equiv 1 \bmod 4$, then d is square free, that is, it is the product of distinct primes. If $d \equiv 0 \bmod 4$, the definition means that $d/4$ is square free and $d/4 \equiv 2$ or $3 \bmod 4$. Even with this restriction, the statement is slightly complicated.

Theorem. *For a fundamental discriminant $d < 0$, we have*

$$r_d(n) = \begin{cases} 0, & \text{if } p^2 | n \text{ for some } p | d, \\ \displaystyle\prod_{\substack{p | n \\ (p, d) = 1}} \left\{ 1 + \left(\frac{d}{p}\right) \right\}, & otherwise. \end{cases} \tag{13.7}$$

Proof. Notice, first, that the theorem says that $r_d(n) = 0$ if n is divisible by the square of any prime p that divides d. This follows from a congruence argument. First, consider p odd. If $m^2 \equiv d$ mod $4n$, then $m^2 \equiv d$ mod p^2. So, for some k, $m^2 = d + kp^2$. But then if p divides d, it also divides m^2, which means that p divides m and p^2 divides m^2. So, p^2 divides d. This is a contradiction when d is a fundamental discriminant; it is not divisible by any odd prime squared. If $4|n$ and $d \equiv 0$ mod 4, then the equation $m^2 \equiv d$ mod 16 similarly leads to a contradiction.

Similarly, the theorem says that $r_d(n) = 0$ if $(\frac{d}{p}) = -1$ for some prime $p|n$, and this is clear. For if there are solutions $m^2 \equiv d$ mod $4n$, there will be solutions modulo p whenever p divides n.

Otherwise, $(\frac{d}{p}) = 1$ for all primes p dividing n but not d. In this case, the theorem claims that $r_d(n)$ is $2^k = (1 + 1)^k$, where k is the number of such primes. We will not prove the theorem in general, but only for the special case where n is odd and $d \equiv 0$ mod 4. Then, we must count the solutions m, $0 \leq m < 2n$, of the congruence $m^2 \equiv d$ mod $4n$. We know that $d = 4\tilde{d}$ for some $\tilde{d}$. This means that solutions m must be even $m = 2\tilde{m}$. Now, $4\tilde{m}^2 \equiv 4\tilde{d}$ mod $4n$ is the same as

$$\tilde{m}^2 \equiv \tilde{d} \text{ mod } n.$$

Furthermore, the restriction $0 \leq 2\tilde{m} < 2n$ becomes $0 \leq \tilde{m} < n$. In other words, we just count solutions to the congruence modulo n. This will be an application of Hensel's Lemma and the Chinese Remainder Theorem discussed in Interlude 4. If some prime q divides n and q also divides d, there is a unique solution $\tilde{m} \equiv 0$ mod q to the congruence. Furthermore, according to the hypothesis, no higher power of q divides n. The remaining primes p dividing n have $(\frac{d}{p}) = 1$. There are two solutions $\tilde{m} \neq -\tilde{m}$ mod p to the congruence $\tilde{m}^2 \equiv d$ mod p. According to Hensel's Lemma, there are exactly two solutions modulo p^j for each exponent j. Factoring n as $q_1 q_2 \ldots q_i p_1^{j_1} p_2^{j_2} \ldots p_k^{j_k}$, the Chinese Remainder Theorem says that the solutions can be combined in 2^k ways to get a solution $\tilde{m}$ modulo n. □

13.2. Analytic Class Number Formula

The following is the opening sentence of Davenport's book (Davenport, 2000):

Analytic number theory may be said to begin with the work of Dirichlet, and in particular with Dirichlet's memoir of 1837 on the existence of primes in a given arithmetic progression.

Dirichlet proved that for any n, and for any a relatively prime to n, there are infinitely many primes $p \equiv a$ mod n. A key ingredient is the nonvanishing of his Dirichlet L-functions $L(s, \chi)$ at $s = 1$, and in the course of proving this, he discovered the Analytic Class Number Formula, which relates the value of the L-function to the class number h of the previous section.

We will not follow Dirichlet's proof. Instead, we will use a later argument that recycles some of Riemann's ideas about theta functions, as discussed in Section 9.2. First, we extend the definition (12.3) of the Legendre symbol to get the Jacobi symbol. For n odd and positive, factor n as $p_1^{k_1} p_2^{k_2} \ldots p_m^{k_m}$ and define

$$\left(\frac{d}{n}\right) = \left(\frac{d}{p_1}\right)^{k_1} \left(\frac{d}{p_2}\right)^{k_2} \ldots \left(\frac{d}{p_m}\right)^{k_m}.$$

By definition, the Jacobi symbol is multiplicative in the "denominator," but it no longer keeps track of squares modulo n. For example,

$$\left(\frac{7}{65}\right) = \left(\frac{7}{5}\right)\left(\frac{7}{13}\right) = (-1)(-1) = 1,$$

but 7 is not a square modulo 65. Also, the Jacobi symbol only depends on d modulo n, according to the Chinese Remainder Theorem discussed in Interlude 4, because each Legendre symbol only depends on d modulo p. For odd positive d and n, Quadratic Reciprocity still works, $\left(\frac{d}{n}\right) = \left(\frac{n}{d}\right)$, unless both d and n are 3 modulo 4, in which case they have opposite signs. The proof consists of factoring d and n and using the version for primes. A shorthand way of writing this is

$$\left(\frac{d}{n}\right) = \left(\frac{n}{d}\right)(-1)^{\frac{(n-1)}{2}\frac{(d-1)}{2}}.$$

The point is that the exponent of -1 is even unless both d and n are 3 modulo 4. The auxiliary rules for 2 and -1 also carry over and have concise notations:

$$\left(\frac{-1}{n}\right) = (-1)^{(n-1)/2} \quad \text{and} \quad \left(\frac{2}{n}\right) = (-1)^{(n^2-1)/8}.$$

To avoid the requirement that n be odd in the definition of the Jacobi symbol, we extend the definition one last time, to get the Kronecker symbol, defined for discriminants d as

$$\left(\frac{d}{2}\right) = \begin{cases} 0, & \text{if } 2|d, \\ +1, & \text{if } d \equiv 1 \bmod 8, \\ -1, & \text{if } d \equiv 5 \bmod 8. \end{cases}$$

For $m = 2^k n$ with n odd, we define

$$\left(\frac{d}{m}\right) = \left(\frac{d}{2}\right)^k \left(\frac{d}{n}\right).$$

This is a lot of definition to absorb. The important facts are as follows.

1. For p an odd prime number, we still have the property

$$\left(\frac{d}{p}\right) = \begin{cases} +1, & \text{if } d \equiv \text{ square } \bmod p, \\ -1, & \text{if } d \equiv \text{ nonsquare } \bmod p, \\ 0, & \text{if } d \equiv 0 \bmod p. \end{cases}$$

2. The Kronecker symbol is multiplicative in the denominator,

$$\left(\frac{d}{mn}\right) = \left(\frac{d}{m}\right)\left(\frac{d}{n}\right),$$

by the way we defined it.

3. If we fix d and vary n, the Kronecker symbol depends only on n modulo $|d|$. That is, if $m \equiv n \bmod |d|$, then

$$\left(\frac{d}{m}\right) = \left(\frac{d}{n}\right).$$

We will prove this in a special case in the next lemma.

Lemma. *For $0 > d \equiv 1 \bmod 4$ and $m > 0$,*

$$\left(\frac{d}{m}\right) = \left(\frac{m}{|d|}\right).$$

Proof. As usual, write $m = 2^k n$ with n odd and factor. We have

$$\left(\frac{d}{2^k}\right) = \left(\frac{d}{2}\right)^k = \left(\frac{2}{|d|}\right)^k = \left(\frac{2^k}{|d|}\right),$$

where the middle step compares the definition to one of the special cases of Quadratic Reciprocity. Meanwhile,

$$\left(\frac{d}{n}\right) = \left(\frac{-1}{n}\right)\left(\frac{|d|}{n}\right)$$

$$= (-1)^{(n-1)/2}\left(\frac{|d|}{n}\right)$$

$$= (-1)^{(n-1)/2}\left(\frac{n}{|d|}\right)(-1)^{\frac{(n-1)}{2}\frac{(|d|-1)}{2}},$$

according to Quadratic Reciprocity again. We are done if the two -1 terms cancel out. But $d \equiv 1 \bmod 4$ means that $|d| = -d \equiv 3 \bmod 4$, and so, $(|d| - 1)/2$ is odd. The two exponents are then either both odd or both even.   □

For d a fundamental discriminant, we define χ_d modulo $|d|$ as

$$\chi_d(n) = \left(\frac{d}{n}\right)$$

and consider the corresponding L-function $L(s, \chi_d)$, defined as

$$L(s, \chi_d) = \sum_{n=1}^{\infty} \chi_d(n) n^{-s}, \qquad \text{for } \mathrm{Re}(s) > 1.$$

Run the argument in Section 11.4 in reverse to see that there is an Euler product:

$$L(s, \chi_d) = \prod_p (1 - \chi_d(p) p^{-s})^{-1}.$$

For example, $d = -4$ is a fundamental discriminant. The definitions say that $\chi_{-4}(n) = 0$ if n is even. For odd n, we get that

$$\left(\frac{-4}{n}\right) = \left(\frac{-1}{n}\right)\left(\frac{4}{n}\right) = (-1)^{(n-1)/2},$$

because 4 is a square modulo any odd prime. So,

$$\chi_{-4}(n) = \begin{cases} +1, & \text{if } n \equiv 1 \bmod 4, \\ -1, & \text{if } n \equiv 3 \bmod 4, \end{cases}$$

and the Dirichlet L-function is

$$L(s, \chi_{-4}) = 1 - \frac{1}{3^s} + \frac{1}{5^s} - \frac{1}{7^s} + \frac{1}{9^s} - \frac{1}{11^s} + \frac{1}{13^s} - \cdots.$$

Another simple example is

$$\chi_{-3}(n) = \left(\frac{n}{3}\right) = \begin{cases} +1, & \text{if } n \equiv 1 \bmod 3, \\ -1, & \text{if } n \equiv 2 \bmod 3, \\ 0, & \text{if } n \equiv 0 \bmod 3, \end{cases}$$

according to the above lemma and the definition of the Jacobi symbol. We have

$$L(s, \chi_{-3}) = 1 - \frac{1}{2^s} + \frac{1}{4^s} - \frac{1}{5^s} + \frac{1}{7^s} - \frac{1}{8^s} + \frac{1}{10^s} - \cdots.$$

In Chapter 11, we looked at some examples with positive fundamental discriminants, namely $d = 8, 12$, and 5. The local phenomenon we saw there for the Euler factor at a prime p holds in general. That is, if we plug $s = 1$ into $(1 - \chi_d(p)p^{-s})^{-1}$, we get

$$\frac{1}{1 - \chi_d(p)/p} = \frac{p}{p - \chi_d(p)}.$$

As in the earlier examples, this is counting the number of solutions of Pell's equation $x^2 - dy^2 \equiv 1 \bmod p$, according to (11.1) and (fact 1) above. It does this regardless of whether d is positive or negative. In arithmetic modulo p there is no distinction between positive and negative. The remarkable thing is that the L-function value at $s = 1$ combines these Euler factors in an arithmetically interesting way, even for $d < 0$ when there are only trivial solutions to the integer Pell equation.

Theorem (Analytic Class Number Formula). *For $d < -4$ a fundamental discriminant,*

$$L(1, \chi_d) = \frac{\pi h}{\sqrt{|d|}},$$

where h is the class number.

The discriminants $d = -4$ and -3 are special because there are extra automorphisms, 4 and 6, respectively. We saw in Exercise 12.6.4 that $L(1, \chi_{-4})$ was given by Gregory's series for $\pi/4$.

Exercise 13.2.1. Use the methods discussed in Section 11.4 to compute $L(1, \chi_{-3})$. (Hint: The integral, after completing the square, involves arctan, not log.) Don't forget to choose the constant of integration so the antiderivative is 0 at $x = 0$.

Proof. To connect the L-function to the class number, we need some function defined in terms of an equivalence class of binary quadratic forms. That function is the EPSTEIN ZETA FUNCTION, defined as

$$\zeta(s, F) = \frac{1}{2}\sideset{}{'}\sum_{(x,y)} F(x, y)^{-s}, \qquad \text{for Re}(s) > 1,$$

where the sum is over all pairs of integers $(x, y) \neq (0, 0)$, which is omitted because it leads to division by 0. Because equivalent forms take on the same values, if we change F to another equivalent form, we get the same sum, merely in a different order. This means that $\zeta(s, F)$ only depends on the class of F. Each individual pair (x, y) has a greatest common divisor m, which can be factored out. The sum on pairs (x, y) can be replaced by a sum over all possible greatest common divisors m, and a sum over all pairs of relatively prime integers, say (u, v). Because

$$F(mu, mv)^{-s} = (m^2 F(u, v))^{-s} = m^{-2s} F(u, v)^{-s},$$

the sum over all possible m contributes $\zeta(2s)$. In what is left, we group all terms where $F(u, v)$ takes the same value n. There are $r_F(n)$ of these, by definition. So, we have that

$$\zeta(s, F) = \zeta(2s) \sum_{n=1}^{\infty} r_F(n) n^{-s}.$$

We will sum over all h equivalence classes of forms to get a new function, denoted

$$\zeta(s, d) = \sum_{\text{classes } F} \zeta(s, F) = \zeta(2s) \sum_{n=1}^{\infty} r_d(n) n^{-s},$$

according to the definition of $r_d(n)$ as the sum of the $r_F(n)$. The connection to the Dirichlet L-function will come through a mysterious-looking intermediate step, which is pulled out of thin air.

Lemma (Thin Air Equation, Part I). *We have an Euler product*

$$\sum_{n=1}^{\infty} r_d(n) n^{-s} = \prod_{p} \frac{1 + p^{-s}}{1 - \chi_d(p) p^{-s}}.$$

Proof. We take a single term in the Euler product and expand it, as usual as a Geometric series:

$$\frac{1 + p^{-s}}{1 - \chi_d(p) p^{-s}} = (1 + p^{-s}) \sum_{k=0}^{\infty} \chi_d(p)^k p^{-ks}$$

$$= \sum_{k=0}^{\infty} \chi_d(p)^k p^{-ks} + \sum_{k=0}^{\infty} \chi_d(p)^k p^{-(k+1)s}$$

$$= 1 + (1 + \chi_d(p)) p^{-s} + \sum_{k=2}^{\infty} (\chi_d(p)^{k-1} + \chi_d(p)^k) p^{-ks}.$$

This simplifies in one of three ways, depending on the value of $\chi_d(p)$:

$$= \begin{cases} 1 + p^{-s}, & \text{if } \chi_d(p) = 0, \\ 1 + 2\sum_k p^{-ks}, & \text{if } \chi_d(p) = 1, \\ 1, & \text{if } \chi_d(p) = -1. \end{cases}$$

In the last case, the terms $\chi_d(p)^{k-1}$ and $\chi_d(p)^k$ always have opposite signs. If we now multiply all the Euler factors together, the coefficient of n^{-s} is $r_d(n)$, according to (13.7). $\qquad \square$

Lemma (Thin Air Equation, Part II). *The Euler product simplifies as*

$$\prod_p \frac{1 + p^{-s}}{1 - \chi_d(p)p^{-s}} = \zeta(2s)^{-1}L(s, \chi_d)\zeta(s).$$

Proof. This easily follows from the Euler products for $\zeta(s)$, $L(s, \chi_d)$, and $\zeta(2s)^{-1}$:

$$\prod_p \frac{1 + p^{-s}}{1 - \chi_d(p)p^{-s}} = \prod_p \frac{(1 + p^{-s})(1 - p^{-s})}{(1 - \chi_d(p)p^{-s})(1 - p^{-s})}$$

$$= \prod_p \frac{(1 - p^{-2s})}{(1 - \chi_d(p)p^{-s})(1 - p^{-s})}.$$

$\qquad \square$

Combining the two lemmas, we get the following theorem.

Theorem.

$$\zeta(s, d) = \zeta(2s) \sum_{n=1}^{\infty} r_d(n) n^{-s}$$

$$= \zeta(2s)\zeta(2s)^{-1}L(s, \chi_d)\zeta(s)$$

$$= L(s, \chi_d)\zeta(s).$$

We are halfway to the Analytic Class Number Formula. What remains is to understand the series expansions of the Epstein zeta functions at $s = 1$. To do so, we introduce a generalization of the theta function from Chapter 9. For a binary quadratic form F of discriminant $d < 0$, we define

$$\Theta(t, F) = \sum_{(x,y)} \exp(-2\pi t F(x, y)/\sqrt{|d|}).$$

As usual, $\Theta(t, F)$ depends only on the equivalence class of F. This theta function has a symmetry as well,

$$\Theta(t^{-1}, F) = t\Theta(t, F'), \tag{13.8}$$

where $F' = \{a, -b, c\}$ when $F = \{a, b, c\}$. This is analogous to (9.2), although the method of proof used does not generalize. We define

$$\Lambda(s, F) = 2|d|^{s/2}(2\pi)^{-s}\Gamma(s)\zeta(s, F)$$

and

$$\Lambda(s, d) = \sum_F \Lambda(s, F) = 2|d|^{s/2}(2\pi)^{-s}\Gamma(s)\zeta(s, d)$$

$$= 2|d|^{s/2}(2\pi)^{-s}\Gamma(s)L(s, \chi_d)\zeta(s) \tag{13.9}$$

according to the theorem above.

Exercise 13.2.2. Imitate the results in Section 9.3 to show that

$$\Lambda(s, F) = \int_0^\infty (\Theta(t, F) - 1) t^s \frac{dt}{t}.$$

Then, show that

$$\Lambda(s, F) = \frac{1}{s-1} - \frac{1}{s}$$

$$+ \int_1^\infty (\Theta(t, F) - 1) t^s \frac{dt}{t} + \int_1^\infty \left(\Theta(t, F') - 1\right) t^{1-s} \frac{dt}{t}. \tag{13.10}$$

The proof uses the symmetry (13.8) exactly the same way the proof of (9.9) used (9.2).

Because the right side of (13.10) is symmetric under $F \to F'$, $s \to 1 - s$, the functional equation is

$$\Lambda(s, F) = \Lambda(1 - s, F').$$

These give a functional equation for $\Lambda(s, d)$ and, thus, also for $L(s, \chi_d)$ under $s \to 1 - s$.

An analysis of the integrals on the right side of (13.10), just as in Chapter 9, shows that

$$\Lambda(s, F) = \frac{1}{s-1} + O(1) \quad \text{as } s \to 1.$$

The miracle here is that the answer is independent of the class F. So, if we sum over all h classes, we get

$$\Lambda(s, d) = \frac{h}{s-1} + O(1) \quad \text{as } s \to 1. \tag{13.11}$$

We already know that

$$\zeta(s) = \frac{1}{s-1} + O(1) \quad \text{as } s \to 1.$$

The rest of (13.9) is well behaved at $s = 1$; a Taylor series expansion at $s = 1$ looks like

$$2|d|^{s/2}(2\pi)^{-s}\Gamma(s)L(s, \chi_d) = 2|d|^{1/2}(2\pi)^{-1}\Gamma(1)L(1, \chi_d) + O(s-1)$$

$$= \frac{\sqrt{|d|}}{\pi}L(1, \chi_d) + O(s-1).$$

Thus, the product of the two is

$$\Lambda(s, d) = \frac{\sqrt{|d|}\pi^{-1}L(1, \chi_d)}{s-1} + O(1) \quad \text{as } s \to 1. \tag{13.12}$$

Comparing the right side of (13.11) with the right side of (13.12), we find that

$$h = \frac{\sqrt{|d|}}{\pi}L(1, \chi_d),$$

which is equivalent to the Analytic Class Number Formula. $\qquad\square$

13.3. Siegel Zeros and the Class Number

Gauss computed the class number for thousands of discriminants, by hand, enumerating all the reduced forms that satisfy (13.4), just as you did in Exercise 13.1.8. As you can see in Table 13.1, he discovered that the class number h seems to slowly increase with $|d|$. Based on these computations, he conjectured in his *Disquisitiones Arithmetica*, in 1801, that for any h, there are only finitely many negative discriminants d with class number h.

The GENERALIZED RIEMANN HYPOTHESIS (GRH) conjectures that all the Dirichlet L-functions have all their nontrivial zeros on the symmetry line $\text{Re}(s) = 1/2$ for the functional equation, just as the Riemann Hypothesis conjectures for $\zeta(s)$. It can be shown (Davenport, 2000) that there is a constant C such that in the rectangle $|\text{Im}(s)| \le 1$ and $\text{Re}(s) \ge 1 - C/\log|d|$, there is,

Table 13.1. Class Numbers of Fundamental Discriminants

d	h	d	h	d	h	d	h	d	h	d	h
−3	1	−115	2	−231	12	−339	6	−452	8	−579	8
−4	1	−116	6	−232	2	−340	4	−455	20	−580	8
−7	1	−119	10	−235	2	−344	10	−456	8	−583	8
−8	1	−120	4	−239	15	−347	5	−463	7	−584	16
−11	1	−123	2	−244	6	−355	4	−467	7	−587	7
−15	2	−127	5	−247	6	−356	12	−471	16	−591	22
−19	1	−131	5	−248	8	−359	19	−472	6	−595	4
−20	2	−132	4	−251	7	−367	9	−479	25	−596	14
−23	3	−136	4	−255	12	−371	8	−483	4	−599	25
−24	2	−139	3	−259	4	−372	4	−487	7	−607	13
−31	3	−143	10	−260	8	−376	8	−488	10	−611	10
−35	2	−148	2	−263	13	−379	3	−491	9	−615	20
−39	4	−151	7	−264	8	−383	17	−499	3	−616	8
−40	2	−152	6	−267	2	−388	4	−503	21	−619	5
−43	1	−155	4	−271	11	−391	14	−511	14	−623	22
−47	5	−159	10	−276	8	−395	8	−515	6	−627	4
−51	2	−163	1	−280	4	−399	16	−516	12	−628	6
−52	2	−164	8	−283	3	−403	2	−519	18	−631	13
−55	4	−167	11	−287	14	−404	14	−520	4	−632	8
−56	4	−168	4	−291	4	−407	16	−523	5	−635	10
−59	3	−179	5	−292	4	−408	4	−527	18	−643	3
−67	1	−183	8	−295	8	−411	6	−532	4	−644	16
−68	3	−184	4	−296	10	−415	10	−535	14	−647	23
−71	7	−187	2	−299	8	−419	9	−536	14	−651	8
−79	5	−191	13	−303	10	−420	8	−543	12	−655	12
−83	3	−195	4	−307	3	−424	6	−547	3	−659	11
−84	4	−199	9	−308	8	−427	2	−548	8	−660	8
−87	6	−203	4	−311	19	−431	21	−551	26	−663	16
−88	2	−211	3	−312	4	−435	4	−552	8	−664	10
−91	2	−212	6	−319	10	−436	6	−555	4	−667	4
−95	8	−215	14	−323	4	−439	15	−559	16	−671	30
−103	5	−219	4	−327	12	−440	12	−563	9	−679	18
−104	6	−223	7	−328	4	−443	5	−564	8	−680	12
−107	3	−227	5	−331	3	−447	14	−568	4	−683	5
−111	8	−228	4	−335	18	−451	6	−571	5	−687	12

at most, one counterexample, $\beta + i\gamma$ with

$$L(\beta + i\gamma, \chi_d) = 0,$$

to GRH. Furthermore, such a counterexample, if it exists, must actually lie on the real axis. These potential counterexamples to GRH are called SIEGEL ZEROS.

In this final section we will show how the possibility of Siegel zeros for $L(s, \chi_d)$ is connected to the size of the class number, h. That there is some connection is not too surprising. After all, $L(s, \chi_d)$ is a continuous function. If it is 0 near $s = 1$, it cannot be too far from 0 at $s = 1$. Nonetheless, the subtle interplay between the zeros and the class number is quite remarkable.

In some sense, the most elementary result about the class number h for a discriminant d is that it is at least 1, because there is certainly the class of the principal form. Because of this trivial lower bound, the Analytic Class Number Formula implies that

$$L(1, \chi_d) \geq \frac{\pi}{\sqrt{|d|}}.$$

This, in turn, will give a weak result on Siegel zeros, that $L(s, \chi_d)$ cannot have a zero that is "too close" to $s = 1$. Before we can prove this, we need some lemmas.

Lemma.

$$\sum_{n \bmod d} \chi_d(n) = 0.$$

Let

$$A(x) = \sum_{n < x} \chi_d(n); \quad then, \quad |A(x)| \leq |d|.$$

Proof. There is at least one m modulo d such that $\chi_d(m) = -1$. This can be proved by factoring d and using the Chinese Remainder Theorem. Because multiplication by m just permutes the congruence classes modulo d, we get that

$$\sum_{n \bmod d} \chi_d(n) = \sum_{n \bmod d} \chi_d(m \cdot n)$$

$$= \sum_{n \bmod d} \chi_d(m) \cdot \chi_d(n)$$

$$= \chi_d(m) \cdot \sum_{n \bmod d} \chi_d(n)$$

$$= -1 \cdot \sum_{n \bmod d} \chi_d(n).$$

This shows that the sum is 0. (You might profitably compare the proof of this lemma to the one on p. 222.) Because $\chi_d(n)$ depends only on n modulo $|d|$, the function $\chi_d(n)$ is periodic with period $|d|$. The previous calculation says that the sum over any complete period is 0. So, it suffices to consider the case of $x \leq |d|$. But, then, the triangle inequality says that

$$|A(x)| \leq \sum_{n<x} |\chi_d(n)| \leq \sum_{n<x} 1 = [x] \leq x \leq |d|.$$

$\square$

The lemma says that for $\sigma > 0$, $L(\sigma, \chi_d)$ converges according to Abel's Theorem, Part II (I2.20).

Lemma. *For* $1 - 1/\log|d| \leq \sigma \leq 1$,

$$L'(\sigma, \chi_d) \ll (\log|d|)^2. \tag{13.13}$$

Proof. If we differentiate the series for $L(\sigma, \chi_d)$ term by term, we get

$$L'(\sigma, \chi_d) = -\sum_n \chi_d(n)\log(n)n^{-\sigma}.$$

We will consider first the terms in the sum with $n < |d|$. With the hypothesis about σ, we have that $1 - \sigma \leq 1/\log|d| < 1/\log(n)$. So, $(1-\sigma)\log(n) < 1$, and

$$n^{1-\sigma} < e \quad \text{or} \quad n^{-\sigma} < \frac{e}{n}.$$

We then estimate

$$\left| \sum_{n<|d|} \chi_d(n)\log(n)n^{-\sigma} \right| < \sum_{n<|d|} \log(n)\frac{e}{n}$$

$$\ll \int_1^{|d|} \frac{\log(x)}{x}dx = (\log|d|)^2$$

according to the usual method of comparing a sum to an integral.

The tail of the series will be smaller, but harder to estimate. We have to use Summation by Parts on the finite sum

$$\sum_{|d| \leq k < N} \chi_d(k)\log(k)k^{-\sigma}.$$

We let $\Delta v(k) = \chi_d(k)$, so $v(k) = A(k) = \sum_{n<k} \chi_d(n)$ and $v(|d|) = 0$, according to the previous lemma. Our $u(k)$ is $\log(k)k^{-\sigma}$. We get that the sum above

is equal to

$$A(n)\frac{\log(N)}{N^\sigma} + \sum_{|d|\leq k < N} A(k+1) \left(\frac{\log(k)}{k^\sigma} - \frac{\log(k+1)}{(k+1)^\sigma} \right).$$

Because $\log(x)x^{-\sigma}$ is a decreasing function, the term in parenthesis is positive. So, the triangle inequality estimates the absolute value of the sum as being bounded by

$$|A(n)|\frac{\log(N)}{N^\sigma} + \sum_{|d|\leq k < N} |A(k+1)| \left(\frac{\log(k)}{k^\sigma} - \frac{\log(k+1)}{(k+1)^\sigma} \right)$$

$$\ll |d|\frac{\log(N)}{N^\sigma} + |d| \sum_{|d|\leq k < N} \left(\frac{\log(k)}{k^\sigma} - \frac{\log(k+1)}{(k+1)^\sigma} \right),$$

according to the estimate on $|A(x)|$

$$= |d|\frac{\log(N)}{N^\sigma} + |d| \left(\frac{\log|d|}{|d|^\sigma} - \frac{\log(N)}{(N)^\sigma} \right),$$

because the sum telescopes. Now we consider what happens when $N \to \infty$. The terms with $\log(N)/N^\sigma$ go to zero. The hypothesis about the location of σ implies, as for n above, that $|d|/|d|^\sigma \leq e$. So, the tail is bounded by $e \log |d|$. ☐

We should have first proven that the series for $L'(\sigma, \chi_d)$ actually converges for $\sigma > 0$. One estimates the tail

$$\left| \sum_{M\leq k < N} \chi_d(k) \log(k)k^{-\sigma} \right|$$

of the series by using this same Summation by Parts trick and shows that it tends to 0 as M, N go to ∞.

Lemma. *For* $1 - 1/\log |d| \leq \sigma \leq 1$,

$$L(\sigma, \chi_d) \ll \log |d|.$$

Exercise 13.3.1. Prove this by imitating the proof of the preceding lemma. It is easier, as there are no $\log(n)$ terms. Split the sum into two parts, $n < |d|$ and $|d| \leq n$. The first part compares to an integral that gives $\log |d|$. Summation by parts does the second, and you find the bound $|d|/|d|^\sigma < e$ for the given range of σ.

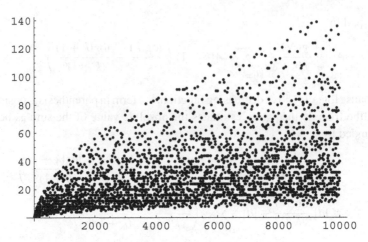

Figure 13.1. Discriminants vs. class numbers.

In fact, if you did the previous exercise, you actually proved the following.

Theorem. *The class number h satisfies*

$$h \ll \log|d|\sqrt{|d|} \quad as \ d \to -\infty.$$

Proof. This follows immediately from the estimate of the lemma at $\sigma = 1$, and from the Analytic Class Number Formula.   □

Figure 13.1 shows pairs $(|d|, h)$ for fundamental discriminants below 10000. You can make out a parabola opening to the right in the large values of h. This follows from the theorem, as

$$h \ll \log|d||d|^{1/2} \ll |d|^{1/2+\epsilon}$$

for any $\epsilon > 0$. It is much, much harder to get a nontrivial lower bound on the class number.

We mentioned previously that a zero near $s = 1$ must be on the real axis. This next theorem, attributable to Page, shows that the positivity of the class number prevents a zero from being too close to $s = 1$.

Theorem. *There is a constant $C > 0$ such that $L(s, \chi_d)$ has no zero β in the region*

$$1 - \frac{C}{\sqrt{|d|}(\log|d|)^2} < \beta < 1.$$

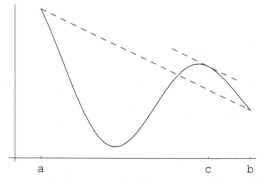

Figure 13.2. Mean Value Theorem.

The C of the theorem is a universal constant independent of d. (In fact, from the proof of (13.13), I expect the $C = 1/(2e)$ will work.) You should think of the theorem as giving a sequence of zero-free intervals, one for each d. We have less information as $|d|$ increases; the intervals shrink.

Proof. The preceding lemmas did all the work; we just need to recall the Mean Value Theorem from calculus, which says that if a function $f(x)$ is differentiable on an interval $a \le x \le b$, then there is some point c in the interval where

$$f(b) - f(a) = f'(c)(b - a).$$

Geometrically, this says that there is some point c where the tangent line at c is parallel to the secant line connecting $(a, f(a))$ to $(b, f(b))$ (see Figure 13.2).

Now, suppose that $L(\beta, \chi_d) = 0$. We may as well assume that $1 - 1/\log|d| < \beta$; the theorem makes no claim to the left of this point, because

$$1 - \frac{1}{\log|d|} < 1 - \frac{C}{\sqrt{|d|}(\log|d|)^2}.$$

We apply the Mean Value Theorem on the interval $\beta \le x \le 1$. Then, we have

$$\begin{aligned}
L(1, \chi_d) &= L(1, \chi_d) - 0 \\
&= L(1, \chi_d) - L(\beta, \chi_d) \\
&= L'(\sigma, \chi_d)(1 - \beta)
\end{aligned}$$

for some σ with $\beta \le \sigma \le 1$. Because

$$\frac{1}{\sqrt{|d|}} < \frac{\pi h}{\sqrt{|d|}} = L(1, \chi_d),$$

we see that

$$\frac{1}{\sqrt{|d|}} < L'(\sigma, \chi_d)(1 - \beta) < C^{-1}(\log |d|)^2(1 - \beta)$$

for some C^{-1}, according to (13.13). This is algebraically equivalent to

$$\beta < 1 - \frac{C}{\sqrt{|d|}(\log |d|)^2}.$$

$\square$

In 1916, Hecke showed a result in the other direction, that if there is no zero too close to $s = 1$, then the class number cannot be too small. More specifically,

Theorem (Hecke). *Suppose that there is some constant C such that no $L(s, \chi_d)$ has a zero β with*

$$1 - \frac{C}{\log |d|} < \beta < 1.$$

Then, there is some other constant $\tilde{C}$ such that

$$h > \tilde{C}\frac{\sqrt{|d|}}{\log |d|}.$$

As in Page's result, the theorem refers to a sequence of zero-free intervals, one for each d, which shrink as $|d|$ increases. Of course, here they are the hypothesis. Because $\sqrt{|d|}/\log |d|$ goes to infinity as $|d|$ does, Hecke's theorem implies Gauss's conjecture but depends on an unproven hypothesis.

Proof. For real $s > 1$, the definition of the Epstein zeta function $\zeta(s, F)$ shows that it is real and positive, and so are the extra terms that go into the definition of $\Lambda(s, F)$. Because of this, they are forced to be real functions for $s < 1$ as well. Recall the expression (13.10) for $\Lambda(s, F)$ as an integral of a theta function $\Theta(t, F)$. Still assuming that s is real, t^s is real and positive, and every term in the sum defining $\Theta(t, F)$ is positive. (Subtracting 1 just removes the contribution of the summand $(x, y) = (0, 0)$ in the definition of $\Theta(t, F)$.) We get an inequality if we throw out positive terms, and Hecke's amazing insight was that there is still useful information if we throw out almost everything in sight. More precisely, for F not the principal class of forms, throw away *every* term in $\Theta(t, F)$. Then, (13.10) implies that

$$\Lambda(s, F) \geq \frac{1}{s-1} - \frac{1}{s}.$$

When $F(x, y)$ is the principal class, throw out everything except the summand $(x, y) = (1, 0)$ in the first integral. Because $F(1, 0) = 1$, (13.10) implies that

$$\Lambda(s, F) \geq \frac{1}{s-1} - \frac{1}{s} + \int_1^\infty \exp(-2\pi t/\sqrt{|d|})t^s\frac{dt}{t}$$

in this case. We now sum these inequalities over all h equivalence classes to get

$$\Lambda(s, d) \geq \frac{h}{s(s-1)} + \int_1^\infty \exp(-2\pi t/\sqrt{|d|})t^s\frac{dt}{t}, \tag{13.14}$$

where we used (13.9) to sum the expressions $\Lambda(s, F)$ and where we put the h expressions $1/(s-1) - 1/s$ over a common denominator. Now, (13.11) says that the left side above is

$$\Lambda(s, d) = \frac{h}{s-1} + O(1) \quad \text{as } s \to 1.$$

So, particularly when s approaches 1 from below, the function is negative, because then $1/(s-1) \to -\infty$. The GRH would then say it must be negative for all s between $1/2$ and 1; it cannot change signs. Otherwise, it is certainly negative for $\beta < s < 1$, where β is the rightmost counterexample to GRH. Now is a good time to point out that any possible counterexamples to GRH must come from $L(s, \chi_d)$ in the factorization (13.9). The term $2|d|^{s/2}(2\pi)^{-s}\Gamma(s)$ is positive for $s > 0$, and we proved in Section 8.1 that $\zeta(s)$ has no real zeros on the interval $0 < s < 1$.

Hecke's idea was simply to investigate the consequences for (13.14) of the inequality $\Lambda(s, d) < 0$. We have

$$0 > \Lambda(s, d) \geq \frac{h}{s(s-1)} + \int_1^\infty \exp(-2\pi t/\sqrt{|d|})t^s\frac{dt}{t},$$

which, on rearranging the terms, says that

$$h > s(1-s)\int_1^\infty \exp(-2\pi t/\sqrt{|d|})t^s\frac{dt}{t}.$$

If we integrate over an even smaller range, $\sqrt{|d|} \leq t$, the integral is yet smaller, and we then change the variables in the integral with $u = t/\sqrt{|d|}$, which gives

$$h > s(1-s)|d|^{s/2} \int_1^\infty \exp(-2\pi u)u^s \frac{du}{u}.$$

The integral is an increasing function of the parameter s because u^s is; so, certainly for $1/2 < s$, we have

$$\int_1^\infty \exp(-2\pi u)u^s \frac{du}{u} > \int_1^\infty \exp(-2\pi u)u^{1/2}\frac{du}{u} = 0.278806\ldots.$$

So, Hecke has deduced that on the interval $1/2 < s < 1$,

$$\Lambda(s,d) < 0 \Rightarrow h > s(1-s)|d|^{s/2}0.278806\ldots.$$

We could stop here and we would have a theorem. Just assume GRH and plug in some value for s. We want to do this in an elegant way, however; getting a good lower bound on h while assuming as little as possible about the zeros. To get the strongest possible lower bound for h is to maximize

$$f(s) = s(1-s)|d|^{s/2}.$$

This is a little calculus problem. We compute

$$f'(s) = (1-2s)|d|^{s/2} + \frac{1}{2}\log|d|s(1-s)|d|^{s/2}.$$

Check that

$$f'(1/2) = \frac{1}{8}\log|d||d|^{1/4} > 0, \qquad f'(1) = -|d|^{1/2} < 0,$$

which means that $f(s)$ has a maximum between $1/2$ and 1. We want to solve

$$(1-2s) + \frac{1}{2}\log|d|s(1-s) = 0$$

for s. This is a quadratic equation. The quadratic formula gives

$$s = \frac{1}{2} - \frac{2}{\log|d|} + \left(\frac{1}{4} + \frac{4}{(\log|d|)^2}\right)^{1/2},$$

because the other choice of sign for the root gives an answer of less than $1/2$. This answer is exact but messy looking. With $x = 2/\log|d|$, we can use the linear approximation $y = 1/2$ to $(1/4 + x^2)^{1/2}$, valid near $x = 0$. This was

Exercise I1.1.2. Now, we have instead, the simpler answer:

$$s = \frac{1}{2} - \frac{2}{\log|d|} + \frac{1}{2} = 1 - \frac{2}{\log|d|}.$$

Having done this calculation, we can now think about what hypothesis we need to make in the theorem. As Hecke's idea shows, if there is no counterexample β to GRH on the interval $1 - 2/\log|d| < \beta < 1$, then we get a lower bound on h. In fact, the bound is just as good if we make the weaker hypothesis that there is some small constant C independent of d such that

$$L(\beta, \chi_d) \neq 0 \quad \text{for} \quad 1 - \frac{C}{\log|d|} < \beta < 1.$$

Then, the maximum for $f(s)$ on this interval occurs at the left endpoint. The value of $f(s)$ at this point is

$$\left(1 - \frac{C}{\log|d|}\right)\frac{C}{\log|d|}\exp\left(\frac{\log|d|}{2} - \frac{C}{2}\right).$$

The term $1 - C/\log|d|$ approaches 1 from below as $|d|$ increases and so is eventually bigger than, say, $1/2$. This $1/2$, the C, and the $\exp(-C/2)$ can all be absorbed into a constant $\tilde{C}$, along with the $0.278806\ldots$ to get that

$$h > \tilde{C}\frac{\sqrt{|d|}}{\log|d|}.$$

$\square$

Later, Heilbronn, extending the work of Deuring and Mordell, proved a lower bound if the generalized Riemann Hypothesis is *false*. This would seem to cover all the possibilities. But their work has the defect that the constant implied by the $\ll$ depends on finding a counterexample to the generalized Riemann Hypothesis. Since no counter example is known, their work showed only that a constant exists. That is, it was known that $h \to \infty$, but nothing was proved about how fast. In 1935, Siegel proved that

$$\log(h) \sim \frac{1}{2}\log|d| \quad \text{as } |d| \to \infty.$$

Here, of course, $\sim$ means asymptotic again. Figure 13.3 shows a plot of data points $(\log|d|, \log(h))$ for $|d| < 10000$. You can see that they converge, slowly, to a line with slope $1/2$.

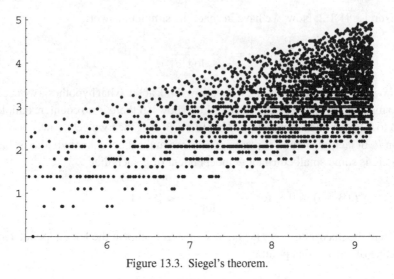

Figure 13.3. Siegel's theorem.

Siegel's theorem implies that for every $\epsilon > 0$, there is a constant $C(\epsilon)$ such that

$$h > C(\epsilon)|d|^{1/2-\epsilon}.$$

Again, the constant $C(\epsilon)$ is *ineffective*; that is, it cannot be computed because it may depend on counterexamples to GRH.

An unconditional estimate on the site of the class number with a effective constant was finally obtained by Goldfeld, Gross, and Zagier in the 1980s. Goldfeld showed that the existence of an elliptic curve with a zero of order exactly 3 at $s = 1$ would give a lower bound on class numbers of the form

$$h > \frac{1}{55} \cdot \prod_{p|d} \left(1 - \frac{2}{\sqrt{p}}\right) \log |d|.$$

Gross and Zagier, in their work on the Birch Swinnerton-Dyer conjecture, showed that such an elliptic curve exists.

Much more might be said about the perfection of the number seven, but this book is already too long . . .

St. Augustine

Solutions

(1.1.1) One possible solution is given in Figure S.1 for the cases $n = 3$ and 4.

Figure S.1. A solution to Exercise 1.1.1.

(1.1.2)

$$8t_n + 1 = 8\frac{n(n+1)}{2} + 1 = 4n^2 + 4n + 1 = (2n+1)^2 = s_{2n+1}.$$

A geometric proof in the case of $n = 2$ is given in Figure S.2.

Figure S.2. A solution to Exercise 1.1.2.

(1.1.3)

$$\frac{n(n+1)}{2} = m^2 \Leftrightarrow$$
$$n(n+1) = 2m^2 \Leftrightarrow$$
$$4n(n+1) = 8m^2 \Leftrightarrow$$
$$4n^2 + 4n + 1 - 1 = 8m^2 \Leftrightarrow$$
$$(2n+1)^2 - 1 = 8m^2.$$

(1.1.4) For the $n = 1$ case, we have $T_1 = t_1 = 1 = 1 \cdot 2 \cdot 3/6$ is true. Now, we can assume that $T_{n-1} = (n-1)n(n+1)/6$. Then, $T_n = T_{n-1} + t_n$, because the nth tetrahedron is formed by adding an extra layer. This is $T_{n-1} + n(n+1)/2$, according to the formula for t_n. According to the induction hypothesis, this is

$$\frac{(n-1)n(n+1)}{6} + \frac{n(n+1)}{2} = \frac{(n-1)n(n+1) + 3n(n+1)}{6},$$

and $(n-1)n(n+1) + 3n(n+1) = (n-1+3)n(n+1)$, which is $(n+2)n(n+1)$.

(1.1.5) What seems to be the pattern when you add up the terms in the kth gnomon? In the second, $2 + 4 + 2 = 8 = 2^3$. In the third, $3 + 6 + 9 + 6 + 3 = 27 = 3^3$. It seems you get k^3 in the kth gnomon. Here's a more careful proof.

(a) 2 is the common factor in the second; 3 is the common factor in the third. The first entry, k, is the common factor in the kth gnomon

(b) Factoring 3 out of the third, we get $1 + 2 + 3 + 2 + 1 = t_3 + t_2$. Factoring 4 out of the fourth we get $1 + 2 + 3 + 4 + 3 + 2 + 1 = t_4 + t_3$. We get $t_k + t_{k-1}$ from the kth.

(c) $t_k + t_{k-1} = s_k$, according to (1.2).

(d) Putting the k back in, $k \cdot s_k = k^3$.

So, the sum of the entries in the first n gnomons is

$$1^3 + 2^3 + 3^3 + \cdots + (n-1)^3 + n^3.$$

On the other hand, the sum of all these entries is the sum of all possible products of a row number from the set $\{1, 2, \ldots, n\}$ with a column number, also from the set $\{1, 2, \ldots, n\}$. According to the distributive law, this is just

$$(1 + 2 + \cdots + n) \cdot (1 + 2 + \cdots + n) = t_n \cdot t_n.$$

(1.2.1) We expect that $\Delta^2 h(n) = 4$. So, $\Delta h(n) = 4n + C$ for some constant C, and

$$h(n) = 4t(n-1) + Cn + D = 2(n-1)n + Cn + D$$

for some other constant D, using the same reasoning that was used for pentagonal numbers. With $h(1) = 1$ and $h(2) = 6$, we get equations

$$0 + C + D = 1,$$
$$4 + 2C + D = 6,$$

with solution $C = 1$, $D = 0$. This gives

$$h(n) = 2(n - 1)n + n = 2n^2 - n.$$

(1.2.2)

$$\Delta(n^{\underline{-2}}) = \frac{1}{(n + 2)(n + 3)} - \frac{1}{(n + 1)(n + 2)}$$
$$= \frac{(n + 1) - (n + 3)}{(n + 1)(n + 2)(n + 3)} = -2 \cdot n^{\underline{-3}}.$$

(1.2.3)

$$n^{\underline{2}}(n - 2)^{\underline{-3}} = n(n - 1) \cdot \frac{1}{(n - 1)n(n + 1)} = n^{\underline{-1}}.$$

(1.2.4) With $\Delta^2 f(n) = a - 2$, $\Delta f(n) = (a - 2)n + C$ for some constant C, and

$$f(n) = (a - 2)t(n - 1) + Cn + D = \frac{a - 2}{2}(n - 1)n + Cn + D$$

for some other constant D. With $f(1) = 1$ and $f(2) = a$, we get equations

$$0 + C + D = 1,$$
$$a - 2 + 2C + D = a,$$

with solution $C = 1$, $D = 0$. This gives

$$f(n) = \frac{a - 2}{2}(n - 1)n + n = \frac{(a - 2)n^2 + (4 - a)n}{2}.$$

(1.2.5)

$$n^{\underline{1}} + 3n^{\underline{2}} + n^{\underline{3}} = n + 3n(n - 1) + n(n - 1)(n - 2)$$
$$= n + 3n^2 - 3n + n^3 - 3n^2 + 2n = n^3.$$

According to the Fundamental Theorem,

$$\sum_{0 \le k < n+1} k^3 = \sum_{0 \le k < n+1} k^{\underline{1}} + 3k^{\underline{2}} + k^{\underline{3}} = \left. \frac{1}{2}k^{\underline{2}} + k^{\underline{3}} + \frac{1}{4}k^{\underline{4}} \right|_{k=0}^{k=n+1}$$

$$= (n+1)n \left(\frac{1}{2} + (n-1) + \frac{1}{4}(n-1)(n-2) \right)$$

$$= (n+1)n \left(\frac{2 + 4(n-1) + (n-1)(n-2)}{4} \right)$$

$$= \frac{(n+1)^2 n^2}{4}.$$

(1.2.6) We want

$$\sum_{0 \le k < n+1} k^4 = \sum_{0 \le k < n+1} k^{\underline{1}} + 7k^{\underline{2}} + 6k^{\underline{3}} + k^{\underline{4}}$$

$$= \left. \frac{1}{2}k^{\underline{2}} + \frac{7}{3}k^{\underline{3}} + \frac{6}{4}k^{\underline{4}} + \frac{1}{5}k^{\underline{5}} \right|_{k=0}^{n+1}$$

$$= (n+1)n \left(\frac{1}{2} + \frac{7}{3}(n-1) + \frac{6}{4}(n-1)(n-2) \right.$$

$$\left. + \frac{1}{5}(n-1)(n-2)(n-3) \right).$$

After some messy algebra, we get

$$= \frac{n(n+1)(2n+1)(3n^2 + 3n - 1)}{30}.$$

(1.2.7) The next row is

$$\left\{ {6 \atop 1} \right\} = 1, \left\{ {6 \atop 2} \right\} = 31, \left\{ {6 \atop 3} \right\} = 90, \left\{ {6 \atop 4} \right\} = 65, \left\{ {6 \atop 5} \right\} = 15, \left\{ {6 \atop 6} \right\} = 1.$$

The theorem that follows this exercise says what the pattern is, but here is the hint again, more explicitly. The 31 is computed from the 1 above and the *second* entry in the row above, the 15. The 90 is computed from the 15 above and the *third* entry in the row above, the 25.

(1.2.8) We want

$$\sum_{0 \le k < n+1} k^5 = \sum_{0 \le k < n+1} k^{\underline{1}} + 15k^{\underline{2}} + 25k^{\underline{3}} + 10k^{\underline{4}} + k^{\underline{5}}$$

$$= \left. \frac{1}{2}k^{\underline{2}} + 5k^{\underline{3}} + \frac{25}{4}k^{\underline{4}} + 2k^{\underline{5}} + \frac{1}{6}k^{\underline{6}} \right|_{k=0}^{n+1}.$$

After some messy algebra, we get

$$= \frac{(2n^2 + 2n - 1)(n + 1)^2 n^2}{12}.$$

(1.2.9) $\Delta 2^n = 2^{n+1} - 2^n = 2^n \cdot (2 - 1) = 2^n$, and $2^0 = 1$ is clear. As a consequence, $\Sigma 2^n = 2^n$ also. (This is meant to remind you of the function $\exp(t)$ in calculus, with the property that $\frac{d}{dt} \exp(t) = \exp(t)$ and $\int \exp(t)dt = \exp(t)$.) According to the Fundamental Theorem,

$$\sum_{0 \le k < n+1} 2^k = 2^k \Big|_{k=0}^{k=n+1} = 2^{n+1} - 1.$$

(1.2.10) $\Delta f(n) = x^{n+1} - x^n = x^n(x - 1) = (x - 1)f(n)$, so $\Sigma(x-1)f(n) = f(n)$, or $\Sigma f(n) = f(n)/(x - 1)$. According to the Fundamental Theorem,

$$\sum_{0 \le k < n+1} f(k) = \frac{f(k)}{x - 1} \Big|_{k=0}^{k=n+1} = \frac{x^{n+1} - 1}{x - 1}.$$

(1.2.11) The question asks for the sum of a Geometric series (missing only the first term):

$$7 + 7^2 + 7^3 + 7^4 + 7^5 = \frac{7^6 - 1}{7 - 1} - 1 = 19607.$$

(1.2.12) With $f(k) = 1/t_k = 2(k - 1)^{\underline{-2}}$, we have $\Sigma f(k) = -2(k-1)^{\underline{-1}}$. The Fundamental Theorem says that

$$\sum_{1 \le k < n+1} f(k) = -2(k-1)^{\underline{-1}} \Big|_{k=1}^{k=n+1} = \frac{-2}{n + 1} + 2.$$

With $T_k = k(k + 1)(k + 2)/6$, you can write $1/T_k = 6(k - 1)^{\underline{-3}}$ to sum the reciprocals of tetrahedral numbers.

(1.2.13) With $u(k) = H_k$ and $\Delta v(k) = k^{\underline{0}}$, we have $\Delta u(k) = k^{\underline{-1}}$, $v(k) = k^{\underline{1}} = k$, and $Ev(k) = k + 1$. Summation by Parts says that

$$\Sigma H_k \cdot k^{\underline{0}} = k \cdot H_k - \Sigma \frac{1}{k + 1} \cdot (k + 1)$$

$$= k \cdot H_k - \Sigma 1 = k \cdot H_k - k.$$

The Fundamental Theorem then says that

$$\sum_{0 \le k < n} H_k = k \cdot H_k - k \Big|_{k=0}^{k=n}$$

$$= (n \cdot H_n - n) - (0 \cdot H_0 - 0)$$

$$= n \cdot H_n - n.$$

(1.2.14) The function 2^k is unchanged under both Δ and Σ, so it can be either $u(k)$ or $\Delta v(k)$. On the other hand, the function k becomes simpler under Δ and more complicated under Σ. So, we choose $u(k) = k$, $\Delta u(k) = 1$. Then, $\Delta v(k) = 2^k$, $v(k) = 2^k$, and $Ev(k) = 2^{k+1}$. Summation by Parts says that

$$\Sigma k 2^k = k 2^k - \Sigma 2^{k+1} = k 2^k - 2^{k+1}.$$

The Fundamental Theorem then says that

$$\sum_{0 \le k < n} k 2^k = k 2^k - 2^{k+1} \Big|_{k=0}^{k=n} = (n 2^n - 2^{n+1}) - (0 - 2).$$

(2.1.1) The divisors of 60 are

$$1, 2, 3, 4, 5, 6, 10, 12, 15, 20, 30, \text{ and } 60.$$

So, $\tau(60) = 12$. The answer is the number of divisors in the list, not the list of divisors itself.

(2.1.2) Observe that

$$\tau(2) = \tau(3) = \tau(5) = \tau(7) = \cdots = 2,$$
$$\tau(4) = \tau(9) = \tau(25) = \tau(49) = \cdots = 3,$$
$$\tau(8) = \tau(27) = \tau(125) = \tau(343) = \cdots = 4,$$

and for any k

$$\sigma(2^k) = \frac{2^{k+1} - 1}{2 - 1}, \quad \sigma(3^k) = \frac{3^{k+1} - 1}{3 - 1}, \quad \sigma(5^k) = \frac{5^{k+1} - 1}{5 - 1}, \ldots.$$

(2.1.3) $\tau(m \cdot n)$ is not equal to $\tau(m) \cdot \tau(n)$ if m and n are not relatively prime. For example, $3 = \tau(4) \ne \tau(2) \cdot \tau(2) = 2 \cdot 2$.

(2.1.4) $\sigma(m \cdot n)$ is not equal to $\sigma(m) \cdot \sigma(n)$ if m and n are not relatively prime. For example, $7 = \sigma(4) \ne \sigma(2) \cdot \sigma(2) = 3 \cdot 3$.

(2.1.5) These are given in (2.2) and (2.3) in the next section.

(2.1.6) Even when m and n are relatively prime, $s(m \cdot n) \ne s(m) \cdot s(m)$. For example, $6 = s(6) \ne s(2) \cdot s(3) = 1 \cdot 1$. For this reason, the function $s(n)$ is really only of historical interest. The modern theory of arithmetic functions is built around functions with multiplicative properties, such as $\tau(n)$ and $\sigma(n)$.

(2.1.7) Observe that 3, 7, 31, and 127 are all prime numbers. Furthermore, they are all one less than a power of two: $3 = 262 - 1, 7 = 2^3 - 1$, $31 = 2^5 - 1, 127 = 2^7 - 1$.

(2.1.8) 1.

$$\sigma(2^8 \cdot 7 \cdot 73) = \sigma(2^8) \cdot \sigma(7) \cdot \sigma(73)$$

$$= \left(\frac{2^9 - 1}{2 - 1} \right) \left(\frac{7^2 - 1}{7 - 1} \right) \left(\frac{73^2 - 1}{73 - 1} \right)$$

$$= (2^9 - 1)(7 + 1)(73 + 1) = 302512.$$

So, $s(130816) = 302512 - 130816 = 171696$, and we do not have a perfect number.

2.

$$\sigma(2^{10} \cdot 23 \cdot 89) = (2^{11} - 1)(23 + 1)(89 + 1) = 4421520.$$

So, $s(2096128) = 4421520 - 2096128 = 2325392$, and again, we do not have a perfect number.

3.

$$\sigma(2^{12} \cdot 8191) = (2^{13} - 1)(8191 + 1) = 8191 \cdot 2^{13}$$
$$= 2 \cdot 2^{12} \cdot 8191.$$

So, $s(2^{12} \cdot 8191) = 2^{12} \cdot 8191$, and we *do* have a perfect number.

(2.1.9) $2^7 - 1 = 127$ is a divisor, and so is $2^{11} - 1 = 2047$. We observed earlier that $2047 = 23 \cdot 89$. After dividing out the primes $23, 89$, and 127, what is left is 581283643249112959, which also happens to be prime. In this case, the Cataldi-Fermat theorem gives a complete factorization into primes.

(2.1.10) Observe that $2^{17} - 1 = 131071$ is a prime. The corresponding perfect number is 8589869056.

(2.1.11) The theorem is proved in the next section.

(2.1.12) Notice that $496 = 1^3 + 3^3 + 5^3 + 7^3$.

(2.1.13) From (1.8),

$$1^3 + 2^3 + \cdots + (2N)^3 = \frac{(2N)^2 (2N + 1)^2}{4},$$

and

$$2^3 + 4^3 + \cdots + (2N)^3 = 8 \cdot (1^3 + 2^3 + \cdots + N^3)$$
$$= 8 \cdot \frac{N^2 (N + 1)^2}{4}.$$

Subtraction gives

$$1^3 + 3^3 + \cdots + (2N-1)^3 = N^2(2N+1)^2 - 2N^2(N+1)^2$$
$$= N^2(2N^2 - 1).$$

If this is equal to $2^{p-1}(2^p - 1)$, then $N^2 = 2^{p-1}$, or $N = 2^{(p-1)/2}$. For example, to get the perfect number $8128 = 2^6(2^7 - 1)$, we will need $N = 2^3 = 8$ and $2N - 1 = 15$:

$$8128 = 1^3 + 3^3 + \cdots + 15^3.$$

(2.1.14) All odd integers below 100 are deficient, whereas the even integers seem to be roughly split between abundant and deficient. Even integers have a much better chance of being abundant, because $n/2$ is already a divisor. Multiples of 6 do even better; because $n/2$, $n/3$, and $n/6$ are all distinct divisors, $s(n) \geq n/2 + n/3 + n/6 = n$. Jordanus de Nemore discovered in the thirteenth century that every multiple of a perfect or abundant number is abundant. Try to prove this.

The parity modulo 2 is almost always preserved. That is, if n is even, then $s(n)$ is even, and if n is odd, then $s(n)$ is odd, *unless n* is a square or twice a square, in which case the parity is reversed. This follows immediately from the three obvious rules "odd plus odd is even, odd plus even is odd, even plus even is even," and the following more subtle fact about $\sigma(n)$. The function $\sigma(n)$ is always even, unless $n = m^2$ or $2m^2$. Any power of two that divides n cannot change the parity of $\sigma(n)$, because $\sigma(2^k) = 2^{k+1} - 1$ is always odd. On the other hand, for an odd prime p,

$$\sigma(p^j) = \underbrace{1 + p + p^2 + \cdots + p^j}_{j+1 \text{ odd summands}}.$$

So, $\sigma(p^j)$ is odd if j is even and even if j is odd. The product over all primes is even as soon as a single factor is even. In other words, the only way $\sigma(n)$ can fail to be even is if every exponent j is even, which says that n is a power of two times a square. This is equivalent to being a square or twice a square.

(2.1.15) Notice that $s(284) = 220$ and $s(220) = 284$. We say that m and n are an AMICABLE PAIR if $s(m) = n$ and $s(n) = m$. This is a generalization of perfect numbers, which are fixed points of the function $s(n)$ under iteration. Amicable pairs are cycles of length two. Cycles of length

4, 5, 6, 8, and 9 are known, but no cycles of length 3 are known to exist.

(2.1.16) If $s(m) = n$ and $s(n) = m$, then $\sigma(m) = s(m) + m = n + m$ and also $\sigma(n) = s(n) + n = m + n$. Conversely, if $\sigma(n) = n + m$, then $s(n) = \sigma(n) - n = m$. If also $\sigma(m) = n + m$, the same reasoning shows that $s(m) = n$.

(2.1.17) With $p = 3 \cdot 2^{k-1} - 1$, $q = 3 \cdot 2^k - 1$, and $r = 9 \cdot 2^{2k-1} - 1$ all primes, we have

$$\sigma(p \cdot q \cdot 2^k) = \sigma(p)\sigma(q)\sigma(2^k) = (p+1)(q+1)(2^{k+1} - 1)$$
$$= (3 \cdot 2^{k-1})(3 \cdot 2^k)(2^{k+1} - 1) = 9 \cdot 2^{2k-1} \cdot (2^{k+1} - 1)$$

and

$$\sigma(r \cdot 2^k) = \sigma(r)\sigma(2^k) = (r+1)(2^{k+1} - 1) = 9 \cdot 2^{2k-1} \cdot (2^{k+1} - 1).$$

Finally, we compute $p \cdot q \cdot 2^k + r \cdot 2^k$ as

$$(3 \cdot 2^{k-1} - 1)(3 \cdot 2^k - 1)2^k + (9 \cdot 2^{2k-1} - 1)2^k$$
$$= ((9 \cdot 2^{2k-1} - 3 \cdot 2^k - 3 \cdot 2^{k-1} + 1) + (9 \cdot 2^{2k-1} - 1))2^k$$
$$= ((9 \cdot 2^{2k-1} - 3 \cdot 2 \cdot 2^{k-1} - 3 \cdot 2^{k-1} + 1) + (9 \cdot 2^{2k-1} - 1))2^k$$
$$= ((9 \cdot 2^{2k-1} - 9 \cdot 2^{k-1} + 1) + (9 \cdot 2^{2k-1} - 1))2^k$$
$$= (9 \cdot 2 \cdot 2^{2k-1} - 9 \cdot 2^{k-1})2^k$$
$$= 9 \cdot (2^{k+1} - 1)2^{2k+1}.$$

This last condition was tedious to check but necessary. The equality $\sigma(m) = \sigma(n)$ alone does not guarantee an amicable pair. For example, $\sigma(6) = 12 = \sigma(11)$, but $s(6) = 6$ and $s(11) = 1$.

(2.1.18) This is just algebra. Plug in the values $p = 3 \cdot 2^{k-1} - 1$, $q = 3 \cdot 2^k - 1$, and $r = 9 \cdot 2^{2k-1} - 1$.

(2.1.19) The sequences you get when iterating $s(n)$ are often called ALIQUOT SEQUENCES. There are only three possibilities, depending on the initial value.

(a) The sequence may eventually reach some prime number p. Then, in the next stage, $s(p) = 1$ and $s(1) = 0$. The sequence terminates.

(b) The sequence may eventually reach a perfect number, a member of an amicable pair, or a member of some longer cycle. The sequence becomes stuck in a loop. Stuck in a loop.

(c) The sequence may grow to infinity.

The Catalan-Dickson conjecture is the claim that this last possibility never happens. It is known to be true for all integers $n < 276$ and many $n > 276$ as well. After 1,284 iterations, the sequence that starts at 276 reaches a number with 117 digits.

Despite this evidence, modern researchers believe that Catalan-Dickson is false, and that 'most' sequences starting with an even integer go to infinity. It follows from Exercise 2.1.14 that if n is even, it is likely that $s(n)$ is too, because squares tend to be very scarce. The techniques of Chapter 4 can be used to show that for an even integer $2n$, the average value of $s(2n)$ is $(5\pi^2/24 - 1) \cdot 2n$. This says that on average, the output $s(2n)$ is $1.05617\ldots$ times as big as the input $2n$.

The problem is that Catalan-Dickson can never be disproved by mere computation. A theorem is needed. Even computation is difficult, because computing $s(n)$ requires computing $\sigma(n)$, which in turn requires factoring n. We can factor a number with 117 digits, slowly, but the next term in the sequence will likely be just as big.

For more on aliquot sequences, see Guy (1994).

(2.1.20) Here're a couple of little theorems. Try to prove them.
  (a) There is no n such that $s(n) = 5$.
  (b) Suppose that the Goldbach conjecture is true; that is, every even integer ≥ 6 is the sum of two distinct odd primes p and q. Then, for every odd integer $n \geq 7$, there is an m such that $s(m) = n$.
  (c) Suppose that $m > n^2$. Show that $s(m) > n$. This implies that the *number* of m such that $s(m) = n$ is bounded by n^2.

(2.1.21) For primes p, q, and r,

$$\tau(p^7) = \tau(p^3 q) = \tau(pqr) = 8,$$

and every integer D with $\tau(D) = 8$ must fall into one of these three cases, because the only ways to factor 8 are with 8, $4 \cdot 2$, or $2 \cdot 2 \cdot 2$. The smallest primes are 2, 3, and 5. The smallest integer that can be formed this way is $2^3 \cdot 3 = 24$. Similar arguments show that $D(12) = 60$, $D(16) = 120$, $D(24) = 360$, and $D(60) = 5040$.

(2.1.22) For any n,

$$\left(\sum_{k|n} \tau(k) \right)^2 = \sum_{k|n} \tau(k)^3.$$

Try to prove this for $n = p^j$, a power of a single prime. The general case requires the techniques discussed in Section 2.3.

(2.2.1) The sum of the row entries is

$$1 \cdot c + \cdots + m \cdot c = (1 + \cdots + m) \cdot c = \sigma(m) \cdot c,$$

because the columns of the table are indexed by all the divisors of m. If we now sum this over all rows, we get

$$\sigma(m) \cdot 1 + \cdots + \sigma(m) \cdot c + \cdots + \sigma(m) \cdot n$$
$$= \sigma(m)(1 + \cdots + c + \cdots + n) = \sigma(m)\sigma(n).$$

On the other hand, the entries of the table are all the divisors of mn. So, the sum of them all is $\sigma(mn)$.

(2.2.2) An even perfect number is of the form

$$2^{p-1}(2^p - 1) = \frac{2^p(2^p - 1)}{2} = t_{2^p-1}.$$

(2.3.1)

$$u * u(n) = \sum_{d|n} 1 \cdot 1 = \tau(n).$$

(2.3.2)

$$f * e(n) = \sum_{d|n} f(d)e(n/d) = 0 + 0 + \cdots + f(n)e(1) = f(n).$$

(2.3.3) See the theorem immediately following this exercise.

(2.3.4) Suppose that $g = f * \mu$. Then,

$$g * u = (f * \mu) * u = f * (\mu * u) = f * e = f.$$

(2.3.5) In the notation of convolutions, the exercise claims that $\tau * \mu = u$. Exercise 2.3.1 says that $\tau = u * u$, and Möbius inversion applies.

(3.1.1) All are true.

(a) $2x + 1 \le Cx$ for, say, $C = 3$ and $x \ge 1$.

(b) This is most easily done in stages. First, show that $100 \ll \exp(x)$. But this is true with $C = 1$ for $x \ge \log(100) = 4.60517$. Next, show that $10x \ll \exp(x)$. Observe that $e^4 = 54.5982 \ge 10 \cdot 4 = 40$. Furthermore, because the graph of $y = \exp(x)$ is concave up, we will get that $\exp(x) \ge 10x$ for all $x \ge 4$. We've shown that

$$100 \le \exp(x) \quad \text{for } x \ge 4.60517,$$
$$10x \le \exp(x) \quad \text{for } x \ge 4.$$

Adding the inequalities in the range where both are true gives

$$100 + 10x \le 2\exp(x) \quad \text{for } x \ge 4.60517.$$

(c) Again, it is easiest to first treat the pieces separately, then to add. With $C = 2$, we certainly have $2 \leq 2 \cdot 1$ for all x. And $\sin(x) \leq 1 = 1 \cdot 1$ for all x. Adding gives $2 + \sin(x) \leq 3 \cdot 1$ for all x, so the bound is true with $C = 3$.

(d) We will have $\exp(-x) \leq C/x$ exactly when $x \leq C \exp(x)$. Can we take $C = 1$? At $x = 1$, the inequality is true: $1 \leq e$. And the concavity argument works again; $x \leq \exp(x)$ for all $x \geq 1$.

(e) Simplify $\log(e^3 x) = 3 + \log(x)$. With $C = 1, 3 \leq Cx$ for $x \geq 3$. Can we also show that $\log(x) \leq 1 \cdot x$? Because $\exp(x)$ is an increasing function, the inequality is the same as $x \leq \exp(x)$, which we just showed is true for $x \geq 1$. Adding the two inequalities in the range where they are both true gives $\log(e^3 x) \leq 2x$ for $x \geq 3$.

Of course, there are other ways to do these problems. The C you get does not have to be the smallest possible constant; it merely needs to give a true inequality.

(3.1.2) The divisors d of n are a subset of the set of all integers less than or equal to n. So,

$$\sigma(n) = \sum_{d \mid n} d \leq \sum_{d=1}^{n} d = t_n = \frac{n(n+1)}{2}.$$

We must show that $n(n+1)/2 \ll n^2$. But dividing by n shows that $n(n+1)/2 \leq Cn^2$ will be true when $n + 1 \leq 2Cn$. This is true for $C = 1$ if $n \geq 1$, because $n + 1 \leq 2n$ is the same as (subtracting n now) $1 \leq n$.

(3.1.3) As the hint says, the divisors of n come in pairs d and n/d, one of which must be less than or equal to $\sqrt{n}$. (Why?) Thus, the number of divisors is certainly less than twice the number of integers less than or equal to $\sqrt{n}$. In other words, $\tau(n) \leq 2\sqrt{n}$ for all n; so, taking $C = 2, \tau(n) \ll \sqrt{n}$.

(3.1.4) From the definitions, $|x/(x+1) - 1| = |-1/(x+1)| = 1/(x+1)$. We have $1/(x+1) \leq C/x$ exactly when $x \leq C(x+1)$. So, we can take $C = 1$; the inequality is always true. Because $|\cosh(x) - \exp(x)/2| = \exp(-x)/2$, we can take $C = 1/2$.

(3.1.5) From the exact formula (1.7), we know that

$$\sum_{k=1}^{n} k^2 = \frac{n(n+1)(2n+1)}{6} = \frac{n^3}{3} + \frac{n^2}{2} + \frac{n}{6}.$$

So, it suffices to show that the difference $n^2/2 + n/6$ is $\ll n^2$. But $n^2/2 + n/6 \le Cn^2$ is equivalent to $n/3 \le (2C - 1)n^2$. Dividing by n shows that equivalent to $1/3 \le (2C - 1)n$. So, we can take $C = 1$; the inequality holds for all $n \ge 1$.

(3.2.1) Just add $\log(n)$ everywhere in the inequality

$$0 < H_n - \log(n) < 1$$

at the end of the previous lemma.

(3.2.2) H_{1000} should be about $\log(1000) + \gamma = 7.48497$, with an error of about $1/1000$, that is, accurate to about three digits. H_{10000} should be about $\log(10000) + \gamma = 9.78756$, accurate to four digits. H_{100000} should be about $\log(100000) + \gamma = 12.09014$, accurate to five digits.

(3.2.3) With $n = 10^k$, it looks line H_n is $\log(n) + \gamma + 1/(2n)$ to within about $2k$ digits of accuracy. For example, 10^2 give four digits (10^{-4}), and 10^3 gives six (10^{-6}). So, you might conjecture that

$$H_n = \log(n) + \gamma + 1/(2n) + O(1/n^2).$$

(3.2.4) Here, each increase by one in the exponent gives an extra four digits of accuracy. So, you might conjecture that

$$H_n = \log(n) + \gamma + 1/(2n) - 1/(12n^2) + O(1/n^4).$$

(3.3.2) From (3.4), we know that for some C,

$$|\log(n!) - (n\log(n) - n)| \le C\log(n).$$

It helps to write

$$n\log(n) - n = \log(n^n) - \log(e^n) = \log((n/e)^n).$$

So, (3.4) says that

$$\left|\log(n!) - \log((n/e)^n)\right| \le C\log(n).$$

By the definition of absolute value, this is equivalent to

$$-C\log(n) \le \log(n!) - \log((n/e)^n) \le C\log(n).$$

Exponentiating, and using the inequality on the right, gives

$$\frac{n!}{(n/e)^n} \le n^C \quad \text{or} \quad n! \le n^C(n/e)^n.$$

This is weaker than what was claimed, but we actually know more than what (3.4) says. The last line of the lemma proved the inequality

$$\log(n!) - (n \log(n) - n) \le \log(n) + 1.$$

(This is a case where it is better not to use $\ll$ notation to hide messy constants.) Imitate the calculations above to get the estimate $n! \ll n(n/e)^n$. What does the "+1" in the above inequality tell you the implied constant C will be?

(3.4.1) Because $1 \le \log(n)$ for $n \ge 3$, it follows that $n \le n \log(n)$. Therefore,

$$n \log(n) + n \le 2n \log(n) \ll n \log(n).$$

(3.4.2) Subtract n from both sides of (3.7) to get $s(n) \le n \log(n)$ for all n.

(3.5.1) The derivative of $(t + 1)p^{-t/3}$ is

$$p^{-t/3} - \log(p)(t + 1)p^{-t/3}/3 = p^{-t/3}(1 - \log(p)(t + 1)/3).$$

This is 0 only when $1 - \log(p)(t + 1)/3 = 0$, or $t = 3/\log(p) - 1$. Plugging back in, we get the value

$$C(p) = \frac{3p^{1/3}}{e \log(p)}.$$

For the primes $p = 2, 3, 5, \ldots 19$, we get the values

$$2.00606,\ 1.44885,\ 1.17258,\ 1.08493,$$
$$1.02359,\ 1.01172,\ 1.00161,\ 1.00017$$

for $C(p)$. The product of all eight is 3.83597.

(3.5.2) For $\log(p) > 4$, that is, for primes greater than or equal to 59,

$$\tau(p^t) = t + 1 \le p^{t/4} \quad \text{for all } t \ge 0.$$

The sixteen primes $2, 3, \ldots, 53$ need to be treated separately. The maximum of the function $(t + 1) \exp(-t \log(p)/4)$ is

$$C(p) = \frac{4p^{1/4}}{e \log(p)}.$$

The product of all sixteen constants happens to be 10.6767.

(3.6.1) If $\sigma(n) > Cn$, then $s(n) > (C - 1)$.

(3.6.2) To get $s(n) > n$ via the construction of the theorem, we want $\sigma(n) > 2n$, or $C = 2$. Because $e^2 = 7.38906$, we take $N = 8$ and

$n = N! = 40320 = 2^7 \cdot 3^2 \cdot 5 \cdot 7$. The factorization is useful in computing

$$\sigma(40320) = \sigma(2^7)\sigma(3^2)\sigma(5)\sigma(7) = 159120.$$

So, $s(40320) = 159120 - 40320 = 118800$, and 40320 is abundant.

The point of this exercise is to show that the theorem guarantees that certain integers exist but does not claim to find the smallest examples. Just to drive this point home, we get $s(n) > 2n$ when $\sigma(n) > 3n$. Because $e^3 = 20.0855$, we need $N = 21$ and

$$n = 21! = 51090942171709440000$$
$$= 2^{18} \cdot 3^9 \cdot 5^4 \cdot 7^3 \cdot 11 \cdot 13 \cdot 17 \cdot 19.$$

(It is easy to factor a factorial integer.) You can then compute $s(21!) = 241369473970791936000$.

(3.6.3)

$$
\begin{array}{llll}
7!! = 105, & \sigma(7!!) = 192, & s(7!!) = 87, \\
9!! = 945, & \sigma(9!!) = 1920, & s(9!!) = 975, \\
11!! = 10395, & \sigma(11!!) = 23040, & s(11!!) = 12645, \\
13!! = 135135, & \sigma(13!!) = 322560, & s(13!!) = 187425.
\end{array}
$$

(3.6.4) We get $\log(N)/2 > 2$ when $N > e^4 = 54.5982$. So, we take $N = 55$, then $2N + 1 = 111$. Observe that $9!!$ is already abundant. You computed this in the previous exercise.

(3.7.1) You can take the integers $2^{100}, 2^{101}, 2^{102}, 2^{103}, \ldots$.

(3.7.2) Nothing, of course. In relation to the previous exercise, the integers $3^{100}, 3^{101}, 3^{102}, 3^{103}, \ldots$ all have more than 100 divisors each.

(3.7.3) The integers $6^{16}, 6^{17}, 6^{18}, 6^{19}, \ldots$ all satisfy $\tau(n) > 10\log(n)$.

(3.7.4) Take $n = 30^m = (2 \cdot 3 \cdot 5)^m$. Then,

$$\tau(n) = (m + 1)^3 = (\log(n)/\log(30) + 1)^3.$$

Now, show that $(\log(n)/\log(30) + 1)^3$ is not $\ll \log(n)^2$.

(4.0.1) Divide both sides of (3.3) by $\log(n)$ and use the fact that $\gamma/\log(n)$ and $1/n\log(n)$ will go to 0 as n gets big.

(4.0.2) In Chapter 3, $t_n = n^2/2 + O(n)$ was proved. Dividing through, we get that

$$\frac{t_n}{n^2/2} = 1 + O(1/n)$$

and $1/n$ goes to 0 as n goes to ∞. The limit is 1.

(4.0.3) Divide both sides of (3.4) by $n \log(n)$ to get

$$\frac{\log(n!)}{n \log(n)} = 1 - \frac{1}{\log(n)} + O(1/n).$$

Because $1/\log(n)$ and $1/n$ go to 0, the limit is 1. Geometrically, this says that the points $(n, \log(n!)/\log(n))$ will tend toward the line $y = x$ with slope 1.

(4.1.1) (a) $\tau(8) = 4$, corresponding to the lattice points $(1, 8)$, $(2, 4)$, $(4, 2)$, $(8, 1)$.

(b) $\tau(6) = 4$, corresponding to $(1, 6)$, $(2, 3)$, $(3, 2)$, $(6, 1)$.

(c) $\tau(4) = 3$, corresponding to $(1, 4)$, $(2, 2)$, $(4, 1)$.

(d) $\tau(7) = 2$, corresponding to $(1, 7)$, $(7, 1)$.

(e) $\tau(5) = 2$, corresponding to $(1, 5)$, $(5, 1)$.

(f) $\tau(3) = 2$, corresponding to $(1, 3)$, $(3, 1)$.

(g) $\tau(2) = 2$, corresponding to $(1, 2)$, $(2, 1)$.

(h) $\tau(1) = 1$, corresponding to just $(1, 1)$.

There are 20 lattice points.

(4.1.2) And

$$\sum_{k=1}^{8} \tau(k) = 4 + 4 + 3 + 2 + 2 + 2 + 2 + 1 = 20,$$

which is the same.

(4.1.3) For general n, $\sum_{k=1}^{n} \tau(k)$ will be equal to the number of lattice points under the hyperbola $xy = n$.

(4.1.4) The point here is that the squares roughly fill in the area under the hyperbola.

(4.1.5)

$$\int_{1}^{n} \frac{n}{x} \, dx = n \int_{1}^{n} \frac{1}{x} \, dx = n \log(x)|_{1}^{n} = n \log(n).$$

(4.1.6) These exercises suggest that

$$\sum_{k=1}^{n} \tau(k) \sim n \log(n) \sim \log(n!) = \sum_{k=1}^{n} \log(k).$$

So, the conjecture is that the average order of $\tau(n)$ is $\log(n)$.

(4.2.2)

$$\frac{[t]([t]+1)}{2} = \frac{(t+O(1))(t+O(1)+1)}{2}$$

$$= \frac{(t+O(1))(t+O(1))}{2}$$

$$= \frac{1}{2}(t^2 + 2tO(1) + O(1)O(1))$$

$$= \frac{1}{2}(t^2 + O(t) + O(1))$$

$$= \frac{1}{2}(t^2 + O(t))$$

$$= \frac{t^2}{2} + O(t).$$

(4.2.3)

$$\log(t) - \log([t]) = \int_{[t]}^{t} \frac{1}{x} dx.$$

Because $t - [t] < 1$ and $1/x < 1/[t]$ for $[t] \le x \le t$, the area under $y = 1/x$ is less than $1 \cdot 1/[t]$.

(4.3.1) $H_1^{(2)} = 1$, $H_2^{(2)} = 1 + 1/4 = 5/4$, and $H_3^{(2)} = 1 + 1/4 + 1/9 = 49/36$.

(4.3.2) For n, a power of 10, say 10^k, $1/n^2 = 10^{-2k}$. The error is bounded by (a constant times) this, so about $2k$ digits should be correct. For example, with $n = 10^2$, $H_n^{(2)} = 1.63498\ldots$ and $\zeta(2) - 1/n = 1.63493\ldots$, four digits of accuracy.

(4.3.3) A natural choice is $H_n^{(3)} = \sum_{k=1}^{n} 1/k^3$. The theorem is that

$$H_n^{(3)} = -\frac{1}{2n^2} + \zeta(3) + O\left(\frac{1}{n^3}\right)$$

or, equivalently,

$$\zeta(3) - H_n^{(3)} = \frac{1}{2n^2} + O\left(\frac{1}{n^3}\right).$$

The term $1/2n^2$ comes from the integral

$$\int_n^\infty \frac{1}{x^3} dx = \frac{-1}{2x^2}\Big|_n^\infty = \frac{1}{2n^2}.$$

(4.3.4) Using the given value $\zeta(3) = 1.2020569031595942854\ldots$, we see that

$$-\frac{1}{2 \cdot 10^2} + \zeta(3) = 1.1970569031595942854\ldots,$$

$$-\frac{1}{2 \cdot 10^4} + \zeta(3) = 1.2020069031595942854\ldots,$$

$$-\frac{1}{2 \cdot 10^6} + \zeta(3) = 1.2020564031595942854\ldots,$$

$$-\frac{1}{2 \cdot 10^8} + \zeta(3) = 1.2020568981595942854\ldots,$$

$$-\frac{1}{2 \cdot 10^{10}} + \zeta(3) = 1.2020569031095942854\ldots,$$

$$-\frac{1}{2 \cdot 10^{12}} + \zeta(3) = 1.2020569031590942854\ldots.$$

As expected, with $n = 10$ (the first example), we get about three digits of accuracy. The second example gives about six digits correctly.

(4.4.1) In $\ll$ notation, (4.7) says that

$$\left| \sum_{k=1}^{n} \sigma(k) - \zeta(2)\frac{n^2}{2} \right| \ll n \log(n).$$

So, for some C,

$$\left| \sum_{k=1}^{n} \sigma(k) - \zeta(2)\frac{n^2}{2} \right| \leq C n \log(n).$$

This means that

$$-Cn \log(n) \leq \sum_{k=1}^{n} \sigma(k) - \zeta(2)\frac{n^2}{2} \leq Cn \log(n).$$

(4.4.2) We already know the stronger estimate,

$$\sum_{k=1}^{n} k = \frac{n^2}{2} + O(n)$$

So, certainly an estimate that allows a larger error is true:

$$\sum_{k=1}^{n} k = \frac{n^2}{2} + O(n \log(n)).$$

So, for some constant D,

$$-Dn \log(n) \le \frac{n^2}{2} - \sum_{k=1}^{n} k \le Dn \log(n).$$

(We need to use a different name for the constant, because we will compare this estimate to the previous one, which has a C in it.)

(4.4.3) Adding the two inequalities and using $\sigma(k) - k = s(k)$, we get

$$-(C + D)n \log(n) \le \sum_{k=1}^{n} s(k) - (\zeta(2) - 1)\frac{n^2}{2} \le (C + D)n \log(n),$$

which says that

$$\sum_{k=1}^{n} s(k) = (\zeta(2) - 1)\frac{n^2}{2} + O(n \log(n)).$$

(4.4.4)

$$\sum_{k=1}^{n} (\zeta(2) - 1)k = (\zeta(2) - 1)\sum_{k=1}^{n} k$$

$$= (\zeta(2) - 1)\left(\frac{n^2}{2} + O(n \log(n))\right)$$

$$= (\zeta(2) - 1)\frac{n^2}{2} + O(n \log(n)).$$

(4.4.5) Dividing both sides of Exercise 4.4.3 by $(\zeta(2) - 1)n^2/2$ and using the fact that $\log(n)/n \to 0$, we see that

$$\sum_{k=1}^{n} s(k) \sim (\zeta(2) - 1)\frac{n^2}{2}.$$

Similarly, Exercise 4.4.4 says that

$$\sum_{k=1}^{n} (\zeta(2) - 1)k \sim (\zeta(2) - 1)\frac{n^2}{2}.$$

So,

$$\sum_{k=1}^{n} s(k) \sim \sum_{k=1}^{n} (\zeta(2) - 1)k.$$

The average order of $s(n)$ is $(\zeta(2) - 1)n$. An "average" integer is deficient by about 65%.

(II.1.1) With $f(t) = \log(1 - t)$, $f'(t) = -1/(1 - t)$ and $f'(0) = -1$. Because $f(0) = 0$, the line is $y - 0 = -(t - 0)$, or just $y = -t$. At

$t = 1/17, y = -1/17 \approx -0.059$. The exact answer is $\log(16/17) \approx -0.061$

(I1.1.2) $f'(x) = 1/2 \cdot (1/4 + x^2)^{-1/2} \cdot 2x = x(1/4 + x^2)^{-1/2}$. So, $f'(0) = 0$. Because $f(0) = 1/2$, the line is $y - 1/2 = 0(x - 0)$, or just $y = 1/2$. This linear approximation happens to be a horizontal line.

(I1.2.1) For $x < 0$, $\exp(x) < e^0 = 1$. So, $\exp(x) - 1 < 0$ is certainly less than the (positive) number $-2x$. This is the trivial half. The other half is equivalent to showing that $2x + 1 \le \exp(x)$ for $x \le 0$, which again follows from the slope of the tangent line.

(I1.2.2) Because $\sin(x)$ is concave down at $x = 0$, it lies under its tangent line, whose slope is 1. A line with slope 2 is always above it.

(I1.3.1) The point is that according to property 4,

$$\int_a^b f(t)dt = \int_a^p f(t)dt + \int_p^q f(t)dt + \int_q^b f(t)dt,$$

and analogously for $\int_a^b |f(t)|dt$. The two integrals from a to p are equal, because when $f(t) > 0$, $f(t) = |f(t)|$. Similarly for the two integrals from q to b. But on the interval from p to q, $f(t) < 0$. So, $f(t) = -|f(t)|$, and on this piece the two integrals have opposite signs according to property 2.

(I1.3.2) Because n is constant in the sum,

$$\sum_{k=1}^n \left(\frac{k}{n}\right)^2 \frac{1}{n} = \frac{1}{n^3} \sum_{k=1}^n k^2 = \frac{n(n+1)(2n+1)}{6n^3} = \frac{1}{3} + \frac{1}{3n} + \frac{1}{6n^2}.$$

(I1.3.3) Observe, for example, that the derivative $f(x)$ is positive where $F(x)$ is increasing; $f(x)$ is negative where $F(x)$ is decreasing. Also, $f(x)$ crosses the axis at the maxima and minima of $F(x)$.

(I1.3.4) According to property 4,

$$\log(xy) = \int_1^{xy} \frac{1}{t} dt = \int_1^x \frac{1}{t} dt + \int_x^{xy} \frac{1}{t} dt$$
$$= \log(x) + \int_x^{xy} \frac{1}{t} dt.$$

With $u = t/x$, $du = 1/x dt$ or $x du = dt$ and $1/t = u/x$. So, $1/t dt = 1/u du$. The change of variables does nothing to the function being integrated, but the limits of integration change: When $t = x, u = 1$, and when $t = xy, u = y$. So, we have

$$= \log(x) + \int_1^y \frac{1}{u} du$$
$$= \log(x) + \log(y).$$

(5.1.2) The smallest integer divisible by both 11 and 17 is $11 \cdot 17 = 187$; this is much bigger than 52. So, no integer near 52 that is divisible by 17 can possibly be divisible by 11 also. These are not independent events; Hypothesis I is not justified.

(5.1.3) Table 5.1 shows seven primes (namely 9511, 9521, 9533, 9539, 9547, 9551, and 9587) between 9500 and 9600. So, $\pi(9600) = \pi(9500) + 7 = 1{,}177 + 7 = 1184$.

(5.1.4) The ratios are 1.20505, 1.0572, 1.01394, 1.00394, 1.00165, 1.00051, 1.00013, 1.00003, and 1.00001.

(5.1.5) Assuming (5.2), we have

$$\prod_{\substack{p \text{ prime} \\ p < x^a}} (1 - 1/p) \sim \frac{\exp(-\gamma)}{\log(x^a)} = \frac{\exp(-\gamma)}{a\log(x)}.$$

If this is supposed to be $\sim 1/\log(x)$, we must have $\exp(-\gamma) = a$.

(5.2.2) We have

$$0 < x - \frac{3.44}{4}(x + \log(2)) \quad \Leftrightarrow \quad \frac{3.44}{4}(x + \log(2)) < x.$$

Each function defines a line, but $3.44/4(x + \log(2))$ has slope $3.44/4 < 1$, and at $x = 4.6$,

$$\frac{3.44}{4}(4.6 + \log(2)) = 4.55 < 4.6.$$

Because the line with the smaller slope is already underneath at $x = 4.6$, it always is.

(I2.1.1)

$$\exp(x) \approx 1 + x,$$
$$\exp(x) \approx 1 + x + x^2/2.$$

(I2.1.2)

$$\log(x) \approx (x - 1) - 1/2(x - 1)^2,$$
$$1/(1 - x) \approx 1 + x + x^2.$$

(I2.1.3)

$$\sin(x) \approx x.$$

(I2.1.4)

$$3x^2 + 7x - 4 \approx -4 + 7x + 3x^2,$$
$$3x^2 + 7x - 4 \approx 6 + 13(x - 1) + 3(x - 1)^2.$$

Observe that after multiplying out, this is $3x^2 + 7x - 4$.

(I2.1.5)

$$\exp(x) \approx 1 + x + x^2/2 + x^3/6,$$
$$\log(x) \approx (x - 1) - 1/2(x - 1)^2 + 1/2(x - 1)^3,$$
$$1/(1 - x) \approx 1 + x + x^2 + x^3,$$
$$\sin(x) \approx -1/6x^3,$$
$$3x^2 + 7x - 4 \approx -4 + 7x + 3x^2 + 0x^3.$$

(I2.1.6)

$$x^2 \cos(3x) \approx x^2.$$

(I2.2.1)

$$\sum_{n=0}^{\infty} (-1)^n x^n, \qquad \sum_{n=0}^{\infty} x^{2n}, \qquad \sum_{n=0}^{\infty} (-1)^n x^{2n}.$$

(I2.2.2)

$$\sum_{n=0}^{\infty} (-1)^n 3^{2n} x^{2n+2}/(2n)!.$$

(I2.2.3)

$$\sum_{n=0}^{\infty} (-1)^n x^{2n}/(2n + 1)!,$$

$$\sum_{n=1}^{\infty} x^{n-1}/n! = \sum_{n=0}^{\infty} x^n/(n + 1)!.$$

(I2.2.4)

$$\sum_{n=0}^{\infty} (-1)^n x^{2n+1}/(2n + 1).$$

(I2.2.5)

$$-x/(1 + x)^2, \qquad \sum_{n=1}^{\infty} (-1)^n n x^n.$$

(I2.2.6)

$$\sum_{n=2}^{\infty} (-1)^{n-1} x^n/n.$$

(I2.2.7)

$$\sum_{n=1}^{\infty} x^{n-1}/n \quad \text{or} \quad \sum_{n=0}^{\infty} x^n/(n+1).$$

(I2.2.8)

$$\sum_{n=1}^{\infty} x^n/(n \cdot n!).$$

(I2.2.9)

$$\sum_{n=1}^{\infty} x^n/n^2.$$

(I2.2.10)

$$1 + x^2/2 + 5x^4/24 + \ldots,$$
$$1 - x/2 + x^2/12 - x^4/720 + \ldots.$$

(I2.2.11) $2n + 2 = 10$ for $n = 4$. So, the tenth derivative is

$$((-1)^4 3^8/8!) \cdot 10! = 590490.$$

(I2.3.1) $2/(x^2 - 1) = 1/(x - 1) - 1/(x + 1)$. One term,

$$\frac{-1}{x+1} = \frac{-1}{x--1},$$

is already a Laurent series at -1. Meanwhile,

$$\frac{1}{x-1} = \frac{-1}{2-(x+1)} = \frac{-1/2}{1-(x+1)/2} = \sum_{n=0}^{\infty}(-1/2)^{n+1}(x+1)^n.$$

So, the expansion at -1 is

$$2/(x^2 - 1) = -1/(x+1) + \sum_{n=0}^{\infty}(-1/2)^{n+1}(x+1)^n.$$

At 0, we have an ordinary Taylor series:

$$-2/(1-x^2) = 2\sum_{n=0}^{\infty}(-1)^n x^{2n}.$$

(I2.3.2) -2.

(I2.3.3) The residue is the number N itself, the order of the zero.

(I2.4.2) $\sum_{n=0}^{\infty}(1/10)^n = 10/9$. At 1 kilometer per 3 minutes, it takes $10/3$ minutes, or 3 minutes 20 seconds.

(I2.5.1) $1 + k/2 = 10$ for $k = 18$. So, $H_{2^{18}} = H_{262144} > 10$.

(I2.5.2) The estimate (I2.11) says that $H_n > 1 + \log_2(n)/2 = 1 + \log(n)/(2\log(2))$.

(I2.6.1) 1, 1/2, 5/6, 7/12, 47/60. Some partial sums that you need a computer to see are $S_{10} = 0.645635\ldots$, $S_{10^2} = 0.688172\ldots$, $S_{10^3} = 0.692647\ldots$, $S_{10^4} = 0.693097\ldots$, $S_{10^5} = 0.693142\ldots$.

(I2.6.2) From Exercise 1.2.12, the Nth partial sum S_N of $\sum_n 1/t_n$ is just $2 - 2/(N+1)$. The sequence $\{2 - 2/(N+1)\}$ converges to 2, because $\{1/(N+1)\}$ converges to 0. Similarly, the other series converges to $3/2$.

(I2.6.3) 1. The $N = 1$ case merely says that $1 - 1/2 = 1/2$, which is true. Now, we assume the $N - 1$ case, that

$$1 - \frac{1}{2} + \frac{1}{3} - \cdots + \frac{1}{2N-2} = \frac{1}{N} + \cdots + \frac{1}{2N-2}.$$

Add $1/(2N-1) - 1/(2N)$ to both sides, so that the left side is in the correct format. On the right, combine the $1/N - 1/(2N)$ to get $+1/(2N)$. So, the left side is also what we want.

2. It is just $H_{2N} - H_N$.

3. We have $S_{2N} = H_{2N} - H_N$, and according to (3.3), we get

$$\begin{aligned}S_{2N} &= (\log(2N) + \gamma + O(1/2N)) - (\log(N) + \gamma + O(1/N)) \\ &= \log(2N) - \log(N) + O(1/N) \\ &= \log(2) + O(1/N).\end{aligned}$$

Because the sequence $\{1/N\}$ converges to 0, the sequence $\{S_{2N}\}$ converges to $\log(2)$.

(I2.6.4) 1. To show that

$$\frac{1}{1+t^2} + \frac{t^{4N+2}}{1+t^2} = \frac{1 + t^{4N+2}}{1+t^2} = \sum_{n=0}^{2N}(-1)^n t^{2n},$$

cross multiply by $1 + t^2$ and observe that all but the first and last terms cancel out.

2. Observe that

$$\int_0^1 \left(\sum_{n=0}^{2N}(-1)^n t^{2n}\right) dt = \sum_{n=0}^{2N}\int_0^1 (-1)^n t^{2n} dt$$

$$= \sum_{n=0}^{2N}\frac{(-1)^n}{2n+1}$$

and

$$\int_0^1 \frac{1}{1+t^2} = \arctan(1) - \arctan(0) = \pi/4 - 0.$$

3. Because $1 + t^2 > 1$, we deduce that $t^{4N+2}/(1+t^2) < t^{4N+2}$. So, according to property v,

$$\int_0^1 \frac{t^{4N+2}}{1+t^2} dt < \int_0^1 t^{4N+2} dt.$$

4.

$$\int_0^1 t^{4N+2} dt = \frac{t^{4N+3}}{4N+3} \Big|_0^1 = \frac{1}{4N+3},$$

which goes to 0.

(I2.6.7) $R = 1$.

(I2.6.11) The hints outline the proof. But observe that $\log(2) = 0.693147\ldots$. The partial sums S_{10^n}, given in the solution to Exercise I2.6.1, are accurate to about n digits, as predicted.

(6.1.1) $\zeta(4) = \pi^4/90$.

(6.2.1)

$$3S_2(n) = \binom{3}{0} B_0 n^3 + \binom{3}{1} B_1 n^2 + \binom{3}{2} B_2 n$$

$$= n^3 - \frac{3}{2} n^2 + \frac{1}{2} n$$

$$= \frac{2n^3 - 3n^2 + n}{2} = 3P_{n-1}.$$

(6.2.2)

$$(\exp(x) - 1) \sum_{k=0}^{\infty} \frac{B_k x^k}{k!} = \sum_{j=1}^{\infty} x^j/j! \sum_{k=0}^{\infty} \frac{B_k x^k}{k!}$$

$$= \sum_{j=1}^{\infty} \sum_{k=0}^{\infty} \frac{B_k}{j!k!} x^{j+k}.$$

Let $j + k = n$. Then, $0 \le k \le n - 1$ because $j \ge 1$; now, $j = n - k$

$$= \sum_{n=0}^{\infty} \left(\sum_{k=0}^{n-1} \frac{B_k}{(n-k)!k!} \right) x^n.$$

Because this is just x^1, we compare powers of x to get that

$$\sum_{k=0}^{n-1} \frac{B_k}{(n-k)!k!} = \begin{cases} 1, & \text{in the case of } n = 1, \\ 0, & \text{in the case of } n > 1. \end{cases}$$

Multiply both sides above by $n!$ and let $m = n - 1$.

(6.2.3) $B_4 = -1/30$; $B_6 = 1/42$.

(6.2.4)

$$S_4(n) = \frac{6n^5 - 15n^4 + 10n^3 - n}{30},$$

$$S_5(n) = \frac{2n^6 - 6n^5 + 5n^4 - n^2}{12},$$

$$S_6(n) = \frac{6n^7 - 21n^6 + 21n^5 - 7n^3 + n}{42}.$$

(6.3.1) From the conjectured formula

$$\sin(\pi z) \overset{?}{=} \pi z \prod_{n=1}^{\infty} \left(1 - \frac{z^2}{n^2}\right),$$

taking logarithmic derivatives gives

$$\frac{\pi \cos(\pi z)}{\sin(\pi z)} \overset{?}{=} \frac{0}{\pi} + \frac{1}{z} + \sum_{n=1}^{\infty} \frac{-2z/n^2}{1 - z^2/n^2},$$

$$\pi \cot(\pi z) \overset{?}{=} \frac{1}{z} + \sum_{n=1}^{\infty} \frac{-2z}{n^2 - z^2},$$

$$\pi z \cot(\pi z) \overset{?}{=} 1 + \sum_{n=1}^{\infty} \frac{-2z^2}{n^2 - z^2}.$$

With $w = \pi z$, we have

$$w \cot(w) \overset{?}{=} 1 + \sum_{n=1}^{\infty} \frac{-2w^2/\pi^2}{n^2 - w^2/\pi^2}$$

$$= 1 + \sum_{n=1}^{\infty} \frac{-2w^2}{\pi^2 n^2 - w^2},$$

which is (6.14).

(7.1.1) $1 - p^{-\epsilon} = 1 - \exp(-\epsilon \log(p)) \approx \log(p) \cdot \epsilon$ for ϵ near 0. The second derivative of $1 - \exp(-\epsilon \log(p))$, with respect to ϵ, is $-\log(p)^2 \exp(-\epsilon \log(p)) < 0$ for all ϵ. Thus, the function is always concave down and lies under its tangent line.

(7.1.2) According to Exercise 7.1.1,

$$\sum_{p<\exp(1/\epsilon)} \{1 - p^{-\epsilon}\} \left\{\frac{1}{p-1} - \frac{1}{p}\right\} < \sum_{p<\exp(1/\epsilon)} \{\epsilon \log(p)\} \left\{\frac{1}{p-1} - \frac{1}{p}\right\}$$

$$= \epsilon \sum_{p<\exp(1/\epsilon)} \frac{\log(p)}{(p-1)p}$$

$$< \epsilon \sum_{\text{all } p} \frac{\log(p)}{(p-1)p}.$$

The sum is finite according to (7.5).

(7.2.1) $n = 2, 3, 4, 5, 6, 8, 10, 12, 18$, and 24. Ironically, even though they satisfy the desired inequality, 2, 3, 4, 5, 6, and 10 are not abundant. The paradox is resolved when you realize that $\log(\log(n)) < 1$ for these very small values of n.

(7.2.2) $n = 9, 16, 20, 30, 36, 48, 60, 72, 84, 120, 180, 240, 360, 720, 840, 2520$, and 5040. Not only is whether there are any more an open problem, it is a *deep* problem. In 1984, Robin proved (Robin, 1984) that it is equivalent to the Riemann Hypothesis. More precisely,

Theorem. *If the Riemann Hypothesis is true,*

$$\sigma(n) \le \exp(\gamma)n\log(\log(n)) \quad \text{for all } n > 5040.$$

If the Riemann Hypothesis is false,

$$\sigma(n) \ge \exp(\gamma)n\log(\log(n)) \quad \text{for infinitely many } n.$$

(7.3.1) With the value $C_2 \approx 0.6601618$ given in the text, $2C_2/\log(10000)^2 \approx 0.0155643$. We expect about 1.5 Sophie Germain primes in intervals of length 100 near 10000.

(7.4.1) With x the 50th Mersenne prime, we have

$$\log(\log(x)) \sim 50 \cdot \log(2)/\exp(\gamma).$$

So, $\log_{10}(x) = \log_{10}(e)\log(x)$ will be approximately

$$\log_{10}(e)\exp(50 \cdot \log(2)/\exp(\gamma)) \approx 1.2 \times 10^8,$$

or about a hundred million digits. This is only a rough estimate, because $f \sim g$ does not imply that $\exp(f) \sim \exp(g)$; look at the definitions.

If $\log_{10}(x) > 10^9$, then

$$n > \frac{\exp(\gamma)}{\log(2)}(\log(10^9) - \log(\log_{10}(2))) \approx 55.3924.$$

So, the 56th Mersenne prime should have a billion digits.

(13.0.1) $(0, 1) = [1, \pi/2].\,[1, \pi/2]\cdot[1, \pi/2] = [1\cdot 1, \pi/2 + \pi/2] = [1, \pi] = (-1, 0).$

(13.0.2) $(1, 0) = [1, 0].\,[1, 0]\cdot[R, \theta] = [1\cdot R, 0 + \theta] = [R, \theta].$

(13.0.3) $\mathrm{Re}(z^2) = x^2 - y^2.\ \mathrm{Im}(z^2) = 2xy.$

(13.0.4) $z\bar{z} = (x + iy)(x - iy) = x^2 + ixy - ixy - i^2y^2 = x^2 + y^2 + 0i = R^2.$

(13.0.5) $1/i = -i.$

(13.0.6) $e^{\pi i} = \cos(\pi) + i\sin(\pi) = -1 + i0 = -1.$

(13.0.7) With $-y/(x^2 + y^2) = c,\ c = 0$ is $y = 0$, the x-axis. Otherwise, we have $x^2 + y/c + y^2 = 0$, which is

$$x^2 + \left(y + \frac{1}{2c}\right)^2 = \frac{1}{4c^2}.$$

These are circles of radius $1/(2c)$ centered at $(0, -1/c)$.

(13.0.8) The real part of $z^3 = (x + iy)^3$ is $x^3 - 3xy^2$; the imaginary part is $3x^2y - y^3$. The real part of $\exp(z) = \exp(x)(\cos(y) + i\sin(y))$ is $\exp(x)\cos(y)$; the imaginary part is $\exp(x)\sin(y)$.

(8.1.1) Using *Mathematica*, you might try

```
slice[sigma_] := ParametricPlot[
{Re[Zeta[sigma + I*t]], Im[Zeta[sigma + I*t]]},
{t, 0, 40}, AspectRatio -> Automatic,
PlotRange -> {{-1, 3}, {-2, 2}},
PlotLabel -> "sigma=" <> ToString[sigma]]
```

This creates a function that takes a real number sigma as input, and returns a -Graphics- object. The *Mathematica* command

```
Table[slice[sigma], {sigma, .1, .9, .01}]
```

will then make a table of -Graphics- objects for the various values of sigma. You can read about how to Animate these in the Help menu.

(8.2.1) Because $\exp(-\log(2)z) = 1 - \log(2)z + O(z^2)$ as $z \to 0$, with $z = s - 1$, we have

$$1 - 2^{1-s} = 1 - \exp(-\log(2)(s - 1))$$
$$= \log(2)(s - 1) + O((s - 1)^2) \quad \text{as } s \to 1.$$

So,

$$(1 - 2^{1-s})^{-1} = \frac{1}{\log(2)} \frac{1}{s-1} + O(1) \quad \text{as } s \to 1.$$

With $\phi(s) = \log(2) + O(s-1)$, we get

$$\zeta(s) = (1 - 2^{1-s})^{-1}\phi(s) = \frac{1}{s-1} + O(1) \quad \text{as } s \to 1.$$

(8.2.2) Applying the Euler operator $\mathcal{E}$ to both the function $-x/(1+x)^2$ and the series expansion $\sum_n (-1)^n n x^n$, we get

$$\frac{x^2 - x}{(1+x)^3} = \sum_n (-1)^n n^2 x^n, \quad \phi(-2) \overset{A}{=} 0,$$

$$\frac{-x^3 + 4x^2 - x}{(1+x)^4} = \sum_n (-1)^n n^3 x^n, \quad \phi(-3) \overset{A}{=} -\frac{1}{8},$$

$$\frac{x^4 - 11x^3 + 11x^2 - x}{(1+x)^5} = \sum_n (-1)^n n^4 x^n, \quad \phi(-4) \overset{A}{=} 0,$$

$$\frac{-x^5 + 26x^4 - 66x^3 + 26x^2 - x}{(1+x)^6} = \sum_n (-1)^n n^5 x^n, \quad \phi(-5) \overset{A}{=} \frac{1}{4},$$

$$\frac{x^6 - 57x^5 + 302x^4 - 302x^3 + 57x^2 - x}{(1+x)^7} = \sum_n (-1)^n n^6 x^n,$$

$$\phi(-6) \overset{A}{=} 0.$$

Finally,

$$\frac{-x^7 + 120x^6 - 1191x^5 + 2416x^4 - 1191x^3 + 120x^2 - x}{(1+x)^8}$$

$$= \sum_n (-1)^n n^7 x^n,$$

and so $\phi(-7) \overset{A}{=} -\frac{17}{16}$. The extra minus sign comes from the definition of ϕ. With $\zeta(-n) = (1 - 2^{1+n})^{-1}\phi(n)$, we get that

$$\zeta(-2) \overset{A}{=} 0 \quad \zeta(-3) \overset{A}{=} \frac{1}{120},$$

$$\zeta(-4) \overset{A}{=} 0 \quad \zeta(-5) \overset{A}{=} -\frac{1}{252},$$

$$\zeta(-6) \overset{A}{=} 0 \quad \zeta(-7) \overset{A}{=} \frac{1}{240}.$$

(8.2.3) Observe that

$$\zeta(-1)/B_2 = -\frac{1}{2},$$

$$\zeta(-3)/B_4 = -\frac{1}{4},$$

$$\zeta(-5)/B_6 = -\frac{1}{6},$$

$$\zeta(-7)/B_8 = -\frac{1}{8}.$$

(8.2.4) The proof is outlined already. The conclusion is that

$$\phi(1-m) = -\frac{B_m(1-2^m)}{m}.$$

So, for integer $m \geq 1$,

$$\zeta(1-m) = (1-2^m)^{-1}\phi(1-m) = -\frac{B_m}{m}.$$

(8.3.1)

$$\Gamma(1) = \int_0^\infty \exp(-t)dt = -\exp(-t)\,|_0^\infty = 0 - (-1) = 1.$$

(8.3.2) With $x = t^s$, $dx = st^{s-1}dt$. So, $t^{s-1}dt = 1/sdx$ and $t = x^{1/s}$. Also, $x = 0$ when $t = 0$ and $x = \infty$ when t does. This gives

$$\Gamma(s+1) = s\Gamma(s) = \int_0^\infty \exp\left(x^{1/s}\right)dx.$$

This is true for all $s > 0$. So, we can change s to $1/s$ everywhere:

$$\Gamma(1/s+1) = \int_0^\infty \exp(x^s)dx.$$

With $s = 2$, we have $\Gamma(3/2) = \sqrt{\pi}/2$ according to Exercise 9.1.2. The recursion gives

$$\Gamma(5/2) = 3/2\Gamma(3/2) = 3\sqrt{\pi}/4,$$
$$\Gamma(7/2) = 5/2\Gamma(5/2) = 15\sqrt{\pi}/8,$$
$$\Gamma(9/2) = 7/2\Gamma(7/2) = 105\sqrt{\pi}/16.$$

For integer $n \geq 1$,

$$\Gamma(n+1/2) = \Gamma((2n+1)/2) = (2n-1)!!/2^n\sqrt{\pi}.$$

Working backward, $\Gamma(3/2) = 1/2\Gamma(1/2)$ implies that $\Gamma(1/2) = 2\Gamma(3/2) = \sqrt{\pi}$. Similarly, $\Gamma(-1/2) = -2\sqrt{\pi}$.

(8.3.3) According to (8.5), $\Gamma(s+1)$ is now given by

$$\int_1^\infty \exp(-t)t^s \, dt + \sum_{k=0}^\infty \frac{(-1)^k}{k!} \frac{1}{s+1+k}.$$

In the integral, integrate by parts with $u = t^s$, $dv = \exp(-t)dt$. Then, $du = st^{s-1}dt$ and $v = -\exp(-t)$. This gives

$$\int_1^\infty \exp(-t)t^s \, dt = -t^s \exp(-t) \big|_1^\infty + s \int_1^\infty \exp(-t)t^{s-1} \, dt$$

$$= (0 + e^{-1}) + s \int_1^\infty \exp(-t)t^{s-1} \, dt.$$

The change of variables given in the hint makes the sum

$$\sum_{j=1}^\infty \frac{(-1)^{j-1}}{j-1!} \frac{1}{s+j} = -\sum_{j=1}^\infty \frac{(-1)^j(s+j-s)}{j!} \frac{1}{s+j}$$

$$= -\sum_{j=1}^\infty \frac{(-1)^j(s+j)}{j!} \frac{1}{s+j} - \sum_{j=1}^\infty \frac{(-1)^j(-s)}{j!} \frac{1}{s+j}$$

$$= -\sum_{j=1}^\infty \frac{(-1)^j}{j!} + s \sum_{j=1}^\infty \frac{(-1)^j}{j!(s+j)}$$

$$= 1 - e^{-1} + s \sum_{j=1}^\infty \frac{(-1)^j}{j!(s+j)}$$

$$= -e^{-1} + s \sum_{j=0}^\infty \frac{(-1)^j}{j!(s+j)}.$$

Combining the sum and the integral gives $s\Gamma(s) + e^{-1} - e^{-1}$.

(8.4.1) Near $s = -2n + 1$,

$$\Gamma(s) = \frac{-1}{2n-1!} \frac{1}{(s+2n-1)} + O(1) \quad \text{according to (8.6); so,}$$

$$\Gamma(s)^{-1} = -(2n-1!)(s+2n-1) + O((s+2n-1)^2).$$

$$F(s) + G(s) = \frac{B_{2n}}{2n!} \frac{1}{(s+2n-1)} + O(1) \quad \text{according to (8.7); so,}$$

$$\zeta(s) = (-(2n-1!)(s+2n-1) + O((s+2n-1)^2))$$

$$\times \left(\frac{B_{2n}}{2n!} \frac{1}{(s+2n-1)} + O(1) \right)$$

$$= -\frac{B_{2n}}{2n} + O(s+2n-1) \quad \text{near } s = -2n + 1.$$

Thus, $\zeta(-2n+1) = -B_{2n}/(2n)$.

(8.4.2) For $x > 0$, the series for $\exp(x)$ contains only positive terms. So, for any positive integer n, if we omit all but the first term and the term $x^n/n!$, we get an inequality $1 + x^n/n! < \exp(x)$, which says that $1/(\exp(x) - 1) < n!/x^n$. Now, for fixed $s > 0$, choose n such that $n > s$. Then,

$$\int_1^\infty \frac{x^{s-1}}{\exp(x) - 1}\, dx < n! \int_1^\infty x^{s-n-1}\, dx$$

according to the comparison test for integrals,

$$= n! \left. \frac{x^{s-n}}{s-n} \right|_1^\infty = \frac{n!}{n-s} < \infty,$$

because $n - s < 0$.

(8.4.3) According to the theorem discussed in Section 6.3, we have

$$\zeta(2k) = (-1)^{k-1} \frac{(2\pi)^{2k} B_{2k}}{2(2k)!} = \frac{(2\pi)^{2k} |B_{2k}|}{2(2k)!},$$

because $\zeta(2k)$ is positive. The comparison test shows that $\zeta(2k) > \zeta(2)$, which implies that

$$\frac{|B_{2k}|}{2k!} < \frac{2\zeta(2)}{(2\pi)^{2k}}.$$

For fixed s not a negative odd integer, there is a closest negative odd integer $-2K + 1$, and $|s + 2K - 1|$ is some constant $C > 0$. Then, for all k, $|s + 2k - 1| \geq C$ and

$$\frac{|B_{2k}|}{2k!} \frac{1}{|s + 2k - 1|} \leq \frac{C^{-1} 2\zeta(2)}{(2\pi)^{2k}}.$$

The series converges by comparison to the Geometric series $\sum_k (2\pi)^{-2k}$.

(8.4.4) The inequality

$$\frac{|B_{2k}|}{2k!} < \frac{2\zeta(2)}{(2\pi)^{2k}}$$

from the previous problem implies that

$$\sum_{k=1}^\infty \frac{B_{2k}}{2k!} z^{2k} \text{ converges absolutely if } \sum_{k=1}^\infty \frac{2\zeta(2)}{(2\pi)^{2k}} z^{2k}$$

does, by the comparison test. Now, the ratio test gives the radius of convergence of this series as 2π.

(9.1.1) The derivatives are $-2x$ and $-1 + 1/(1 + x) = -x/(1 + x)$, respectively. Both derivatives are negative for $x > 0$.

(9.1.2)

$$\int_{-\infty}^{\infty}\int_{-\infty}^{\infty} \exp(-x^2 - y^2)dxdy = \int_0^{\infty}\int_0^{2\pi} r\exp(-r^2)d\theta dr$$

$$= 2\pi \int_0^{\infty} r\exp(-r^2)dr$$

$$= \pi \exp(-r^2)\big|_0^{\infty} = \pi.$$

There is a story about this exercise.

Once when lecturing in class [Kelvin] used the word "mathematician" and then interrupting himself asked his class: "Do you know what a mathematician is?" Stepping to his blackboard he wrote upon it:

$$\int_{-\infty}^{\infty} \exp(-x^2)dx = \sqrt{\pi}$$

Then putting his finger on what he had written, he turned to his class and said, "A mathematician is one to whom that is as obvious as that twice two makes four is to you."

S. P. Thompson, Life of Lord Kelvin

(9.1.3)

$$\frac{10!}{(10/e)^{10}\sqrt{2\pi 10}} = 1.00837\ldots,$$

$$\frac{20!}{(20/e)^{20}\sqrt{2\pi 20}} = 1.00418\ldots,$$

$$\frac{30!}{(30/e)^{30}\sqrt{2\pi 30}} = 1.00278\ldots,$$

$$\frac{40!}{(40/e)^{40}\sqrt{2\pi 40}} = 1.00209\ldots.$$

(9.1.4) The $m = 2$ approximation is

$$n! \approx (n/e)^n\sqrt{2\pi n}\exp\left(\frac{1}{12n} - \frac{1}{360n^3}\right)$$

and

$$(50/e)^{50}\sqrt{2\pi 50}\exp\left(\frac{1}{12\cdot 50} - \frac{1}{360\cdot 50^3}\right) =$$

$$3.0414093201636159061694647316522177\ldots \times 10^{64}.$$

(10.2.1) With $x = 12.99$, the sum is

$$\log(2) + \log(3) + \log(2) + \log(5) + \log(7)$$
$$+ \log(2) + \log(3) + \log(11) \approx 10.2299\ldots,$$

with the terms corresponding to $2, 3, 2^2, 5, 7, 2^3, 3^2$, and 11. With $x = 13.01$, add $\log(13)$ to get $12.7949\ldots$.

(10.2.2) $(10.2299\ldots + 12.7949\ldots)/2 = 11.5124\ldots$.

(10.2.3)

$$\int_1^\infty 1 \cdot x^{-s-1} dx = \left.\frac{x^{-s}}{-s}\right|_1^\infty = \frac{1}{s}.$$

(10.2.4)

$$\int_1^\infty x \cdot x^{-s-1} dx = \int_1^\infty x^{-s} dx = \left.\frac{x^{1-s}}{1-s}\right|_1^\infty = \frac{1}{s-1}.$$

(10.2.5)

$$\int_1^\infty -\frac{x^\rho}{\rho} \cdot x^{-s-1} dx = -\frac{1}{\rho}\int_1^\infty x^{\rho-s-1} dx$$
$$= -\frac{1}{\rho}\left.\frac{x^{\rho-s}}{\rho-s}\right|_1^\infty = -\frac{1}{\rho(s-\rho)}.$$

(10.2.6) From (I2.8),

$$-\log(1-x) = \sum_{n=1}^\infty \frac{x^n}{n} \quad \text{for } 0 \le x < 1.$$

So,

$$-\frac{1}{2}\log(1 - \frac{1}{x^2}) = \frac{1}{2}\sum_{n=1}^\infty \frac{x^{-2n}}{n} \quad \text{for } x > 1.$$

(10.2.7) Integrate term by term in the series expansion of Exercise (10.2.6); an individual summand contributes

$$\int_1^\infty \frac{x^{-2n}}{2n} x^{-s-1} dx = \left.\frac{x^{-2n-s}}{2n(-2n-s)}\right|_1^\infty = \frac{1}{2n(s+2n)}.$$

(10.2.8) In *Mathematica*, first make a table of as many zeros from Table 10.1 as you want to use (five are used for illustration).

```
gammas =
{14.13472, 21.02203, 25.01085, 30.42487, 32.93506};
```

The following defines vmef, a function of x, which includes the contribution of the first n zeros to the right side of the Von Mangoldt explicit formula (10.5) for $\Psi(x)$.

```
vmef[x_, n_] := x - Log[1 - 1/x^2]/2 - Log[2Pi] -
Sum[2x^(1/2)Cos[gammas[[k]]Log[x] - Arc-
Tan[2*gammas[[k]]]]/
Abs[1/2 + I*gammas[[k]]], {k, 1, n}]
```

The function plotvm takes n as input and returns a -Graphics-object, the plot of vmef[x,n].

```
plotvm[n_] := Plot[vmef[x, n], {x, 2, 20},
AxesOrigin -> {0, 0}, PlotPoints -> 50,
PlotRange -> {0, 20}, AspectRatio -> Automatic,
PlotLabel -> ToString[n] <> " zeros"]
```

Finally, you can make a Table of these for increasing values of n.

```
Table[plotvm[n], {n, 1, 5}]
```

The bigger n is, the more work *Mathematica* has to do, so it may run slowly. The result can then be Animate-d to make a movie if you wish. You will need more than 5 zeros to see a good approximation to $\Psi(x)$. Figure S.3 shows the contribution of the first zero, the first 10, and the first 100.

(10.3.1) This is routine. With $u = 1/\log(t)^2$, $du = -2/(\log(t)^3 t)dt$, $dv = dt$, and $v = t$.

(10.4.1) First, compute

$$x \cdot \frac{d}{dx}\left(\text{Li}(x^\alpha) \cdot \log(x)\right)$$

$$= x \cdot \left(\frac{d}{dx}\text{Li}(x^\alpha)\right) \cdot \log(x) + x \cdot \text{Li}(x^\alpha) \cdot \frac{d}{dx}(\log(x))$$

$$= x \cdot \frac{x^{\alpha-1}}{\log(x)} \cdot \log(x) + x \cdot \text{Li}(x^\alpha) \cdot \frac{1}{x} = x^\alpha + \text{Li}(x^\alpha).$$

Subtracting the $\text{Li}(x^\alpha)$ as instructed leaves x^α. If we start from the other end,

$$x \cdot \frac{d}{dx}\left(\frac{x^\alpha}{\alpha}\right) = x \cdot \frac{\alpha x^{\alpha-1}}{\alpha} = x^\alpha.$$

(10.4.2) Using *Mathematica*, first make a table of as many zeros from Table 10.1 as you want to use, five in this case.

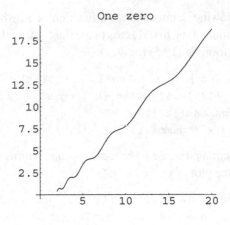

One zero

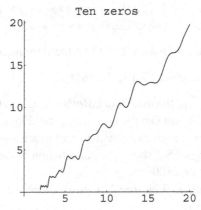

Ten zeros

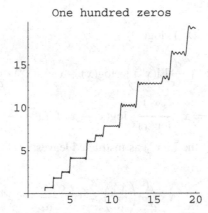

One hundred zeros

Figure S.3. Solutions to Exercise 10.2.8.

```
gammas =
{14.13472, 21.02203, 25.01085, 30.42487, 32.93506};
```

The following loads the `Audio` package.

```
<< Miscellaneous`Audio`
```

The default values of the parameters make a nice sound, but for the highest quality sound, you can reset them as follows:

```
SetOptions[ListWaveform, SampleDepth -> 16,
SampleRate -> 48000]
```

The following creates a table of frequencies $\gamma = \mathrm{Im}(\rho)$ and amplitudes $2/|\rho|$, as in the explicit formula. All the low-lying ρ that we will use have $\beta = \mathrm{Re}(\rho) = 1/2$.

```
musiclist =
Transpose[{gammas, 2/Abs[1/2 + I*gammas]}];
```

Ignore the phase shift $-\arctan(1/2\beta)$, because *Mathematica* can't produce it and we probably can't hear it. We have factored out the "universal amplitude" $x^{1/2}/\log(x)$ and imagined an exponential timescale to get rid of the $\log(x)$ inside the cos. Because the explicit formula is "dimensionless," we have to choose a timescale to play it on, a fundamental frequency the others are multiples of. This is a subjective choice. Try, first, 10 cycles per second. (You can change this later.) The command

```
primemusic=
Table[ListWaveform[Take[musiclist, n], 10, .2],
{n, 1, 5}]
```

creates (but does not play) a table of -Sound- files. The first is the contribution of just the first zero for 0.2 seconds, then the contribution of the first two zeros for 0.2 seconds, then the first three, and so on. You should now save your work before going on. If the command

```
Show[primemusic]
```

does not crash your machine, it will actually play the sound file. Mathematica has to do a lot of work to play the sound in real time; if it crashes your machine, you can instead Export the sound to a file, which can then be played in some other application. For example,

```
Export["primemusic.aiff",
Table[ListWaveform[Take[musiclist, n], 10, .2],
{n, 1, 5}]]
```

will produce a sound file `primemusic.aiff` in the .aiff format. You can also produce a sound in the .wav format. Try this also with the fundamental frequency 5 cycles per second, or 20. Try it also with more zeros.

(I4.0.1)   1. $a - a = 0 = 0 \cdot n$ is a multiple of n. So, $a \equiv a$.

2. If $a \equiv b \bmod n$, then for some k, $b - a = k \cdot n$. So, $a - b = (-k) \cdot n$. So, $b \equiv a$.

3. If $b - a = k \cdot n$ and $c - b = m \cdot n$, then $c - a = (c - b) + (b - a) = m \cdot n + k \cdot n = (m + k) \cdot n$. So, $a \equiv c$.

(I4.0.2)

×	0	1	2	3	4	5
0	0	0	0	0	0	0
1	0	1	2	3	4	5
2	0	2	4	0	2	4
3	0	3	0	3	0	3
4	0	4	2	0	4	2
5	0	5	4	3	2	1

(I4.0.3) Because $2 \cdot 4 \equiv 1 \bmod 7$, $2 = 4^{-1} = 1/4$ and $4 = 2^{-1} = 1/2$. Similarly, $3 = 5^{-1}$ and $5 = 3^{-1}$. Of course, 1 is its own inverse, and 0 has no inverse.

(I4.0.4) $3^2 \equiv 2 \bmod 7, 3^3 \equiv 3 \cdot 3^2 \equiv 6, 3^4 \equiv 4, 3^5 \equiv 5, 3^6 \equiv 1$. Thus, $3^7 \equiv 3$ and the cycle repeats. The powers of 3 give every nonzero class modulo 7. $2^2 \equiv 4 \bmod 7$, $2^3 \equiv 1$, and $2^4 \equiv 2$ again. The powers of 2 do not give every nonzero class modulo 7.

(I4.0.5) We have $0^2 \equiv 0 \bmod 4$, $1^2 \equiv 1$, $2^2 \equiv 0$, and $3^2 \equiv 1$. The only squares modulo 4 are 0 and 1. The possible sums of squares are $0 + 0 \equiv 0, 0 + 1 \equiv 1 + 0 \equiv 1$, and $1 + 1 \equiv 2$.

(I4.0.6) Imitating the argument in Exercise I4.0.5 but using arithmetic modulo 3, one sees that x^2 is always congruent to 0 or 1 modulo 3 and, so is $x^2 + 3y^2$ as $3y^2 \equiv 0$. So, no prime $p \equiv 2 \bmod 3$ can be written as $p = x^2 + 3y^2$. The converse, that every prime $p \equiv 1 \bmod 3$ can be written as $p = x^2 + 3y^2$, is true but harder to prove. Examples are $7 = 2^2 + 3 \cdot 1^2$, $13 = 1^2 + 3 \cdot 2^2$, and $19 = 4^2 + 3 \cdot 1^2$.

(I4.0.7) The converse is the following.

+	0	1	2	3	4	5
0	0	1	2	3	4	5
1	1	2	3	4	5	0
2	2	3	4	5	0	1
3	3	4	5	0	1	2
4	4	5	0	1	2	3
5	5	0	1	2	3	4

×	0	1	2	3	4	5	6
0	0	0	0	0	0	0	0
1	0	1	2	3	4	5	6
2	0	2	4	6	1	3	5
3	0	3	6	2	5	1	4
4	0	4	1	5	2	6	3
5	0	5	3	1	6	4	2
6	0	6	5	4	3	2	1

+	0	1	2	3	4	5	6
0	0	1	2	3	4	5	6
1	1	2	3	4	5	6	0
2	2	3	4	5	6	0	1
3	3	4	5	6	0	1	2
4	4	5	6	0	1	2	3
5	5	6	0	1	2	3	4
6	6	0	1	2	3	4	5

Lemma. *Suppose that e is the least positive integer such that $a^e \equiv 1 \bmod q$. If e divides k, then $a^k \equiv 1 \bmod q$.*

Proof. If e divides k, then $k = e \cdot d$ for some d and $a^k \equiv a^{e \cdot d} \equiv (a^e)^d \equiv 1^d \equiv 1 \bmod q$. $\square$

(I4.0.8) 59 is prime but does not divide M_{29}. $117 = 58 \cdot 2 + 1 = 9 \cdot 13$ is composite. $175 = 58 \cdot 3 + 1$ is, of course, $25 \cdot 7$, composite. $233 = 54 \cdot 4 + 1$ is prime, and we have found a divisor of the Mersenne number: $M_{29} = 233 \cdot 2304167$. So, M_{29} is composite.

(I4.0.9) $18 = 3 + 3 \cdot 5$ is congruent to 3 mod 5, and $18^2 = 324 = 13 \cdot 25 - 1$. So, $18^2 \equiv -1 \bmod 25$. $68 = 18 + 2 \cdot 25$ is congruent to 18 mod 25 and $68^2 = 4624 = 37 \cdot 125 - 1$, so $68^2 \equiv -1 \bmod 125$.

(I4.0.10) We have $m = 3$ and $n = 5$, $a = 2$, and $b = 4$. Because $2 \cdot 3 - 1 \cdot 5 = 1$, we can take $c = 2$ and $d = -1$. Then, $x = 2(-1)5 + 4(2)3 = 14$ works; $14 \equiv 2 \bmod 3$ and $14 \equiv 4 \bmod 5$.

(11.2.1) By expanding and grouping terms, one sees that

$$(ax + Nby)^2 - N(bx + ay)^2 = (a^2 - Nb^2)x^2 - N(a^2 - Nb^2)y^2$$
$$= x^2 - Ny^2 = 1.$$

(11.2.2)

$$(1, 0) \cdot (x, y) = (x + N \cdot 0, 0 \cdot x + y) = (x, y),$$
$$(x, y) \cdot (x, -y) = (x^2 - Ny^2, xy - xy) = (1, 0).$$

(11.2.3)

$$(2 + \sqrt{3})^6 = \sum_{k=0}^{6} \binom{6}{k} 2^k \sqrt{3}^{6-k} = 1351 + 780\sqrt{3}.$$

(11.4.1)

$$f_8(x) = \sum_{k=0}^{\infty} \left\{ \frac{x^{8k+1}}{8k+1} - \frac{x^{8k+3}}{8k+3} - \frac{x^{8k+5}}{8k+5} + \frac{x^{8k+7}}{8k+7} \right\}.$$

So,

$$f_8'(x) = \sum_{k=0}^{\infty} \left\{ x^{8k} - x^{8k+2} - x^{8k+4} + x^{8k+6} \right\}$$

$$= (1 - x^2 - x^4 + x^6) \sum_{k=0}^{\infty} x^{8k}$$

$$= \frac{1 - x^2 - x^4 + x^6}{1 - x^8} = \frac{(1 - x^4)(1 - x)(1 + x)}{(1 - x^4)(1 + x^4)}$$

$$= \frac{(1 - x)(1 + x)}{1 + x^4}.$$

(11.4.2) This is just messy algebra. Put

$$\frac{2x + \sqrt{2}}{x^2 + \sqrt{2}x + 1} - \frac{2x - \sqrt{2}}{x^2 - \sqrt{2}x + 1}$$

over a common denominator equal to

$$(x^2 + \sqrt{2}x + 1)(x^2 - \sqrt{2}x + 1) = x^4 + 1.$$

(11.4.3) The two rational functions can each be integrated by u-substitution, with

$$u = x^2 + \sqrt{2}x + 1, \quad du = (2x + \sqrt{2})dx,$$

$$u = x^2 - \sqrt{2}x + 1, \quad du = (2x - \sqrt{2})dx,$$

respectively.

(11.4.4) The first part is just algebra, and the second is just like Exercise 11.2.1.

(11.4.5) Use (11.4), with $d = 5$. For odd primes $p \neq 5$, there are $p - 1$ solutions if 5 is congruent to a square modulo p, and $p + 1$ solutions if 5 is not congruent to a square modulo p. Because $5 \equiv 1 \bmod 4$, Quadratic Reciprocity says that these two cases are exactly the same as p congruent to a square (respectively, not congruent to a square) modulo 5. The squares modulo 5 are 1 and 4: $1^2 \equiv 1, 2^2 \equiv 4, 3^2 \equiv 4$, $4^2 \equiv 1$ modulo 5. So, we have $p - 1$ solutions if $p \equiv 1$ or 4 mod 5, and $p + 1$ solutions if $p \equiv 2$ or 3 mod 5.

For $p = 2$, one finds, by considering all possibilities, that $(0, 1)$, $(1, 0)$, and $(1, 1)$ are all the solutions.

(11.4.6)

$$f_5(x) = \sum_{k=0}^{\infty} \left\{ \frac{x^{5k+1}}{5k+1} - \frac{x^{5k+2}}{5k+2} - \frac{x^{5k+3}}{5k+3} + \frac{x^{5k+4}}{5k+4} \right\}.$$

So,

$$f_5'(x) = \sum_{k=0}^{\infty} \left\{ x^{5k} - x^{5k+1} - x^{5k+2} + x^{5k+3} \right\}$$

$$= (1 - x - x^2 + x^3) \sum_{k=0}^{\infty} x^{5k}$$

$$= \frac{1 - x - x^2 + x^3}{1 - x^5} = \frac{(1 - x)(1 + x)(1 - x)}{(1 - x)(1 + x + x^2 + x^3 + x^4)}$$

$$= \frac{(1 - x)(1 + x)}{1 + x + x^2 + x^3 + x^4}.$$

(11.4.7) Just as in Exercise 11.4.2, put the two rational functions over a common denominator.

(11.4.8) Just as in Exercise 11.4.3, use two u-substitution integrals, with

$$u = x^2 + \epsilon_+ x + 1, \quad du = (2x + \epsilon_+)dx,$$

$$u = x^2 + \epsilon_- x + 1, \quad du = (2x + \epsilon_-)dx,$$

respectively.

(12.1.1) 1. The line through $P = (-2, 1)$ and $-Q = (0, 1)$ has slope 0. So, it is $y = 1$. Plugging into the equation of the curve gives

$$1^2 = x^3 - 4x + 1 \quad \text{or} \quad x^3 - 4x = 0,$$

which has solutions $x = 0, \pm 2$. Because $x = 0$ comes from $-Q$ and $x = -2$ comes from P, the third point of intersection has $x = 2$, and of course $y = 1$. So, $P - Q = (2, -1)$.

2. As in the text, implicit differentiation shows that the slope of the tangent line at $(0, -1)$ is $(3 \cdot 0^2 - 4)/(2 \cdot -1) = 2$. So, the tangent line at Q is $y + 1 = 2(x - 0)$, or $y = 2x - 1$. Plugging into the equation of the curve gives

$$(2x - 1)^2 = x^3 - 4x + 1 \quad \text{or} \quad x^3 - 4x^2 = 0.$$

So, as expected, $x = 0$ is a double root corresponding to the point Q, and the other root is $x = 4$. The point on the line has $y = 2 \cdot 4 - 1 = 7$. So, $2Q = (4, -7)$.

3. $-P = (-2, -1)$ and $2Q = (4, -7)$ is computed above. $-P + 2Q = (-1, 2)$.

4. $2P = (20, -89)$ is computed in the text and $-Q = (0, 1)$. $2P - Q = (1/4, 1/8)$.

5. $3Q = Q + 2Q = (0, -1) + (4, -7) = (-7/4, -13/8)$.

(12.2.1) We have $a = 3/2, b = 20/3, c = 41/6, n = ab/2 = 5, t = c^2/4 = (41/12)^2 = 1{,}681/144, t - n = 961/144 = (31/12)^2$, and $t + n = 2{,}401/144 = (49/12)^2$. So, $(t - n)t(t + n) = ((31/12)(41/12)(49/12))^2$ and the rational point is

$$\left(\frac{1681}{144}, \frac{31}{12} \cdot \frac{41}{12} \cdot \frac{49}{12} \right) = \left(\frac{1681}{144}, \frac{62279}{1728} \right).$$

(12.3.1) For $p = 11$, there are 12 points: $(0, \pm 4)$, $(4, \pm 5)$, $(5, \pm 3)$, $(6, \pm 1)$, $(8, 0)$, $(10, \pm 2)$, and the point at infinity. For $p = 13$, there are 16 points: $(2, 0)$, $(4, \pm 2)$, $(5, 0)$, $(6, 0)$, $(7, \pm 6)$, $(8, \pm 6)$, $(10, \pm 2)$, $(11, \pm 6)$, $(12, \pm 2)$, and the point at infinity. For $p = 17$, there are 18 points: $(2, \pm 8)$, $(3, \pm 7)$, $(4, \pm 1)$, $(6, 0)$, $(7, \pm 5)$, $(10, \pm 6)$, $(12, \pm 4)$, $(13, \pm 3)$, $(16, \pm 2)$, and the point at infinity.   Observe that 11 and 17 are congruent to 2 modulo 3; these examples are in agreement with (12.4).

(12.3.2) $124 = 4^2 + 27 \cdot 2^2$, and $4 \equiv 1 \bmod 3$. $5^2 \equiv 25 \equiv -6 \bmod 31$. $5^4 \equiv (-6)^2 \equiv 5 \bmod 31$. $5^8 \equiv 5^2 \equiv -6$, and $5^{10} \equiv 5^8 \cdot 5^2 \equiv 5 \bmod 31$.

Now, $5 \cdot 4 > 31/2$. So, we choose the representative $a = -31 + 20 = -11 \equiv 20$ mod 31. The theorem says that the number of points is $31 - 11 - 1 = 19$.

(12.4.1) For small primes $p \leq 13$, you can define a function nsmall, which counts the number of points on the curve $y^2 = x^3 - 432d^2$ modulo p by brute force and ignorance, testing all possible x and y values.

```
nsmall[p_, d_] :=
(pointcount = 1;
Do[
    Do[
        If[Mod[x^3 - 432*d^2, p] == Mod[y^2, p],
            pointcount = pointcount + 1],
        {x, 0, p - 1}],
{y, 0, p - 1}];
pointcount)
```

The function rep takes a prime $p \equiv 1$ mod 3 as input and returns the unique $L \equiv 1$ mod 3 such that $4p = L^2 + 27M^2$.

```
rep[p_] :=
If[Mod[p, 3] == 1,
    Do[M = Sqrt[(4*p - L^2)/27];
        If[IntegerQ[M],
            If[Mod[L, 3] == 1, Return[L], Return[-L]]
        ],
    {L, 1, Sqrt[4*p]}]
]
```

For primes $p > 13$, the function nbig, which uses (12.4) and (12.5), will be more efficient.

```
nbig[p_, d_] :=
If[Mod[p, 3] == 2,
    p + 1,
    L = rep[p];a = Mod[d^((p - 1)/3)*L, p];
    If[a > p/2, a = a - p]; p + a + 1
]
```

The function bsd takes x and d as input and returns the product over primes $p < x$ of the ratio N_p/p for the curve $y^2 = x^3 - 432d^2$. It is very inefficient if you want many different values of x, because it recomputes previously used values of N_p. Try to think of another way to do this.

```
bsd[x_, d_] :=
Product[p = Prime[j];
        If[p <= 13,
            N[nsmall[p, d]/p],
            N[nbig[p, d]/p]
           ],
{j, 1, PrimePi[x]}]
```

E_5 has rank 0, and so the product should grow like $\log(x)^0$, that is, not at all. E_7 has rank 1, so the product should grow like $\log(x)^1$.

(13.1.1) 1. Take $M = \left[\begin{smallmatrix} 1 & 0 \\ 0 & 1 \end{smallmatrix}\right]$.

2. If M makes $F \sim G$, then M^{-1} makes $G \sim F$.

3. If M makes $F \sim G$ and M' makes $G \sim H$, then $M'M$ makes $F \sim H$.

(13.1.2) The determinant is $ac - (b/2)^2 = -d/4$. This is unchanged under multiplication by M and ^{tr}M, because M has determinant 1. (The determinant of a product of two matrices is the product of the determinants.)

(13.1.3) $0^2 - 4 \cdot 2 \cdot 3 = -24 = 0^2 - 4 \cdot 1 \cdot 6$. $G(1, 0) = 1$, but if $(x, y) \neq (0, 0)$, then

$$F(x, y) = 2x^2 + 3y^2 \geq 2x^2 + 2y^2 = 2(x^2 + y^2) \geq 2,$$

and of course $F(0, 0) = 0$.

(13.1.4) Suppose that $c < |b|$. The matrix

$$\begin{bmatrix} 1 & -\text{sgn}(b) \\ 0 & 1 \end{bmatrix}$$

changes (x, y) to $(x, y - \text{sgn}(b)x)$. The corresponding form is

$$ax^2 + bx(y - \text{sgn}(b)x) + c(y - \text{sgn}(b)x)^2$$
$$= (a + c - |b|)x^2 + (b - 2\text{sgn}(b)c)xy + cy^2.$$

We have $(a + c - |b|) + c < a + c$ because $c < |b|$.

(13.1.5) Because $\sqrt{23/3} = 2.76877\ldots$, we know that $|b| \leq 2$ and $1 \leq a \leq 2$. Again, b is odd because -23 is; so, $b = \pm 1$. Then, $b^2 - d = 24$. So, the choice $a = 1$ gives the forms $\{1, \pm 1, 6\}$. The choice $a = 2$ gives the forms $\{2, \pm 1, 3\}$. The class number is, at most, 4.

(13.1.6)

$$F = \{12, 11, 3\} \quad G = \{39, 43, 12\} \quad H = \{93, 109, 32\}$$
$$\sim \{4, 5, 3\} \qquad \sim \{39, -35, 8\} \qquad \sim \{93, -77, 16\}$$
$$\sim \{4, -3, 2\} \qquad \sim \{12, -19, 8\} \qquad \sim \{32, -45, 16\}$$
$$\sim \{3, 1, 2\} \qquad \sim \{12, 5, 1\} \qquad \sim \{32, 19, 3\}$$
$$\sim \{2, -1, 3\} \qquad \sim \{8, 3, 1\} \qquad \sim \{16, 13, 3\}$$
$$\qquad\qquad \sim \{6, 1, 1\} \qquad \sim \{6, 7, 3\}$$
$$\qquad\qquad\qquad\qquad \sim \{6, -5, 2\}$$
$$\qquad\qquad\qquad\qquad \sim \{3, -1, 2\}$$
$$\qquad\qquad\qquad\qquad \sim \{2, 1, 3\}.$$

(13.1.7) The function `red` takes a list $\{a, b, c\}$ as input carries out one stage of the reduction.

```
red[list_] :=
(a = list[[1]]; b = list[[2]]; c = list[[3]];
If[Abs[b] > a,
    Return[{a, b - 2Sign[b]a, a + c - Abs[b]}],
    If[Abs[b] > c,
        Return[{a + c - Abs[b], b - 2Sign[b]c, c}],
        If[a > c,
            Return[{c, -b, a}],
            Return[{a, b, c}]]]])
```

To carry out the complete reduction on a list $\{a, b, c\}$, use *Mathematica*'s `FixedPoint` function:

```
FixedPoint[red,{a,b,c}]
```

(13.1.8) See Table 13.1

(13.1.9) You might do something like

```
classno[d_] :=
(h = 0;
Do[
    If[Mod[b, 2] == Mod[d, 2],
        Do[c = (b^2 - d)/(4a);
            If[IntegerQ[c],
                If[b == 0 || b == a || a == c,
                    h = h + 1,
```

Table S.1. *Examples of Representation Numbers*

	−159	−163	−164	−424	−427	−431
2	2	0	1	1	0	2
3	1	0	2	0	0	2
5	2	0	2	2	0	2
7	2	0	2	0	1	0
11	0	0	2	2	0	2
13	2	0	0	0	0	0
17	0	0	0	2	2	0
19	0	0	2	0	0	2
23	2	0	0	2	0	2
29	0	0	0	0	0	2
31	0	0	0	2	2	0
37	2	0	2	0	0	0

```
            h = h + 2
          ]
        ],
      {a, Max[b, 1], Floor[Sqrt[(b^2 - d)/4]]}]
    ],
  {b, 0, Floor[Sqrt[Abs[d]/3]]}];
  h)
```

(13.1.10) Modulo $4 \cdot 5 = 20$, the only squares are $0, 1, 4, 5, 9$, and 16. $-23 \equiv -3 \equiv 17 \bmod 20$ is not a square. Modulo 12, $-23 \equiv 1$ is a square. And modulo 24,

$$- 23 \equiv 1 \equiv 1^2 \equiv 5^2 \equiv 7^2 \equiv 11^2.$$

So, 6 has 4 representations.

(13.1.11) The representation numbers are in Table S.1. Observe that the class numbers of the discriminants are, respectively, 10, 1, 8 and 6, 2, 21 (taken from Table 13.1). The point here is that discriminants with relatively smaller class numbers tend to represent fewer small primes than those with larger class numbers.

(13.1.12) You might do something like

```
repno[n_, d_] :=
(count = 0;
Do[
    If[Mod[m^2, 4n] == Mod[d, 4n], count = count + 1],
{m, 0, 2n - 1}];
count)
```

(13.2.1) Take

$$f_{-3}(x) = \sum_{k=0}^{\infty} \left\{ \frac{x^{3k+1}}{3k+1} - \frac{x^{3k+2}}{3k+2} \right\}$$

thus $f_{-3}(1) = L(1, \chi_{-3})$. Then,

$$f'_{-3}(x) = \sum_{k=0}^{\infty} x^{3k} - x^{3k+1} = (1-x) \sum_{k=0}^{\infty} x^{3k}$$

$$= \frac{1-x}{1-x^3} = \frac{1}{1+x+x^2}.$$

Completing the square, we can write

$$f'_{-3}(x) = \frac{1}{(x+1/2)^2 + 3/4}.$$

Then, via u-substitution (or looking in a table of integrals),

$$f_{-3}(x) = \frac{2}{\sqrt{3}} \arctan((2x+1)/\sqrt{3}) + C.$$

Because we require that $f_{-3}(0) = 0$ from the series expansion, the usual choice of $C = 0$ won't do. Instead, $\arctan(1/\sqrt{3}) = \pi/6$ forces us to pick $C = -\pi/(3\sqrt{3})$. Then,

$$f_{-3}(x) = \frac{2}{\sqrt{3}} \arctan((2x+1)/\sqrt{3}) - \pi/(3\sqrt{3})$$

has $f_{-3}(0) = 0$, and $f_{-3}(1) = \pi/(3\sqrt{3}) = L(1, \chi_{-3})$.

(13.3.1) Consider, first, the terms in the sum with $n < |d|$. With the hypothesis about σ, we still have that $n^{-\sigma} < e/n$, as before. Then, we can estimate

$$\left| \sum_{n<|d|} \chi_d(n) n^{-\sigma} \right| < \sum_{n<|d|} \frac{e}{n} \ll \int_1^{|d|} \frac{1}{x} dx = \log |d|.$$

For the tail of the series, we again use summation by parts. Consider

$$\sum_{|d| \le k < N} \chi_d(k) k^{-\sigma}.$$

We let $\Delta v(k) = \chi_d(k)$. So, $v(k) = A(k) = \sum_{n<k} \chi_d(n)$ and $v(|d|) = 0$, according to the previous lemma. Our $u(k)$ is $k^{-\sigma}$. We

get that the sum above is equal to

$$A(n)\frac{1}{N^\sigma} + \sum_{|d|\leq k<N} A(k+1)\left(\frac{1}{k^\sigma} - \frac{1}{(k+1)^\sigma}\right).$$

Because $x^{-\sigma}$ is a decreasing function, the term in parenthesis is positive. So, the triangle inequality estimates the absolute value of the sum as being bounded by

$$|A(n)|\frac{1}{N^\sigma} + \sum_{|d|\leq k<N} |A(k+1)|\left(\frac{1}{k^\sigma} - \frac{1}{(k+1)^\sigma}\right)$$

$$\ll |d|\frac{1}{N^\sigma} + |d| \sum_{|d|\leq k<N} \left(\frac{1}{k^\sigma} - \frac{1}{(k+1)^\sigma}\right),$$

according to the estimate on $|A(x)|$,

$$= |d|\frac{1}{N^\sigma} + |d|\left(\frac{1}{|d|^\sigma} - \frac{1}{(N)^\sigma}\right),$$

because the sum telescopes. Now, we consider what happens when $N \to \infty$. The terms with $1/N^\sigma$ go to zero. The hypothesis about the location of σ implies, just as for n above, that $|d|/|d|^\sigma \leq e$. So, the tail is bounded by $e \ll \log|d|$.

Bibliography

Recreational–Expository

Cipra, B. (1988). Zeroing in on the zeta zeta function (sic), *Science*, **239**, pp. 1241–2; or http://www.jstor.org/

Cipra, B. (1996). Prime formula weds number theory and quantum physics, *Science*, **274**, pp. 2014–5; or http://www.jstor.org/

Cipra, B. (1999). *A Prime Case of Chaos, What's Happening in the Mathematical Sciences*, **4**; AMS or http://www.ams.org/new-in-math/cover/prime-chaos.pdf

Conway, J. Guy, R. (1996). *The Book of Numbers*, Copernicus.

Guy, R. (1994). *Unsolved Problems in Number Theory*, 2nd ed., Springer-Verlag.

Klarreich, E. (2000). Prime Time, *New Sci.*, 11 Nov, no. 2264.

Polya, G. (1966). Heuristic reasoning in the theory of numbers, *Amer. Math. Monthly*, **66**, no. 5, pp. 375–84; or http://www.jstor.org/

Wagon, S. (1986). Where are the zeros of zeta of s?, *Math. Intelligencer*, **8**, no. 4, pp. 57–62.

Wells, D. (1986). *The Penguin Dictionary of Curious and Interesting Numbers*, Penguin Books.

Zagier, D. (1977). The first 50 million prime numbers, *Math. Intelligencer*, **0**, pp. 7–19.

Historical

Ayoub, R. (1974). Euler and the zeta function, *Amer. Math. Monthly*, **81**, pp. 1067–86; or http://www.jstor.org/

Bell, A. (1895). The "cattle problem," by Archimedes 251 B.C., *Amer. Math. Monthly*, **2**, no. 5, pp.140–41; or http://www.jstor.org/

Brown, J. (1897). *The Life and Legend of Michael Scot.*

Charles, R. (1916). *The Chronicle of John, Bishop of Nikiu*, Williams & Norgate.

Cicero, (1928). *Tusculan Disputations* (transl. by J. E. King), Putnam.

Clagett, M. (1968). *Nicole Oresme and the Medieval Geometry of Qualities and Motions*, University of Wisconsin Press.

Davis, P. J. (1959). Leonhard Euler's integral: A historical profile of the gamma function, *Amer. Math. Monthly*, **66**, pp. 849–69; or http://www.jstor.org/

Dickson, L. E. (1999). *History of the Theory of Numbers*, AMS Chelsea.

Dictionary of Scientific Biography, C. C. Gillispie, ed., Scribner. (1970–1980).

Gregorovius, F. (1971). *Rome and Medieval Culture; Selections from History of the City of Rome in the Middle Ages* (trans. by Mrs. G. W. Hamilton), University of Chicago Press.

Hardy, G. H. (1978). *Ramanujan*, Chelsea.

Heath, T. (1981). *A History of Greek Mathematics*, Dover.

Heath, T. (1964). *Diophantus of Alexandria*, Dover.

Hopper, V. (2000). *Medieval Number Symbolism*, Dover.

Iamblichus, (1989). *On the Pythagorean Life* (trans. by G. Clark), Liverpool University Press.

Lévy, T. (1996). L'histoire des nombres amiables: le témoinage des textes hébreux médiévaux, *Arabic Sci. Philos.*, **6**, no. 1, pp. 63–87.

Masi, M. (1983). *Boethian Number Theory, A translation of De Institutione Arithmetica, Studies in Classical Antiquity*, **6** Amsterdam, Rodopi.

Monastyrsky, M. (1987). *Riemann, Topology, and Physics*, Birkhäuser.

Nicomachus of Gerasa, (1938). *Introduction to Arithmetic* (trans. by M. L. D'ooge), University of Michigan Press.

Norwich, J. (1997). *Byzantium: The Apogee*, Knopf.

Oresme, N. (1961). *Questiones Super Geometriam Euclidis, Janus Revue Internationale de l'Histoire des Science Suppléments*, **III**, Dr. H. L. L. Busard, ed., Brill, Leiden.

Private Correspondence and Miscellaneous Papers of Samuel Pepys, 1679–1703, J. Tanner, ed., Harcourt Brace (1925).

Rose, P. L. (1975). The Italian Renaissance of mathematics, in *Travaux D'Humanism et Renaissance*, **CXLV** Librarie Droz.

Stanley, T. (1978). *A History of Philosophy* (1687), facsimile ed. Garland.

Vardi, I. (1998). Archimedes' cattle problem, *Amer. Math. Monthly*, **105**, no. 4, pp. 305-19; or http://www.jstor.org/

Williams, H. German, R. Zarnke, C. (1965). Solution to the cattle problem of Archimedes, *Math. Comp.*, **19**, no. 92, pp. 671–4; or http://www.jstor.org/

Weil, A. (1983). *Number Theory, an Approach through History*, Birkhäuser.

Technical

Apostol, T. (1995). *Introduction to Analytic Number Theory*, 5th ed., Springer-Verlag.

Bellman, R. (1961). *A Brief Introduction to Theta Functions*, Holt, Rinehart and Winston.

Berry, M. V. and Keating, J. P. (1991). The Riemann zeros and eigenvalue asymptotics, *SIAM Rev.*, **41**, no. 2, pp. 236–66.

Cox, D. (1989). *Primes of the Form $x^2 + ny^2$*, Wiley.

Davenport, H. (2000). *Multiplicative Number Theory*, 3rd ed., *Graduate Texts in Mathematics*, **74**, Springer-Verlag.

DeBruijn, N. G. (1981). *Asymptotic Methods in Analysis*, Dover.

Edwards, H. M. (2001). *Riemann's Zeta Function*, Dover.

Gradshteyn, I. and Ryzhik, I. (1979). *Tables of Integrals, Series, and Products*, 4th ed., Academic.

Graham, R., Knuth, D. and Patashnik, O. (1994). *Concrete Mathematics*, Addison-Wesley.

Hardy, G. H. and Littlewood, J. E. (1922). Some problems of 'Partio numerorum.' III. On the expression of a number as a sum of primes, *Acta Math.*, **44**, pp. 1–70.

Hardy, G. H. and Wright, E. M. (1980). *An Introduction to the Theory of Numbers*, 5th ed., Oxford University Press.

Ingham, A. E. (1990). *The Distribution of Prime Numbers*, Cambridge Mathematical Library.

Ireland, K. and Rosen, M. (1990). *A Classical Introduction to Modern Number Theory*, Springer-Verlag.

LeVeque, W. (2002). *Topics in Number Theory*, Dover.

McKean, H. and Moll, V. (1997). *Elliptic Curves*, Cambridge University Press.

Narkiewicz, W. (2000). *The Development of Prime Number Theory*, Springer-Verlag.

Polya, G. (1927). Elementarer Beweis einer Thetaformel, *Sitz. Phys.-Math. Klasse*, pp. 158–61 (reprinted in *Collected Papers*, **1**, pp. 303–6 MIT Press, 1974).

Ram Murty, M. (2001). *Problems in Analytic Number Theory, Graduate Texts in Mathematics*, **206**, Springer-Verlag.

Robin, G. (1984). Grandes valuers de la fonction somme des diviseurs et hypothèse de Riemann, *J. Math. Pure Appl.*, **63**, pp. 187–213.

Schroeder, M. R. (1997). *Number Theory in Science and Communication*, Springer-Verlag.

Shanks, D. (1985). *Solved and Unsolved Problems in Number Theory*, Chelsea.

Stark, H. (1975). The analytic theory of algebraic numbers, *Bull. Amer. Math. Soc.*, **81**, no. 6, pp. 961–72

Tennenbaum, G. and Mendés France, M. (2000). *The Prime Numbers and Their Distribution*, American Mathematical Society.

Wagstaff, S. (1983). Divisors of Mersenne numbers, *Math. Comp.*, **40**, no. 161, pp. 385–97; or http://www.jstor.org/

Index